Bulk and Surface Acoustic Waves

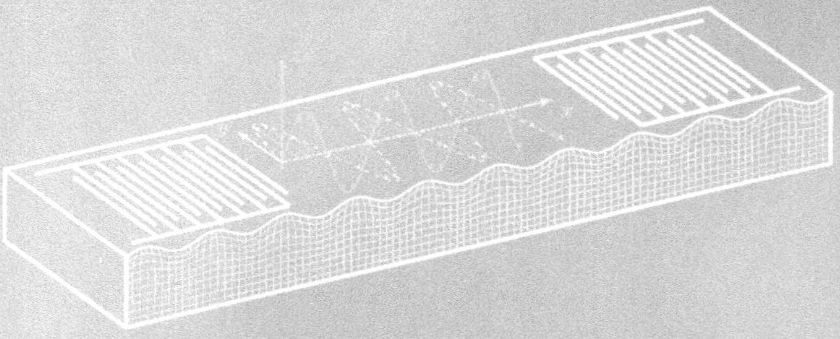

Bulk and Surface Acoustic Waves

Fundamentals, Devices, and Applications

Guigen Zhang

Jenny Stanford Publishing

Published by

Jenny Stanford Publishing Pte. Ltd.
Level 34, Centennial Tower
3 Temasek Avenue
Singapore 039190

Email: editorial@jennystanford.com
Web: www.jennystanford.com

British Library Cataloguing-in-Publication Data
A catalogue record for this book is available from the British Library.

Bulk and Surface Acoustic Waves: Fundamentals, Devices, and Applications

ISBN 978-981-4877-73-2 (Hardcover)
ISBN 978-1-003-25662-5 (eBook)

For Sucheng,
Andrew, and Gordon

And to the memory of
my Mother and Father

Contents

Preface

Acoustic wave devices, especially surface acoustic wave (SAW) devices, have been used for military and telecommunication applications for many decades, and they are regarded as possessing great potential for further applications in mobile communication, earth and ocean exploration, and medical imaging and detection, among others. Yet, unleashing this potential appears to be challenging. One of the signs for such challenges is that even to this day most of SAW devices are still made from crystal cuts identified decades earlier. While it is indeed a challenging task to identify new crystal cuts for desired features to meet the emerging needs, personally I would attribute this lack of advancements to a dearth of well-structured and easy-to-understand texts laying out a solid foundation for engineers to gain insights into the underlying wave theories and connecting these insights to device applications.

On the one hand, although a large number of texts have been written on this subject, most of them have placed an emphasis on applications with only superficial coverage of the underlying sciences and theories. These texts are of good values to experienced engineers in the SAW community, but they do not seem to help much the newcomers to the field in terms of gaining a good fundamental understanding of the SAW technologies. On the other hand, there are also texts available on the subject that dive deep into acoustic theories. One thing I noted is that all theoretical developments and discussions in these texts are based on tensor notions across the board. In other words, these texts need some physics or mathematical prerequisites in tensor notions and manipulations for readers to understand and appreciate.

My intent for this book is to address this dilemma by introducing acoustic wave theories using a reader-friendly matrix-based linear algebra approach. I hope that doing so will enable the reader to take advantage of software tools such as MATLAB (commercial codes) or OCTAVE (open-source codes) to gain better and deeper understanding of the underlying physics quickly. In this aspect, this text can be regarded as a practical introduction of the acoustic wave theories in an easy-to-follow linear algebra format using matrix manipulations instead of an abstract approach relying on tensor manipulations.

I ventured into the acoustic wave field in the early 2000s when my then doctoral student Yeswanth Rao and I were working on developing SAW-based biosensors. Since that time, I have kept exploring further and deep into the basics as well as applications of the acoustic wave technologies. This journey has helped me note the dilemma and challenges mentioned earlier. In my search over the years for a simplified way to introduce and discuss the acoustic wave theories, I have made countless notes, derivations, calculations, and device analyses. This book is a collection of these notes, calculations, and analytic results in an organized form.

This book is structured in nine chapters. It begins with a brief introduction in Chapter 1 on sound waves and acoustics, a brief history of wave theory development, and uniqueness of acoustic devices. The discussion of the fundamental aspects of acoustic waves is laid out in Chapter 2 through Chapter 4. Specifically, in Chapter 2, equations of motion, constitutive relations for isotropic and anisotropic elastic and piezoelectric materials, and governing equations for acoustic waves in elastic and piezoelectric solids are discussed; in Chapter 3, coordinate rotation, definition of Euler angles, transformation of stress and strain tensors and vectors, and piezoelectric material properties are discussed; in Chapter 4, the 32 known classes of symmetry in piezoelectric crystals are presented along with how a symmetric condition dictates the characteristics of the material properties. In-depth discussions of the characteristics of bulk and surface waves are presented in Chapter 5 and Chapter 6, where the propagation of both bulk and surface acoustic waves in various elastic and piezoelectric solids, the relationship between the type of crystal symmetry and the decoupling of different SAWs, and the phase velocity of SAWs traveling piezoelectric materials in different crystal cuts are discussed in detail. Chapter 7 discusses the effect of temperature on SAW propagation, and Chapter 8 looks into the underlying science for SAWs traveling in thin layers, including the Love waves and Lamb waves. In Chapter 9, case studies for the application of BAWs and SAWs in several devices are presented, including BAW devices such as disc resonators, micro-balances, and tuning-fork resonators and SAW devices such as filters, sensors, and resonators.

With this book, I also intend to achieve an important goal of demonstrating how the fundamentals on acoustic waves discussed throughout the book are applied in examples, case studies, and device analyses such that the connections and interdependences between the underlying sciences and the observed behavior and performances can be better appreciated by the reader. To achieve this, all problems for illustrations, examples, case studies, and device analyses are developed and solved based on the mathematical foundations laid out in the book, rather than relying on reference information and results from different sources. Moreover, in my continued efforts to introduce engineering concepts illustratively, I also created my own graphics to provide visualizations to assist the reader to develop a better understanding of important concepts, including wave polarization and propagation, types of piezoelectric crystal cuts, and substrate deformations caused by various types of surface waves, to name just a few. These efforts would not have been possible without the use of some software programs. Among them, MATLAB and COMSOL, commercially available software codes, are used to solve wave equations, find values for the phase velocity of bulk and surface waves in various piezoelectric materials in different crystal orientations and cuts, and perform analyses of various acoustic wave devices; LaTeX, TikZ, and PGF, open-source codes thanks to a community of creators and supporters, are used to create illustrative graphics, plots, and curves. The availability of these software tools has made my tasks a lot easier toward advancing engineering education and technology development.

Finally, a note on finishing the book amid the global COVID-19 pandemic. Although this journey started years ago, completing the book in the midst of the pandemic seems to provide a much-needed solace to distract me from the daily grind and give me the pleasure of getting something done in this difficult time.

Guigen Zhang

Chapter 1

Introduction

It is said that sound was recognized as traveling in waves as early as in the sixth century BC by the Greek philosopher Pythagoras (570–495 BC) but scientific understanding of sound waves came much later in the 16th and 17th centuries. In this chapter, we will conduct a brief overview of human's search in understanding the sound waves, the various types of seismic and acoustic waves, and the development of wave theories, along with a modern day application of acoustic wave technologies.

1.1 Sound Waves and Acoustics

Italian physicist Galileo Galilei (1564–1642) was believed to be the first scientist to demonstrate the relationship between the frequency of a wave and the pitch it produces. French mathematician Marin Mersenne (1588–1648) conducted the first experiments in 1640 to determine the speed of sound traveling in air. Twenty years later, British scientist Robert Boyle (1627–1691) demonstrated that the transmission of sound waves required a medium such as air by using improved vacuum technology to show that sound intensity decreased virtually to zero as air was pumped out.

Bulk and Surface Acoustic Waves: Fundamentals, Devices, and Applications
Guigen Zhang

ISBN 978-981-4877-73-2 (Hardcover), 978-1-003-25662-5 (eBook)
www.jennystanford.com

English physicist Robert Hooke (1635–1703) first produced a sound wave of known frequency. Later, German physicist Ernst Chladni (1756–1827) made extensive analyses of sound-producing vibrations. In 1801 the French mathematician Joseph Fourier (1768–1830) discovered that the sound waves produced by a vibrating string were just a combination of a series of simple periodic waves in different frequencies. Although the attempt to measure the speed of sound started in the 17th century by Mersenne, the earliest precise experimental value for the speed of sound was obtained at the Academy of Sciences in Paris in 1738 with a value of 332 m/s, which is very close to the presently accepted value of 331.29 m/s at 0°C.

French physicist Joseph Sauveur (1653–1716) was believed to be the first scientist to use the term acoustics for the study of sound. Acoustics is also a term we use today to associate with string instruments and audible sounds in a frequency range from 20 Hz to 20 KHz. Waves with frequency higher than 20 KHz are typically referred to as ultrasound. In this book, however, acoustic waves are mechanical waves generated by the displacements of particles in a medium or a solid material having frequency typically in the ultrasound range.

Speaking of ultrasound, Italian physiologist and biologist Lazzaro Spallanzani (1729–1799) was credited for the discovery that bats navigate in the dark through the reflection of high frequency sounds (echolocation), laying the basis of ultrasound physics. While bats use their biological capability to generate and detect ultrasound, an engineering means to produce ultrasound was made possible only with the discovery of direct and reverse piezoelectric effects by the French physicists and brothers, Pierre Curie (1859–1906) and Jacques Curie (1855–1941) in 1880. Their discovery of piezoelectric effects provided a primary means for generating and receiving acoustic waves. The actual realization of ultrasonic technologies, however, was not possible until the development of the electronic oscillator and amplifier to drive the piezoelectric element in the early 20th century. The tragic sinking of the Titanic in 1912 played a role in accelerating scientific efforts to develop a system to visualize underwater structures and the First World War also provided another impetus to speed up the development of high-frequency sound wave technologies to detect submarines. One such successful outcome was the development of SONAR (Sound Navigation and Ranging) by the U.S. Navy in the 1930s.

1.2 Main Types of Acoustic Waves

One commonly known type of acoustic waves is seismic tremors generated from an earthquake event. From the standpoint of spatial concentration of vibrational energy, these waves can be grouped into body (or bulk) waves and surface waves. Apparently, by their definitions, body or bulk waves are waves that propagate through the interior of a medium, and surface waves are those traveling along the surface of the medium. Music that propagates in air is an examples of bulk waves, and ripples traveling along the surface of a body of water are of surface waves.

There are two types of bulk waves, namely, *longitudinal waves* and *transverse waves*. Longitudinal waves (or L waves, for short) are often called compressional, dilatation or irrotational waves. In seismological terms, they are primary waves, or P waves for short. Particles vibration caused by a longitudinal wave occurs in the same direction in which the wave travels. Sound waves are examples of longitudinal waves. By contrast, transverse waves (often termed T waves for short) are shear, rotational or equivoluminal waves. In seismology, they are known as secondary waves, or S waves. Particle vibration caused by transverse waves occurs in a direction perpendicular to the direction in which the wave travels.

In other words, longitudinal waves are waves that polarize and propagate in the same direction and transverse waves are those that polarize in a direction perpendicular to the direction of wave propagation. In reference to a three-dimensional (3D) Cartesian Coordinates system, we can imagine that along each of the three coordinates directions, three waves could possibly exist. For example, when acoustic waves propagate in the X direction, there could be a longitudinal wave (L wave) polarizing in the X direction, and two transverse waves, one polarizing in the Y direction (the YX-transverse wave) and the other in the Z direction (the ZX-transverse wave). The same can be said to waves propagating along the Y direction, the Z direction, or any arbitrary direction.

When the medium is bounded by a free surface, the surface can serves as a guiding boundary for acoustic waves to travel along. For this reason, surface waves are often considered as guided waves. There are a few different types of surface waves, among them, *Rayleigh waves*, *Love waves*, *Lamb waves* and *Bleustein–Gulyaev wave*. Rayleigh waves, discovered by Lord Rayleigh (1842–1919), are surface acoustic waves consisting of a transverse wave component vibrating in the direction normal to the surface and a longitudinal wave component vibrating in the direction in which the waves travel. So Rayleigh waves are elliptically polarized surface waves. Love waves, discovered by Augustus Edward Hough Love (1863–1940), are transverse waves traveling in a thin surface layer bonded to a thick support material of different properties with particle vibration occurring in a direction that is parallel to the surface but perpendicular to the direction in which the waves propagate. Lamb waves, discovered by Horace Lamb (1849–1934), are waves traveling along both the upper and lower surfaces of a plate structure. The connection between Rayleigh waves and Lamb waves is that in the former the medium is bounded by one free surface as the guiding boundary and in the latter the medium is bounded by two free surfaces on both the upper and lower sides to serve as the guiding boundaries. Bleustein–Gulyaev wave is a horizontal shear wave formed due to piezoelectric coupling. We will discuss all these waves in later chapters.

1.3 A Brief History of Wave Theory Development

In 1831 Siméon Denis Poisson (1781–1840) showed that a homogeneous isotropic elastic solid body of unlimited extent can transmit two kinds of waves with different velocities. Of the two waves, the quicker wave is the longitudinal wave and the slower one the transverse wave. In 1849, Sir George Stokes (1819–1903) published his proof that the quicker wave, traveling at the velocity of $\sqrt{(\lambda_L + 2\mu_L)/\rho}$, is a wave consisting of particles polarizing and vibrating in an irrotational dilation manner, and the slower wave, traveling at the velocity of $\sqrt{\mu_L/\rho}$, is a wave of equivoluminal distortion, where λ_L and μ_L are elastic constants (known today as the *Lamé constants*) of the medium, and ρ its density. The theory of the dilatational and distortional waves does not consider the existence of a boundary, which posed some difficulties when it was used to explain the seismic movements because all observations and recordings of seismic tremors were made at the surface of the earth.

In 1887 Lord Rayleigh showed that an irrotational dilation displacement (the longitudinal wave) and an equivoluminal rotation displacement (the transverse wave) can occur in a coupled form near a surface free from traction and that neither of two waves penetrates far beneath the surface. Such coupled acoustic waves that travel along a surface have since been called Rayleigh waves.

The year 1889 marked the beginning of systematic recording of the disturbances caused by earthquakes. Early records showed two distinct stages, namely, a preliminary tremor followed by a main shock. At the time, the two events were regarded as dilational and distortional waves emerging at the earth surface. In 1900, Richard Dixon Oldham (1858–1936) applied the concept of Rayleigh waves to explain the seismic records. He hypothesized that the first and second phases of the preliminary tremors were the results of dilational and distortional waves, respectively, and the main shock was the result of Rayleigh waves. The suggestion that the main shock was the result of Rayleigh waves was not well received because observation had shown that a large part of the motion transmitted in the main shock was a horizontal movement perpendicular to the direction of wave propagation.

In Lord Rayleigh's investigation, the waves were regarded as having already arrived from a distant source traveling along the surface. In 1903 and 1904, Lamb published his theory on surface wave propagation by considering an arbitrary disturbance such as impulsive pressure suddenly applied at a point of the surface of a homogeneous semi-infinite body. Lamb's theory helped explain conveniently some of the most prominent features of seismic records including the first and second phases of the preliminary tremors and the large disturbance of the main shock, and even the gradual decay of the movement, but it cannot account for the presence of horizontal transverse movement.

In a long essay published in 1911 in honoring his receiving the Adams' Prize of 1910, Love summarized many of his notable works, among them the theory of the propagation of seismic waves. In this work, Love showed that a different type of surface waves could exist in a horizontal plane with particle polarization occurring in a direction perpendicular to the direction in which the waves travel. This finding provided the missing linking in the understanding of the seismic records, particularly the existence of the horizontal movement perpendicular to the direction of wave propagation. Such transverse surface waves that travel along a surface with particle polarization occurring in a horizontal plane have since been called Love waves .

1.4 Uniqueness of Acoustic Wave Devices

In this book, we will focus our discussions on acoustic wave propagation in solid piezoelectric materials such as lithium niobate ($LiNbO_3$), lithium tantalate ($LiTaO_3$), zinc oxide (ZnO), cadmium sulphate (CdS), gallium arsenide (GaAs), langasite (LGS) and quartz (SiO_2), etc. For piezoelectric materials, as the Curie brothers demonstrated in 1880, when a mechanical strain is applied, atoms inside the material will be displaced from their original locations, resulting in polarization of dipoles and generation of an electric field. This effect is commonly known as the direct piezoelectric effect. Conversely, when a piezoelectric material becomes strained when placed in an electric field, it is known as the reverse piezoelectric effect. Since the discovery of the piezoelectric effect in 1880 the science of piezoelectricity has gradually been moved from research laboratories to real world applications. The sonar devices developed during and after World War I are first examples of its practical use. To date, piezoelectric devices are ubiquitous, including for the production and detection of sound (or ultrasound), mechanical actuation and excitation, electronics and communications, signal processing, various sensors (e.g., thickness and temperature), and even biosensors, among others.

Bulk acoustic wave or waves (BAW or BAWs) typically travel in solids with a phase velocity in a range on the scale of 10^3 – 10^4 m/s, which is significantly slower than electromagnetic waves. Take an average velocity of 3×10^3 m/s for an acoustic wave, and 3×10^8 m/s for an electromagnetic wave traveling at the speed of light, as an example, by the commonly known relationship between velocity (v), frequency (f) and wavelength (λ): $\lambda = v/f$, it is seen that realizing a predetermined operational frequency (e.g., 100 MHz) would require a wavelength of 3 m for the electromagnetic wave, but only 3×10^{-5} m, or 30 μm, for the acoustic wave, which is 10^5 times shorter than the former. This means that acoustic devices operating at a desirable frequency can be made on a very small piece of piezoelectric material. Surface acoustic wave or waves (SAW or SAWs) typically travels with a phase velocity that is less than that of the slowest bulk acoustic wave. However, the Love type of surface acoustic waves, particularly those traveling in a thin layer bounded to a thicker substrate, typically travels with a phase velocity that is faster than that of the transverse horizontal waves in the layer but slower than that in the substrate.

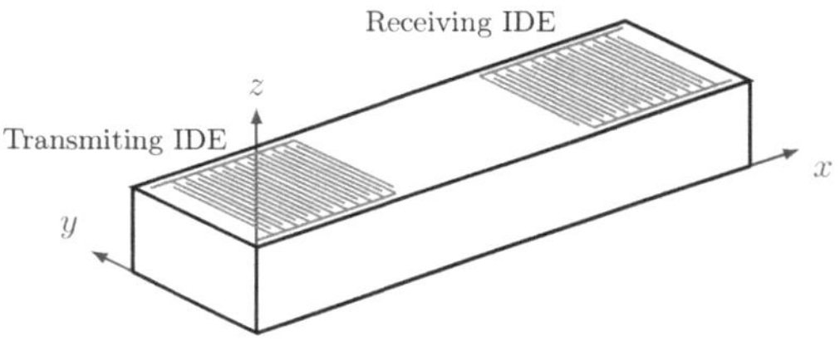

Figure 1.1 A two-port delay-line SAW filter device.

The various techniques created in the 1950s and 1960s for the generation and detection of both BAWs and SAWs have helped accelerate the use of acoustic devices in many applications. In the case of BAWs, quartz resonators were developed with high Q-values as a result of the low acoustic losses. For SAWs, the introduction of planar interdigitated electrodes (IDE) as transducers in the 1960s marked the beginning of its wide-spread applications. Figure 1.1 shows a typical two-port delay-line transversal SAW filter or sensor device with an equal number of finger pairs in both the transmitting and receiving IDEs. In this device, two IDE transducers are placed on the surface of a rectangle-shaped piezoelectric substrate, one to generate acoustic waves as a transmitter and the other to detect the waves as a receiver. The transmitting IDE at the left end of the device converts an input signal into mechanical strains, which in turn generate acoustic waves. The generated acoustic waves will travel along the surface of the piezoelectric material toward the receiving IDE and get converted back to electrical output signal via a reverse piezoelectric effect by the receiving IDE.

In a SAW filter application, the input and output signals can be used to assess the center frequency, bandwidth, cut-off frequencies, attenuation, and insertion loss of the device. As an example, we show here a transversal SAW filter made from a thin slab of 128° rotated Y cut $LiNbO_3$ crystal with wave propagation set in the x direction. Figure 1.2(a) shows the output potential measured at the receiving IDE in response to a bipolar (±0.5V) 1-nanosecond impulse input signal applied at the transmitter IDE and Figure 1.2(b) shows the obtained frequency response of the SAW filter.

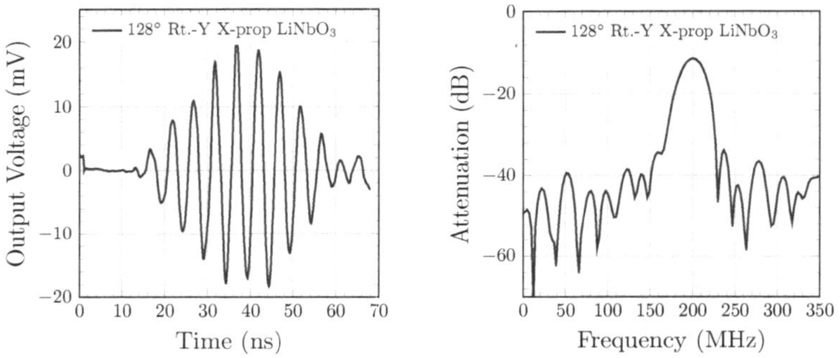

(a) Output potential (b) Frequency response

Figure 1.2 Output potential signal (a) and frequency response curve (b) of a two-port delay-line SAW filter made of 128° Rt.-Y X-propagating $LiNbO_3$ crystal cut.

In this case, the two IDEs are designed to have their finger width and gap spacing set at an equal width of 5 μm. This produces a wavelength of 20 μm for the generated SAW according to the formula: $\lambda = 2\times$(width + spacing). Knowing that the Rayleigh-type SAW travels along the surface of a 128° rotated Y-cut and X-propagating $LiNbO_3$ crystal at v = 3995 m/s, and using the relationship between the phase velocity (v), frequency (f) and wavelength (λ), $f = v/\lambda$, we estimate the center frequency for this SAW filter at 199.75 MHz. From the frequency response curve given in Figure 1.2(b), we can see that this filter is indeed operating at a center frequency of approximately 200 MHz.

In a biosensor application, the SAW waves, while traveling from the transmitting IDE to the receiving IDE, will perturb a sensing area located between the two IDEs. The sensing area is often coated with a molecular layer for target recognition, absorption or adsorption and a change in frequency responses due to molecular addition in the sensing area is quantified. Because of the confinement of the acoustic energy near the surface within a distance only a few wavelengths, SAW biosensors can be highly sensitive for discerning any surface perturbation such as molecular absorption or adsorption by the measurement of frequency shift and/or insertion loss.

1.5 An Overview of the Book

This book intends to provide the reader fundamental information and important underlying science on bulk and surface acoustic waves traveling in various types of piezoelectric crystal materials. Unlike other books on the subjects that almost universally rely heavily on the use of tensor notions and manipulations, this book aims to deliver the "scary" mathematic content in an easy-to-understand and easy-to-follow manner by taking advantage of Linear Algebra and matrix operations in theoretical derivation and development. It is believed that through the use of a software tool that can handle linear algebra, such as open-source program OCTAVE or commercial available program MATLAB, the reader can develop quickly a better understanding of the fundamentals of acoustic waves through hands-on exercises. As an end goal, it is hoped that readers will gain a mastery of acoustic theory and a deep understanding of the operating principles of acoustic devices to equip themselves to take advantage of acoustic technologies and further unleash their untapped potential.

To achieve the set goals, we will start in Chapter 2 with a discussion on free-body diagram, the state of equilibrium, equations of motion, constitutive relations for isotropic and anisotropic elastic and piezoelectric materials, and governing equations for acoustic waves in elastic and piezoelectric solids. In Chapter 3, we will learn coordinate rotation, definition of Euler angles, transformation of stress and strain tensor matrices and vectors, and transformation of matrices of piezoelectric material properties. In Chapter 4, we will discuss the 32 known classes of symmetry in piezoelectric crystals and show how a symmetric condition in a crystal dictates the characteristics of its material properties. In Chapter 5, we will dive in to learn the propagation of bulk acoustic waves in various elastic and piezoelectric solids, study the slowness curves, the piezoelectric coupling effect, as well as the energy velocity and Poynting angle of various bulk acoustic waves. In Chapter 6, we will investigate the propagation of surface acoustic waves in various elastic and piezoelectric solids, examine the relationship between the type of crystal symmetry and the decoupling of different SAWs, and determine the phase velocity of various SAWs traveling in different crystals, as well as the electromechanical coupling coefficient, energy velocity and Poynting angle of various surface acoustic waves. In Chapter 7, we will discuss the effect of temperature on the propagation of SAWs and learn not only the concepts but also how to evaluate the temperature coefficient of delay for various piezoelectric materials. In Chapter 8, we will introduce the underlying science for acoustic waves traveling in thin layers, such as Love waves and Lamb waves. Finally, to close the loop to help the reader bridge the connections from acoustic theory to acoustic devices, Chapter 9 will discuss in detail case studies of various BAW devices including disc resonators and tuning-fork resonator and SAW devices including filters, sensors and resonators. For these case studies, we will take advantage of finite element based computational modeling results to not only provide quantitative analysis of these acoustic devices with visualization of acoustic waves but also relate the observed device performances to the underlying mechanisms governing the propagation of acoustic waves in different anisotropic piezoelectric materials.

1.6 Study Problems

1. Name the various types of bulk waves discussed in this chapter.
2. How is an L wave different from a P wave?
3. How many different types of surface waves are discussed in this chapter?
4. Can you provide any proof to show why we regard that a distortional T wave travels slower than a dilational L wave?

5. Name one key advantage of a BAW or SAW device for high frequency applications.
6. In designing a two-port delay-line SAW filter, knowing that the SAW to be generated will travel at 3600 m/s, what wavelength should be selected in order for the filter to operate at 120 MHz as its center frequency? If two IDEs are of equal width and spacing, determine the finger width and spacing for the IDEs.
7. Do some literature search to find some key differences between Rayleigh waves and Lamb waves, and between Love waves and Bleustein–Gulyaev wave.

Chapter 2

Elasticity and Piezoelectricity

The theory of *Elasticity* deals with the elastic aspects of a material's mechanical properties and the theory of *Piezoelectricity* deals with the coupled mechanical and electrical properties of a material that becomes electrically polarized when it is mechanically strained. A good grasp of some fundamental knowledge about these theories will allow us to deal with complex stress–strain relationships in problems of higher dimensions and in materials of different anisotropy, and at the same time enable us to deal with relationships between the generated electrical field and applied strains, or vice versa. This chapter intends to help the reader to develop such knowledge.

2.1 Elastic and Piezoelectric Behavior of Materials

The term "elastic" describes the ability of a material to regain its original shape after the removal of an applied load. It should not be confused with the word "linear" although they are often used side by

Bulk and Surface Acoustic Waves: Fundamentals, Devices, and Applications
Guigen Zhang

ISBN 978-981-4877-73-2 (Hardcover), 978-1-003-25662-5 (eBook)
www.jennystanford.com

side. The word "linear" is a geometric term often used to describe a relationship between two variables such as in a load-displacement curve. When the relationship is a straight line, it is linear, and when curved it is nonlinear. So a load-displacement curve can be nonlinear even when the material deforms elastically. One such example is when you gently use fingers to pinch an aluminum can and then release it without causing any permanent shape change. In this case, the visible deformation in the can caused by pinching is nonlinear elastic. Another example is the deformation of a diving board when a diver makes a jump. It is therefore prudent to filter out this geometric influence when dealing with a material's elastic properties. This can be done by examining the stress–strain relationship, rather than the load-displacement relationship.

In the simplest form, the stress–strain relationship of an elastic material is often expressed as $\sigma = E\epsilon$, where σ represents stress, ϵ strain and E Young's modulus, or modulus of elasticity. While this equation captures well, conceptually, the stress–strain relationship governed by Hooke's law, it is valid only for situations where the axial stress and strain in a one-dimensional (1D) isotropic structure free of any lateral constraints are of concern. In the sections that follow, we will learn how the theory of elasticity can assist us to deal with complex stress–strain relationships in problems of higher dimensions and in materials of different anisotropy.

The term "piezoelectric," on the other hand, describes the coupled behavior linking a material's mechanical and electrical properties together. There are two forms of piezoelectric effect. *Direct piezoelectric effect* describes the generation of electric dipoles and polarization as a result of displaced atoms and the associated redistribution of electrons inside a solid when it is mechanically strained. This polarization in turn generates an electric field. By contrast, *reverse piezoelectric effect* is a phenomenon in which a solid becomes strained when placed in an electric field. Piezoelectric effect was first discovered and demonstrated in its direct form by the brothers Pierre and Jacques Curie in 1880. To date, piezoelectric effect has found its use in many applications, including the sonar technology for the production and detection of sound, mechanical actuation devices, thickness and temperature sensors, even biosensors, resonators, transmitters, receivers and amplifiers, among others.

2.2 Free Body Diagram and Equations of Motion

2.2.1 Free Body Diagram

In a three-dimensional (3D) Cartesian Coordinates system, to deal with vector-field solid mechanics problems we often use a free-body diagram, as shown in Figure 2.1, to examine its state of equilibrium. This free-body, in the shape of an infinitesimal cube with dimensions *dx*, *dy* and *dz* along the three axes, *x*, *y* and *z*, respectively, represents an arbitrarily small unit of a physical object of interest.

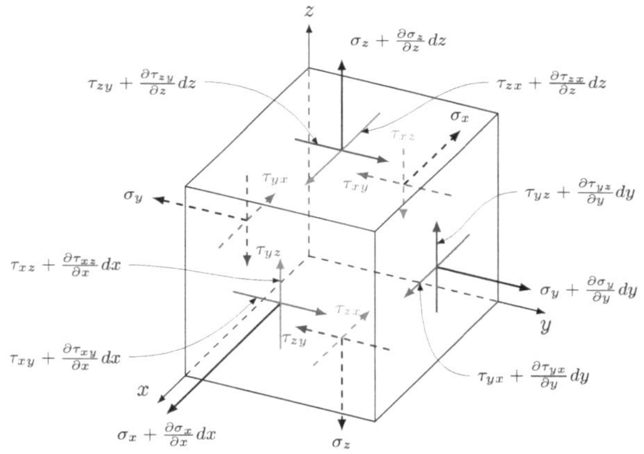

Figure 2.1 A 3D stress cube in Cartesian coordinates system.

Each surface of the cube experiences three stress components as marked in the Figure. To describe the stress state of the cube, we write the following three stress vectors in terms of three directional unit vectors $\vec{x}, \vec{y}$ and $\vec{z}$:

$$\begin{aligned} &\mathrm{x-section}: \sigma_x\vec{x} + \tau_{xy}\vec{y} + \tau_{xz}\vec{z} \\ &\mathrm{y-section}: \tau_{yx}\vec{x} + \sigma_y\vec{y} + \tau_{yz}\vec{z} \\ &\mathrm{z-section}: \tau_{zx}\vec{x} + \tau_{zy}\vec{y} + \sigma_z\vec{z} \end{aligned} \tag{2.1}$$

in which $\tau_{xy} = \tau_{yx}, \tau_{xz} = \tau_{zx}, \tau_{yz} = \tau_{zy}$ because shear stresses occur in pairs. This reflects the fact that of the nine stress components, only six are independent. Knowing these stress components exerting on this infinitesimal cube, we can calculate the net force in each of the three axes by taking the differential forces (differential stresses multiplied by the corresponding area) in the opposite surfaces as

$$\begin{aligned}
\sum F_x &: \left(\sigma_x + \tfrac{\partial \sigma_x}{\partial x} dx - \sigma_x\right) dydz + \left(\tau_{xy} + \tfrac{\partial \tau_{xy}}{\partial y} dy - \tau_{xy}\right) dxdz \\
&\quad + \left(\tau_{xz} + \tfrac{\partial \tau_{xz}}{\partial z} dz - \tau_{xz}\right) dxdy + f_x dxdydz \\
\sum F_y &: \left(\sigma_y + \tfrac{\partial \sigma_y}{\partial y} dy - \sigma_y\right) dxdz + \left(\tau_{xy} + \tfrac{\partial \tau_{xy}}{\partial x} dx - \tau_{xy}\right) dydz \\
&\quad + \left(\tau_{yz} + \tfrac{\partial \tau_{yz}}{\partial z} dz - \tau_{yz}\right) dxdy + f_y dxdydz \\
\sum F_z &: \left(\sigma_z + \tfrac{\partial \sigma_z}{\partial z} dz - \sigma_z\right) dxdy + \left(\tau_{xz} + \tfrac{\partial \tau_{xz}}{\partial x} dx - \tau_{xz}\right) dydz \\
&\quad + \left(\tau_{yz} + \tfrac{\partial \tau_{yz}}{\partial y} dy - \tau_{yz}\right) dydz + f_z dxdydz
\end{aligned}$$

where f_x, f_y, f_z are the three components of the specific volumetric force (i.e., force per unit volume) in a Cartesian coordinates system.

2.2.2 Equations of Motion

By considering Newton's second law of motion (i.e., $F = ma$), we express the following three partial differential equations of motion through free-body equilibrium after dividing both sides of the equations by the unit volume of the cube, *dxdydz*:

$$\begin{aligned}
\sum F_x = ma_x &: \frac{\partial \sigma_x}{\partial x} + \frac{\partial \tau_{xy}}{\partial y} + \frac{\partial \tau_{xz}}{\partial z} + f_x = \rho \frac{\partial^2 u_x}{\partial t^2} \\
\sum F_y = ma_y &: \frac{\partial \tau_{xy}}{\partial x} + \frac{\partial \sigma_y}{\partial y} + \frac{\partial \tau_{yz}}{\partial z} + f_y = \rho \frac{\partial^2 u_y}{\partial t^2} \\
\sum F_z = ma_z &: \frac{\partial \tau_{xz}}{\partial x} + \frac{\partial \tau_{yz}}{\partial y} + \frac{\partial \sigma_z}{\partial z} + f_z = \rho \frac{\partial^2 u_z}{\partial t^2}
\end{aligned} \tag{2.2}$$

where $\rho = m/dxdydz$ is the mass density of the material, and u_x, u_y and u_z are displacements in the x, y and z directions, respectively, and t is time. Since as dx, dy and dz become smaller and approach zero, the cube reduces to a point in space, thus these equations also capture the motion of particles.

2.2.3 The State of Stresses and Strains

The stress vectors given in Eq. (2.1) that depict the state of stresses of the infinitesimal cube or point shown in Figure 2.1 can be expressed in a 3×3 square and symmetric tensor matrix $[\mathbf{T}]$ as

$$[\mathbf{T}] = \begin{bmatrix} \sigma_x & \tau_{xy} & \tau_{xz} \\ \tau_{xy} & \sigma_y & \tau_{yz} \\ \tau_{xz} & \tau_{yz} & \sigma_z \end{bmatrix} \tag{2.3}$$

It contains the six independent stress components, in which x, y and z are indexes representing the three orthogonal directions of the Cartesian Coordinates system. Similarly, the corresponding state of strains is expressed using a 3×3 square and symmetric tensor matrix $[\mathbf{S}]$ containing six independent strain components (with a factor of 1/2 applied to the γ terms per the common convention, mainly for easy matrix operations) to represent the state of strains at the point of interest:

$$[\mathbf{S}] = \begin{bmatrix} \epsilon_x & \frac{\gamma_{xy}}{2} & \frac{\gamma_{xz}}{2} \\ \frac{\gamma_{xy}}{2} & \epsilon_y & \frac{\gamma_{yz}}{2} \\ \frac{\gamma_{xz}}{2} & \frac{\gamma_{yz}}{2} & \epsilon_z \end{bmatrix} \tag{2.4}$$

Example 1: Knowing the six stress components at a point in a physical object as: σ_x = 25 MPa, σ_y = 21 MPa, σ_z = 19 MPa, τ_{xy} = 15 MPa, τ_{xz} = 9 MPa and τ_{yz} = 11 MPa, express the stress tensor matrix for this point.

Answer: Using Eq. (2.3), we obtain the following 3×3 square and symmetric matrix to represent the stress state of the point of interest:

$$[\mathbf{T}] = \begin{bmatrix} 25 & 15 & 9 \\ 15 & 21 & 11 \\ 9 & 11 & 19 \end{bmatrix} \text{(MPa)}$$

Example 2: Comparing the equations of motion given in (2.2) with our familiar Newton's second law of motion, what similarity and difference do you see?

Answer: Newton's second law of motion states

$$F = ma$$

where F is force, m is mass and a is acceleration. In a comparison, we can clearly see a similarity in that equations in (2.2) also contain a force term on the left-hand side and an acceleration term on the right-hand side, albeit both are expressed per unit volume. One obvious difference is the inclusion of stresses in (2.2), accounting for the stress induced particle motion.

2.3 Elastic Constitutive Relations

2.3.1 Generalized Hooke's Law

One of the difficulties in vector-field solid mechanics is dealing with six stress or strain components in a 3D coordinates system. Although all six stress and strain components are conveniently expressed in 3×3 tensor matrices in Eqs. (2.3) and (2.4), the generalized Hooke's law calls for 6×1 column vectors for expressing stresses and strains

$$\{T\} = \left\{\sigma_x\ \sigma_y\ \sigma_z\ \tau_{yz}\ \tau_{xz}\ \tau_{xy}\right\}^T,$$
$$\{S\} = \left\{\epsilon_x\ \epsilon_y\ \epsilon_z\ \gamma_{yz}\ \gamma_{xz}\ \gamma_{xy}\right\}^T \tag{2.5}$$

in writing the elastic constitutive stress–strain relations:

$$\{T\} = \begin{Bmatrix} \sigma_x \\ \sigma_y \\ \sigma_z \\ \tau_{yz} \\ \tau_{xz} \\ \tau_{xy} \end{Bmatrix} = \begin{bmatrix} c_{11} & c_{12} & c_{13} & c_{14} & c_{15} & c_{16} \\ c_{12} & c_{22} & c_{23} & c_{24} & c_{25} & c_{26} \\ c_{13} & c_{23} & c_{33} & c_{34} & c_{35} & c_{36} \\ c_{14} & c_{24} & c_{34} & c_{44} & c_{45} & c_{46} \\ c_{15} & c_{25} & c_{35} & c_{45} & c_{55} & c_{56} \\ c_{16} & c_{26} & c_{36} & c_{46} & c_{56} & c_{66} \end{bmatrix} \begin{Bmatrix} \epsilon_x \\ \epsilon_y \\ \epsilon_z \\ \gamma_{yz} \\ \gamma_{xz} \\ \gamma_{xy} \end{Bmatrix} = [c]\{S\} \tag{2.6}$$

or

$$\{S\} = \begin{Bmatrix} \epsilon_x \\ \epsilon_y \\ \epsilon_z \\ \gamma_{yz} \\ \gamma_{xz} \\ \gamma_{xy} \end{Bmatrix} = \begin{bmatrix} s_{11} & s_{12} & s_{13} & s_{14} & s_{15} & s_{16} \\ s_{12} & s_{22} & s_{23} & s_{24} & s_{25} & s_{26} \\ s_{13} & s_{23} & s_{33} & s_{34} & s_{35} & s_{36} \\ s_{14} & s_{24} & s_{34} & s_{44} & s_{45} & s_{46} \\ s_{15} & s_{25} & s_{35} & s_{45} & s_{55} & s_{56} \\ s_{16} & s_{26} & s_{36} & s_{46} & s_{56} & s_{66} \end{bmatrix} \begin{Bmatrix} \sigma_x \\ \sigma_y \\ \sigma_z \\ \tau_{yz} \\ \tau_{xz} \\ \tau_{xy} \end{Bmatrix} = [s]\{T\} \tag{2.7}$$

Eqs. (2.6) and (2.7) are known as the elastic constitutive relations governed by generalized Hooke's law for solid materials in its stiffness and compliance form, respectively, in which

$$[c] = \begin{bmatrix} c_{11} & c_{12} & c_{13} & c_{14} & c_{15} & c_{16} \\ c_{12} & c_{22} & c_{23} & c_{24} & c_{25} & c_{26} \\ c_{13} & c_{23} & c_{33} & c_{34} & c_{35} & c_{36} \\ c_{14} & c_{24} & c_{34} & c_{44} & c_{45} & c_{46} \\ c_{15} & c_{25} & c_{35} & c_{45} & c_{55} & c_{56} \\ c_{16} & c_{26} & c_{36} & c_{46} & c_{56} & c_{66} \end{bmatrix}$$

and

$$[s] = \begin{bmatrix} s_{11} & s_{12} & s_{13} & s_{14} & s_{15} & s_{16} \\ s_{12} & s_{22} & s_{23} & s_{24} & s_{25} & s_{26} \\ s_{13} & s_{23} & s_{33} & s_{34} & s_{35} & s_{36} \\ s_{14} & s_{24} & s_{34} & s_{44} & s_{45} & s_{46} \\ s_{15} & s_{25} & s_{35} & s_{45} & s_{55} & s_{56} \\ s_{16} & s_{26} & s_{36} & s_{46} & s_{56} & s_{66} \end{bmatrix}$$

The 6×6 $[c]$ matrix is often called the stiffness matrix, and $[s]$ the compliance matrix of a solid material, in which c_{ij} and s_{ij} are property constants. Due to the symmetric nature of a stiffness or compliance matrix, both these 6×6 matrices have only 21 independent constants. From Eqs. (2.6) and (2.7) we can see that the stiffness matrix $[c]$ and compliance matrix $[s]$ are related in an inverse relationship as:

$$[c] = [s]^{-1} \quad \text{or} \quad [s] = [c]^{-1} \tag{2.8}$$

2.3.2 Isotropic Materials

When an elastic material possesses the same homogeneous properties along any arbitrary direction, it is known as an isotropic material. The stiffness matrix of an isotropic material is a much simplified matrix with only two independent constants

$$[c] = \begin{bmatrix} c_{11} & c_{12} & c_{12} & 0 & 0 & 0 \\ c_{12} & c_{11} & c_{12} & 0 & 0 & 0 \\ c_{12} & c_{12} & c_{11} & 0 & 0 & 0 \\ 0 & 0 & 0 & \frac{c_{11}-c_{12}}{2} & 0 & 0 \\ 0 & 0 & 0 & 0 & \frac{c_{11}-c_{12}}{2} & 0 \\ 0 & 0 & 0 & 0 & 0 & \frac{c_{11}-c_{12}}{2} \end{bmatrix} \tag{2.9}$$

It is sometimes expressed in terms of Lamé constants, λ_L and μ_L, so letting $c_{11} = \lambda_L + 2\mu_L$ and $c_{12} = \lambda_L$, we have

$$[c] = \begin{bmatrix} \lambda_L + 2\mu_L & \lambda_L & \lambda_L & 0 & 0 & 0 \\ \lambda_L & \lambda_L + 2\mu_L & \lambda_L & 0 & 0 & 0 \\ \lambda_L & \lambda_L & \lambda_L + 2\mu_L & 0 & 0 & 0 \\ 0 & 0 & 0 & \mu_L & 0 & 0 \\ 0 & 0 & 0 & 0 & \mu_L & 0 \\ 0 & 0 & 0 & 0 & 0 & \mu_L \end{bmatrix} \tag{2.10}$$

The two Lamé constants are related to our familiar terms, E, Young's modulus and, ν, Poisson's ratio, in the following expressions:

$$\lambda_L = \frac{E\nu}{(1+\nu)(1-2\nu)}, \quad \mu_L = \frac{E}{2(1+\nu)}$$

In other words, the $[c]$ matrix of an isotropic material can be written as

$$[c] = \frac{E}{(1+\nu)(1-2\nu)} \begin{bmatrix} 1-\nu & \nu & \nu & 0 & 0 & 0 \\ \nu & 1-\nu & \nu & 0 & 0 & 0 \\ \nu & \nu & 1-\nu & 0 & 0 & 0 \\ 0 & 0 & 0 & 0.5-\nu & 0 & 0 \\ 0 & 0 & 0 & 0 & 0.5-\nu & 0 \\ 0 & 0 & 0 & 0 & 0 & 0.5-\nu \end{bmatrix} \tag{2.11}$$

Per Eq. (2.8), by taking the inverse of $[c]$, we have the compliance matrix $[s]$ for the same isotropic material as

$$[s] = \frac{1}{E}\begin{bmatrix} 1 & -\nu & -\nu & 0 & 0 & 0 \\ -\nu & 1 & -\nu & 0 & 0 & 0 \\ -\nu & -\nu & 1 & 0 & 0 & 0 \\ 0 & 0 & 0 & 2(1+\nu) & 0 & 0 \\ 0 & 0 & 0 & 0 & 2(1+\nu) & 0 \\ 0 & 0 & 0 & 0 & 0 & 2(1+\nu) \end{bmatrix} \quad (2.12)$$

2.3.3 Anisotropic Materials

Materials that are not isotropic are often referred to as anisotropic materials. As the name indicates, the properties of anisotropic materials are direction dependent. Almost all classes of crystal materials are anisotropic, though some are piezoelectric and some are not. Depending on its class of symmetry, the $[c]$ matrix of a crystal material can have different number of nonzero independent constants, typically between 3 and 21. Below we list some anisotropic materials (see more detailed discussions in Chapter 3).

2.3.3.1 Triclinic Crystals

Triclinic crystals do not have any structural symmetry; hence, their $[c]$ matrix is a full 6×6 square matrix:

$$[c] = \begin{bmatrix} c_{11} & c_{12} & c_{13} & c_{14} & c_{15} & c_{16} \\ c_{12} & c_{22} & c_{23} & c_{24} & c_{25} & c_{26} \\ c_{13} & c_{23} & c_{33} & c_{34} & c_{35} & c_{36} \\ c_{14} & c_{24} & c_{34} & c_{44} & c_{45} & c_{46} \\ c_{15} & c_{25} & c_{35} & c_{45} & c_{55} & c_{56} \\ c_{16} & c_{26} & c_{36} & c_{46} & c_{56} & c_{66} \end{bmatrix}$$

While these crystals lack a structural symmetry, their stiffness matrix is always symmetric about the principal diagonal. This means that this stiffness matrix has only 21 independent constants.

2.3.3.2 Monoclinic Crystals

Monoclinic crystals typically have either a two-fold (a rotation about an axis by 180°) or a mirror symmetry (mirroring about a plane perpendicular to an axis), so their $[c]$ matrix has 13 independent constants:

$$[c] = \begin{bmatrix} c_{11} & c_{12} & c_{13} & 0 & c_{15} & 0 \\ c_{12} & c_{22} & c_{23} & 0 & c_{25} & 0 \\ c_{13} & c_{23} & c_{33} & 0 & c_{35} & 0 \\ 0 & 0 & 0 & c_{44} & 0 & c_{46} \\ c_{15} & c_{25} & c_{35} & 0 & c_{55} & 0 \\ 0 & 0 & 0 & c_{46} & 0 & c_{66} \end{bmatrix}$$

2.3.3.3 Orthorhombic Crystals

Orthorhombic crystals have combined two-fold and mirror symmetry, so their $[c]$ matrix has 9 independent constants:

$$[c] = \begin{bmatrix} c_{11} & c_{12} & c_{13} & 0 & 0 & 0 \\ c_{12} & c_{22} & c_{23} & 0 & 0 & 0 \\ c_{13} & c_{23} & c_{33} & 0 & 0 & 0 \\ 0 & 0 & 0 & c_{44} & 0 & 0 \\ 0 & 0 & 0 & 0 & c_{55} & 0 \\ 0 & 0 & 0 & 0 & 0 & c_{66} \end{bmatrix}$$

Example 3: Knowing the stiffness matrix for orthorhombic crystal barium sodium niobate ($Ba_2NaNb_5O_{15}$) given below, use the relationship between $[c]$ and $[s]$ to find its compliance matrix.

$$[c] = 10^{10} \times \begin{bmatrix} 23.9 & 10.4 & 5.0 & 0 & 0 & 0 \\ 10.4 & 24.7 & 5.2 & 0 & 0 & 0 \\ 5.0 & 5.2 & 13.5 & 0 & 0 & 0 \\ 0 & 0 & 0 & 6.5 & 0 & 0 \\ 0 & 0 & 0 & 0 & 6.6 & 0 \\ 0 & 0 & 0 & 0 & 0 & 7.6 \end{bmatrix}$$

Answer: By the inverse relationship given in Eq. (2.8) we take the inverse of the $[c]$ matrix by using MATLAB or any software packages that can handle linear algebra, and obtain

$$[s] = 10^{-12} \times \begin{bmatrix} 5.30 & -1.98 & -1.20 & 0 & 0 & 0 \\ -1.98 & 5.14 & -1.25 & 0 & 0 & 0 \\ -1.20 & -1.25 & 8.33 & 0 & 0 & 0 \\ 0 & 0 & 0 & 15.38 & 0 & 0 \\ 0 & 0 & 0 & 0 & 15.15 & 0 \\ 0 & 0 & 0 & 0 & 0 & 13.16 \end{bmatrix}$$

2.3.3.4 Tetragonal Crystals

Tetragonal crystals have four-fold symmetry (a rotation about an axis by 90°). When such a crystal lacks a two-fold symmetry, its $[c]$ matrix has 7 independent constants:

$$[c] = \begin{bmatrix} c_{11} & c_{12} & c_{13} & 0 & 0 & c_{16} \\ c_{12} & c_{11} & c_{13} & 0 & 0 & -c_{16} \\ c_{13} & c_{13} & c_{33} & 0 & 0 & 0 \\ 0 & 0 & 0 & c_{44} & 0 & 0 \\ 0 & 0 & 0 & 0 & c_{44} & 0 \\ c_{16} & -c_{16} & 0 & 0 & 0 & c_{66} \end{bmatrix}$$

When a tetragonal crystal also has a two-fold symmetry, or a mirror symmetry which is equivalent to a two-fold symmetry, its $[c]$ matrix reduces to one with only 6 independent constants:

$$[c] = \begin{bmatrix} c_{11} & c_{12} & c_{13} & 0 & 0 & 0 \\ c_{12} & c_{11} & c_{13} & 0 & 0 & 0 \\ c_{13} & c_{13} & c_{33} & 0 & 0 & 0 \\ 0 & 0 & 0 & c_{44} & 0 & 0 \\ 0 & 0 & 0 & 0 & c_{44} & 0 \\ 0 & 0 & 0 & 0 & 0 & c_{66} \end{bmatrix}$$

Example 4: Knowing the stiffness matrix for tetragonal crystal lead molybdate ($PbMoO_4$), find its compliance matrix.

$$[c] = 10^{10} \times \begin{bmatrix} 10.92 & 6.83 & 5.8 & 0 & 0 & 1.36 \\ 6.83 & 10.92 & 5.8 & 0 & 0 & -1.36 \\ 5.8 & 5.8 & 9.17 & 0 & 0 & 0 \\ 0 & 0 & 0 & 2.67 & 0 & 0 \\ 0 & 0 & 0 & 0 & 2.67 & 0 \\ 1.36 & -1.36 & 0 & 0 & 0 & 3.37 \end{bmatrix}$$

Answer: By taking the inverse of the $[c]$ matrix we find

$$[s] = 10^{-12} \times \begin{bmatrix} 21.51 & -11.91 & -6.07 & 0 & 0 & -13.49 \\ -11.91 & 21.51 & -6.07 & 0 & 0 & 13.49 \\ -6.07 & -6.07 & 18.59 & 0 & 0 & 0 \\ 0 & 0 & 0 & 37.45 & 0 & 0 \\ 0 & 0 & 0 & 0 & 37.45 & 0 \\ -13.49 & 13.49 & 0 & 0 & 0 & 40.56 \end{bmatrix}$$

Example 5: Knowing the stiffness matrix for tetragonal crystal tellurium oxide (TeO_2), find its compliance matrix.

$$[c] = 10^{10} \times \begin{bmatrix} 5.6 & 5.1 & 2.2 & 0 & 0 & 0 \\ 5.1 & 5.6 & 2.2 & 0 & 0 & 0 \\ 2.2 & 2.2 & 10.6 & 0 & 0 & 0 \\ 0 & 0 & 0 & 2.65 & 0 & 0 \\ 0 & 0 & 0 & 0 & 2.65 & 0 \\ 0 & 0 & 0 & 0 & 0 & 6.6 \end{bmatrix}$$

Answer: By taking the inverse of the $[c]$ matrix we find

$$[s] = 10^{-12} \times \begin{bmatrix} 105.1 & -94.9 & -2.1 & 0 & 0 & 0 \\ -94.9 & 105.1 & -2.1 & 0 & 0 & 0 \\ -2.1 & -2.1 & 10.3 & 0 & 0 & 0 \\ 0 & 0 & 0 & 37.7 & 0 & 0 \\ 0 & 0 & 0 & 0 & 37.7 & 0 \\ 0 & 0 & 0 & 0 & 0 & 15.2 \end{bmatrix}$$

2.3.3.5 Trigonal Crystals

Trigonal crystals have three-fold symmetry (a rotation about an axis by 120°). When a trigonal crystal lacks other types of symmetry, its $[c]$ matrix has 7 non-zero constants:

$$[c] = \begin{bmatrix} c_{11} & c_{12} & c_{13} & c_{14} & -c_{25} & 0 \\ c_{12} & c_{11} & c_{13} & -c_{14} & c_{25} & 0 \\ c_{13} & c_{13} & c_{33} & 0 & 0 & 0 \\ c_{14} & -c_{14} & 0 & c_{44} & 0 & c_{25} \\ -c_{25} & c_{25} & 0 & 0 & c_{44} & c_{14} \\ 0 & 0 & 0 & c_{25} & c_{14} & \frac{c_{11}-c_{12}}{2} \end{bmatrix}$$

When a trigonal crystal also has a two-fold or mirror symmetry, its $[c]$ matrix reduces to one with 6 independent constants:

$$[c] = \begin{bmatrix} c_{11} & c_{12} & c_{13} & c_{14} & 0 & 0 \\ c_{12} & c_{11} & c_{13} & -c_{14} & 0 & 0 \\ c_{13} & c_{13} & c_{33} & 0 & 0 & 0 \\ c_{14} & -c_{14} & 0 & c_{44} & 0 & 0 \\ 0 & 0 & 0 & 0 & c_{44} & c_{14} \\ 0 & 0 & 0 & 0 & c_{14} & \frac{c_{11}-c_{12}}{2} \end{bmatrix}$$

Example 6: Knowing the stiffness matrix for trigonal crystal alumina (Al_2O_3), find its compliance matrix.

$$[c] = 10^{10} \times \begin{bmatrix} 49.7 & 16.3 & 11.1 & -2.35 & 0 & 0 \\ 16.3 & 49.7 & 11.1 & 2.35 & 0 & 0 \\ 11.1 & 11.1 & 49.8 & 0 & 0 & 0 \\ -2.35 & 2.35 & 0 & 14.7 & 0 & 0 \\ 0 & 0 & 0 & 0 & 14.7 & -2.35 \\ 0 & 0 & 0 & 0 & -2.35 & 16.7 \end{bmatrix}$$

Answer: By taking the inverse of the $[c]$ matrix we find

$$[s] = 10^{-12} \times \begin{bmatrix} 2.350 & -0.712 & -0.365 & 0.490 & 0 & 0 \\ -0.712 & 2.350 & -0.365 & -0.490 & 0 & 0 \\ -0.365 & -0.365 & 2.171 & 0 & 0 & 0 \\ 0.490 & -0.490 & 0 & 6.959 & 0 & 0 \\ 0 & 0 & 0 & 0 & 6.959 & 0.979 \\ 0 & 0 & 0 & 0 & 0.979 & 6.126 \end{bmatrix}$$

2.3.3.6 Hexagonal Crystals

Hexagonal crystals have six-fold symmetry (a rotation about an axis by 60°), and their $[c]$ matrix has 5 independent constants:

$$[c] = \begin{bmatrix} c_{11} & c_{12} & c_{13} & 0 & 0 & 0 \\ c_{12} & c_{11} & c_{13} & 0 & 0 & 0 \\ c_{13} & c_{13} & c_{33} & 0 & 0 & 0 \\ 0 & 0 & 0 & c_{44} & 0 & 0 \\ 0 & 0 & 0 & 0 & c_{44} & 0 \\ 0 & 0 & 0 & 0 & 0 & \frac{c_{11}-c_{12}}{2} \end{bmatrix}$$

Example 7: Knowing the stiffness matrix for hexagonal crystal zinc oxide (ZnO), find its compliance matrix.

$$[c] = 10^{10} \times \begin{bmatrix} 20.9 & 12.11 & 10.51 & 0 & 0 & 0 \\ 12.11 & 20.9 & 10.51 & 0 & 0 & 0 \\ 10.51 & 10.51 & 21.09 & 0 & 0 & 0 \\ 0 & 0 & 0 & 4.25 & 0 & 0 \\ 0 & 0 & 0 & 0 & 4.25 & 0 \\ 0 & 0 & 0 & 0 & 0 & 4.4 \end{bmatrix}$$

Answer: By taking the inverse of the $[c]$ matrix we find

$$[s] = 10^{-12} \times \begin{bmatrix} 7.91 & -3.47 & -2.21 & 0 & 0 & 0 \\ -3.47 & 7.91 & -2.21 & 0 & 0 & 0 \\ -2.21 & -2.21 & 6.95 & 0 & 0 & 0 \\ 0 & 0 & 0 & 23.53 & 0 & 0 \\ 0 & 0 & 0 & 0 & 23.53 & 0 \\ 0 & 0 & 0 & 0 & 0 & 22.73 \end{bmatrix}$$

2.3.3.7 Cubic Crystals

Cubic crystals have multiple symmetry, sometimes with a two-fold symmetry and a rotational symmetry (e.g., symmetric after two consecutive rotations, the first about the Z axis by 90° followed by the second about the rotated x axis by 90°). The $[c]$ matrix of cubic crystals has only 3 independent constants:

$$[c] = \begin{bmatrix} c_{11} & c_{12} & c_{12} & 0 & 0 & 0 \\ c_{12} & c_{11} & c_{12} & 0 & 0 & 0 \\ c_{12} & c_{12} & c_{11} & 0 & 0 & 0 \\ 0 & 0 & 0 & c_{44} & 0 & 0 \\ 0 & 0 & 0 & 0 & c_{44} & 0 \\ 0 & 0 & 0 & 0 & 0 & c_{44} \end{bmatrix}$$

Example 8: Knowing the stiffness matrix for cubic crystal silicon (Si), find its compliance matrix.

$$[c] = 10^{10} \times \begin{bmatrix} 16.56 & 6.39 & 6.39 & 0 & 0 & 0 \\ 6.39 & 16.56 & 6.39 & 0 & 0 & 0 \\ 6.39 & 6.39 & 16.56 & 0 & 0 & 0 \\ 0 & 0 & 0 & 7.59 & 0 & 0 \\ 0 & 0 & 0 & 0 & 7.59 & 0 \\ 0 & 0 & 0 & 0 & 0 & 7.59 \end{bmatrix}$$

Answer: By taking the inverse of the $[c]$ matrix we find

$$[s] = 10^{-12} \times \begin{bmatrix} 7.69 & -2.14 & -2.14 & 0 & 0 & 0 \\ -2.14 & 7.69 & -2.14 & 0 & 0 & 0 \\ -2.14 & -2.14 & 7.69 & 0 & 0 & 0 \\ 0 & 0 & 0 & 13.18 & 0 & 0 \\ 0 & 0 & 0 & 0 & 13.18 & 0 \\ 0 & 0 & 0 & 0 & 0 & 13.18 \end{bmatrix}$$

Note that from Examples 3 through 8, a common observation can be made in that all the non-zero terms in both the stiffness and compliance matrices occupy the same locations in their respective matrices. This reflects the fact that examining a solid material problem either by considering the stiffness matrix properties or compliance matrix properties will not affect the isotropy of the material of interest.

As a closure to this material's isotropy discussion, let us return to isotropic solids. Seeing the stiffness matrices for various types of anisotropic materials, it is not difficult to see why the stiffness matrix of an isotropic material has only two independent constants. Since the term isotropic also means that symmetry in material properties exists in any directions, if we include, arbitrarily, either a trigonal or hexagonal type of symmetry to a cubic type of symmetry, the stiffness matrix should remain unchanged. This results in the following relationship:

$$c_{44} = \frac{c_{11} - c_{12}}{2}$$

which leads us to Eq. (2.9) for the stiffness matrix for an isotropic material.

2.4 Simplification of 3D Problems

In daily life we always encounter three-dimensional (3D) objects, yet we often use the terms two-dimensional (2D) and one-dimensional (1D) problems. Technically speaking, all problems are 3D in a geometry sense, but in certain circumstances they can be simplified through approximation into 2D and 1D problems. Here we provide a close look from elastic constitutive relations on how such simplification is done from 3D problems to 2D and 1D problems for isotropic solid materials.

2.4.1 2D Plane-Stress Situation

Because of the approximation nature, the conditions for making the simplification should be considered carefully. For example, when a material is thin in its third dimension (say, along the *z* axis) compared with the other two dimensions, and if the loading and deformations in the material occur only within the thin plane, the stresses in the third dimension, as illustrated in Figure 2.2, σ_z, τ_{yz}, τ_{xz}, are negligible.

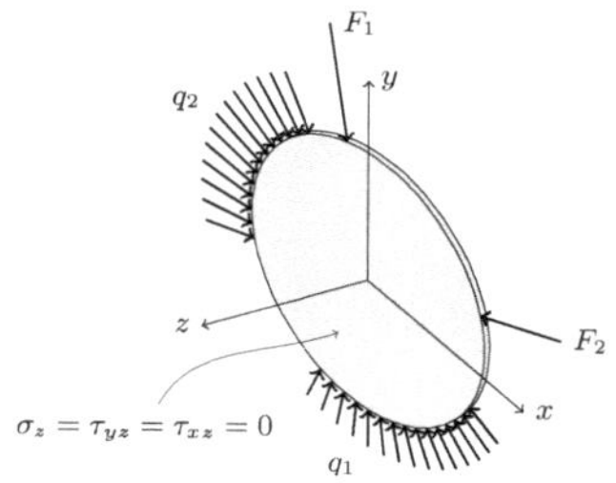

Figure 2.2 2D simplification in a plane-stress situation.

As illustrated in Figure 2.2, all loads and deformations are considered to occur only in the x-y plane, and the two circular surfaces are stress-free. Thus we can treat all the out-of-plane stress components, namely, σ_z, τ_{yz}, τ_{xz}, as zero:

$$\sigma_z = \tau_{yz} = \tau_{xz} = 0$$

For this reason, we regard this situation as 2D plane-stress situation. By plugging these zero terms into Eq. (2.7) and using Eq. (2.12), we obtain

$$\begin{Bmatrix} \epsilon_x \\ \epsilon_y \\ \epsilon_z \\ \gamma_{yz} \\ \gamma_{xz} \\ \gamma_{xy} \end{Bmatrix} = \frac{1}{E} \begin{bmatrix} 1 & -\nu & -\nu & 0 & 0 & 0 \\ -\nu & 1 & -\nu & 0 & 0 & 0 \\ -\nu & -\nu & 1 & 0 & 0 & 0 \\ 0 & 0 & 0 & 2(1+\nu) & 0 & 0 \\ 0 & 0 & 0 & 0 & 2(1+\nu) & 0 \\ 0 & 0 & 0 & 0 & 0 & 2(1+\nu) \end{bmatrix} \begin{Bmatrix} \sigma_x \\ \sigma_y \\ 0 \\ 0 \\ 0 \\ \tau_{xy} \end{Bmatrix}$$

After striking out the rows and columns associated with the zero stress terms, we obtain the reduced strain-stress relationship for a 2D isotropic material in a plane-stress situation

$$\begin{Bmatrix} \epsilon_x \\ \epsilon_y \\ \gamma_{xy} \end{Bmatrix} = \frac{1}{E} \begin{bmatrix} 1 & -\nu & 0 \\ -\nu & 1 & 0 \\ 0 & 0 & 2(1+\nu) \end{bmatrix} \begin{Bmatrix} \sigma_x \\ \sigma_y \\ \tau_{xy} \end{Bmatrix}$$

After taking the inverse of the above equation we have

$$\begin{Bmatrix} \sigma_x \\ \sigma_y \\ \tau_{xy} \end{Bmatrix} = \frac{E}{(1-\nu^2)} \begin{bmatrix} 1 & \nu & 0 \\ \nu & 1 & 0 \\ 0 & 0 & (1-\nu)/2 \end{bmatrix} \begin{Bmatrix} \epsilon_x \\ \epsilon_y \\ \gamma_{xy} \end{Bmatrix}$$

This gives rise to

$$[c] = \frac{E}{(1-\nu^2)} \begin{bmatrix} 1 & \nu & 0 \\ \nu & 1 & 0 \\ 0 & 0 & (1-\nu)/2 \end{bmatrix}$$

While the three stress components associated with the z axis are zero, the corresponding strain components are not necessarily zero, and they can be found by matrix multiplication, as:

$$\epsilon_z = -\nu(\sigma_x + \sigma_y)/E, \quad \gamma_{yz} = \gamma_{xz} = 0$$

Apparently, only the shear strains are zero and the normal strain is not.

2.4.2 2D Plane-Strain Situation

Another situation is when the material is very long in its longitudinal direction (the z direction here) as compared with the lateral dimension (in the xy plane here) and the loading, deformations and the cross section area remain unchanged over the entire longitudinal length, as depicted in Figure 2.3. Since strains are defined as changes in length per unit length, the relatively long length in the z direction will make all the strains associated with the z axis, namely, $\epsilon_z, \gamma_{yz}, \gamma_{xz}$, negligibly small. In this situation, we can simplify the 3D problem into a 2D plane-strain problem.

By taking the same steps as in the plane-stress situation with substitution of the three zero strain terms depicted in Figure 2.3, we

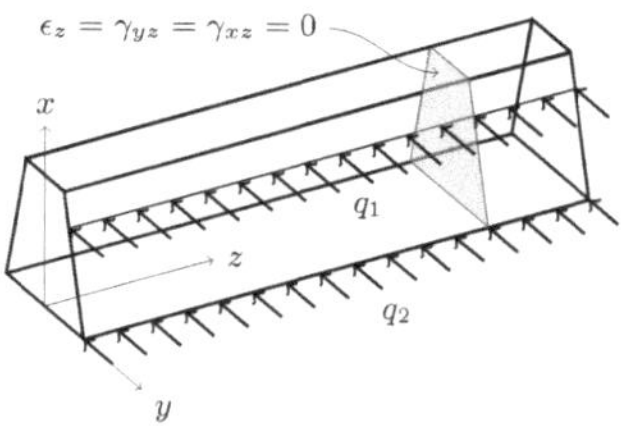

Figure 2.3 2D simplification in a plane-strain situation.

obtain

$$\begin{Bmatrix} \sigma_x \\ \sigma_y \\ \sigma_z \\ \tau_{yz} \\ \tau_{xz} \\ \tau_{xy} \end{Bmatrix} = \frac{E}{(1+\nu)(1-2\nu)} \times \begin{bmatrix} 1-\nu & \nu & \nu & 0 & 0 & 0 \\ \nu & 1-\nu & \nu & 0 & 0 & 0 \\ \nu & \nu & 1-\nu & 0 & 0 & 0 \\ 0 & 0 & 0 & 0.5-\nu & 0 & 0 \\ 0 & 0 & 0 & 0 & 0.5-\nu & 0 \\ 0 & 0 & 0 & 0 & 0 & 0.5-\nu \end{bmatrix} \begin{Bmatrix} \epsilon_x \\ \epsilon_y \\ 0 \\ 0 \\ 0 \\ \gamma_{xy} \end{Bmatrix}$$

After striking out the rows and columns associated with the zero strain terms, we obtain the reduced strain-stress relationship for a 2D isotropic material in a plane-strain situation as follows:

$$\begin{Bmatrix} \sigma_x \\ \sigma_y \\ \tau_{xy} \end{Bmatrix} = \frac{E}{(1+\nu)(1-2\nu)} \begin{bmatrix} 1-\nu & \nu & 0 \\ \nu & 1-\nu & 0 \\ 0 & 0 & 0.5-\nu \end{bmatrix} \begin{Bmatrix} \epsilon_x \\ \epsilon_y \\ \gamma_{xy} \end{Bmatrix}$$

which leads to

$$[c] = \frac{E}{(1+\nu)(1-2\nu)} \begin{bmatrix} 1-\nu & \nu & 0 \\ \nu & 1-\nu & 0 \\ 0 & 0 & 0.5-\nu \end{bmatrix}$$

Taking the inverse of the stress–strain equation, we have

$$\begin{Bmatrix} \epsilon_x \\ \epsilon_y \\ \gamma_{xy} \end{Bmatrix} = \frac{(1+\nu)}{E} \begin{bmatrix} 1-\nu & -\nu & 0 \\ -\nu & 1-\nu & 0 \\ 0 & 0 & 2 \end{bmatrix} \begin{Bmatrix} \sigma_x \\ \sigma_y \\ \tau_{xy} \end{Bmatrix}$$

Similarly, the stress components corresponding to the three zero strains in a 2D plane-stain approximation situation are

$$\sigma_z = E\nu(\epsilon_x + \epsilon_y)/[(1+\nu)(1-2\nu), \quad \tau_{yz} = \tau_{xz} = 0$$

Example 9: For an isotropic material with E = 210 GPa and $\nu =$ 0.33, find the $[c]$ and $[s]$ matrices for 2D simplified plane-stress and plane-strain situations, respectively. What will happen if someone mistakenly treats a plane-stress problem as a plane-strain problem?

Answer: Plane-stress situation: By plugging the given E and ν values into the $[c]$ matrix derived for a 2D plane-stress situation, we have

$$[c] = 10^{11} \times \begin{bmatrix} 2.360 & 0.778 & 0 \\ 0.778 & 2.360 & 0 \\ 0 & 0 & 0.789 \end{bmatrix}$$

Taking its inverse, we find

$$[s] = 10^{-11} \times \begin{bmatrix} 0.475 & -0.157 & 0 \\ -0.157 & 0.4750 & 0 \\ 0 & 0 & 1.267 \end{bmatrix}$$

Plane-strain situation: By doing the same for the $[c]$ matrix derived for a 2D plane-strain situation, we obtain

$$[c] = 10^{11} \times \begin{bmatrix} 3.110 & 1.530 & 0 \\ 1.530 & 3.110 & 0 \\ 0 & 0 & 0.789 \end{bmatrix}$$

and taking its inverse we find

$$[s] = 10^{-11} \times \begin{bmatrix} 0.424 & -0.209 & 0 \\ -0.209 & 0.4750 & 0 \\ 0 & 0 & 1.267 \end{bmatrix}$$

In a close examination of the two $[c]$ matrices, it is apparent that the $[c]$ matrix for the plane-strain situation is stiffened as compared with that for the plane-stress situation. Aside from the shear term at the lower right corner, two distinct terms of the four listed in the first two rows are 3.110×10^{11} and 1.530×10^{11} for the plane-strain situation, which are approximately 1.32 and 1.97 times higher than the corresponding terms in the plane-stress situation (i.e., 2.360×10^{11} and 0.778×10^{11}, respectively). This suggests that treating a plane-stress problem by mistake as a plane-strain problem would lead to stiffening the material significantly. The danger of such a practice is over-estimating the capability of the structure under design and analysis.

2.4.3 1D Situation

The 1D simplified constitutive relation can be derived from the 2D plane-stress situation because a typical 1D situation is often free of lateral constraints. Thus we can assume $\sigma_y = \tau_{xy} = 0$ additionally, thus write.

$$\begin{Bmatrix} \epsilon_x \\ \epsilon_y \\ \gamma_{xy} \end{Bmatrix} = \frac{1}{E} \begin{bmatrix} 1 & -\nu & 0 \\ -\nu & 1 & 0 \\ 0 & 0 & 2(1+\nu) \end{bmatrix} \begin{Bmatrix} \sigma_x \\ 0 \\ 0 \end{Bmatrix}$$

which leads to

$$\sigma_x = E\epsilon_x \text{ with } \epsilon_y = -\nu\sigma_x/E \text{ and } \gamma_{xy} = 0$$

If the 1D material is constrained laterally, we may derive its 1D constitutive relation from a plane-strain situation.

2.5 Governing Equation for Acoustic Waves in Elastic Solids

In this book we mainly concern ourselves to acoustic waves generated by mechanical displacements of atoms in a solid material. In Sections 2.2 and 2.3 we have developed the equations of motion for particles in a solid medium and the constitutive equations relating the induced stresses to strains. We will now utilize these equations to put together a set of governing equations to predict the movement and propagation of acoustic waves.

To do so, we first use the definition of strains to express the six components of strains in terms of displacements. In a matrix form, these relationships can be written as

$$\{S\} = \begin{Bmatrix} \epsilon_x \\ \epsilon_y \\ \epsilon_z \\ \gamma_{yz} \\ \gamma_{xz} \\ \gamma_{xy} \end{Bmatrix} = \begin{Bmatrix} \frac{\partial u_x}{\partial x} \\ \frac{\partial u_y}{\partial y} \\ \frac{\partial u_z}{\partial z} \\ \frac{\partial u_y}{\partial z} + \frac{\partial u_z}{\partial y} \\ \frac{\partial u_x}{\partial z} + \frac{\partial u_z}{\partial x} \\ \frac{\partial u_x}{\partial y} + \frac{\partial u_y}{\partial x} \end{Bmatrix}$$

where u_x, u_y and u_z are displacement components in the x, y and z directions, respectively.

By introducing two matrix-form differential operators, $[\nabla]$ and $[\nabla_s]$, as

$$[\nabla] = \begin{bmatrix} \frac{\partial}{\partial x} & \frac{\partial}{\partial y} & \frac{\partial}{\partial z} \end{bmatrix}, \quad \text{or} \quad [\nabla]^T = \begin{Bmatrix} \frac{\partial}{\partial x} \\ \frac{\partial}{\partial y} \\ \frac{\partial}{\partial z} \end{Bmatrix} \tag{2.13}$$

and

$$[\nabla_s] = \begin{bmatrix} \frac{\partial}{\partial x} & 0 & 0 & 0 & \frac{\partial}{\partial z} & \frac{\partial}{\partial y} \\ 0 & \frac{\partial}{\partial y} & 0 & \frac{\partial}{\partial z} & 0 & \frac{\partial}{\partial x} \\ 0 & 0 & \frac{\partial}{\partial z} & \frac{\partial}{\partial y} & \frac{\partial}{\partial x} & 0 \end{bmatrix}, \text{ or } [\nabla_s]^T = \begin{bmatrix} \frac{\partial}{\partial x} & 0 & 0 \\ 0 & \frac{\partial}{\partial y} & 0 \\ 0 & 0 & \frac{\partial}{\partial z} \\ 0 & \frac{\partial}{\partial z} & \frac{\partial}{\partial y} \\ \frac{\partial}{\partial z} & 0 & \frac{\partial}{\partial x} \\ \frac{\partial}{\partial y} & \frac{\partial}{\partial x} & 0 \end{bmatrix} \tag{2.14}$$

along with expressing the displacement and force vectors in a matrix form

$$\{u\} = \begin{Bmatrix} u_x \\ u_y \\ u_z \end{Bmatrix}, \quad \{f\} = \begin{Bmatrix} f_x \\ f_y \\ f_z \end{Bmatrix}$$

we obtain the following relationships by plugging these expressions into Eq. (2.5):

$$\{S\} = [\nabla_s]^T \{u\} \tag{2.15}$$

and

$$\begin{Bmatrix} \frac{\partial \sigma_x}{\partial x} + \frac{\partial \tau_{xy}}{\partial y} + \frac{\partial \tau_{xz}}{\partial z} \\ \frac{\partial \tau_{xy}}{\partial x} + \frac{\partial \sigma_y}{\partial y} + \frac{\partial \tau_{yz}}{\partial z} \\ \frac{\partial \tau_{xz}}{\partial x} + \frac{\partial \tau_{yz}}{\partial y} + \frac{\partial \sigma_z}{\partial z} \end{Bmatrix} = \nabla_s \begin{Bmatrix} \sigma_x \\ \sigma_y \\ \sigma_z \\ \tau_{yz} \\ \tau_{xz} \\ \tau_{xy} \end{Bmatrix} \tag{2.16}$$

Using these expressions, we rewrite the equations of motion given in (2.2) in a matrix form as follows:

$$[\nabla_s]\{T\} + \{f\} = \rho \frac{\partial^2}{\partial t^2}\{u\} \tag{2.17}$$

Putting them together along with the constitutive equation, $\{T\} = [c]\{S\}$, we obtain the following explicit governing equation linking the acoustics displacements, $\{u\}$, to the mechanical loading condition, $\{f\}$, and materials properties, $[c]$ and ρ:

$$[\nabla_s][c][\nabla_s]^T\{u\} - \rho\frac{\partial^2}{\partial t^2}\{u\} + \{f\} = 0 \tag{2.18}$$

Here we regard this equation as the governing equation for acoustic waves in elastic solids.

Example 10: For a piece of $LiTaO_3$ crystal with ρ = 7450 kg/m^3 and stiffness matrix $[c]$ given below, assuming $\{f\} = 0$, apply Eq. (2.18) to express the governing equation for the longitudinal (L) wave traveling in the x direction, and determine the wave velocity if possible.

$$[c] = 10^{11} \times \begin{bmatrix} 2.33 & 0.47 & 0.80 & -0.11 & 0 & 0 \\ 0.47 & 2.33 & 0.80 & 0.11 & 0 & 0 \\ 0.80 & 0.80 & 2.75 & 0 & 0 & 0 \\ -0.11 & 0.11 & 0 & 0.94 & 0 & 0 \\ 0 & 0 & 0 & 0 & 0.94 & -0.11 \\ 0 & 0 & 0 & 0 & -0.11 & 0.93 \end{bmatrix} (\mathrm{N/m^2})$$

Answer: For the longitudinal (L) wave traveling in the x direction, we consider only the wave component that not only travels but also polarizes in the x direction, so by ignoring the other two components, we can express its particle displacement field as

$$\{u\} = \begin{Bmatrix} {}^\circ u_x e^{ik(vt-x)} \\ 0 \\ 0 \end{Bmatrix}$$

in which ${}^\circ u_x$ is the magnitude of the wave polarization, k is wavenumber, v is wave travel velocity, t is time, and i is the imaginary unit ($i = \sqrt{-1}$).

Knowing the displacement expression we first work on the first term in Eq. (2.18), step by step, as follows. First, we calculate

$$[\nabla_s]^T\{u\} = \begin{bmatrix} \frac{\partial}{\partial x} & 0 & 0 \\ 0 & \frac{\partial}{\partial y} & 0 \\ 0 & 0 & \frac{\partial}{\partial z} \\ 0 & \frac{\partial}{\partial z} & \frac{\partial}{\partial y} \\ \frac{\partial}{\partial z} & 0 & \frac{\partial}{\partial x} \\ \frac{\partial}{\partial y} & \frac{\partial}{\partial x} & 0 \end{bmatrix} \begin{Bmatrix} {}^{\circ}u_x e^{ik(vt-x)} \\ 0 \\ 0 \end{Bmatrix} = \begin{Bmatrix} -ik \\ 0 \\ 0 \\ 0 \\ 0 \\ 0 \end{Bmatrix} {}^{\circ}u_x e^{ik(vt-x)}$$

With multiplication of the $[c]$ matrix to the left side, we have

$$[c][\nabla_s]^T\{u\}$$

$$= 10^{11} \times \begin{bmatrix} 2.33 & 0.47 & 0.80 & -0.11 & 0 & 0 \\ 0.47 & 2.33 & 0.80 & 0.11 & 0 & 0 \\ 0.80 & 0.80 & 2.75 & 0 & 0 & 0 \\ -0.11 & 0.11 & 0 & 0.94 & 0 & 0 \\ 0 & 0 & 0 & 0 & 0.94 & -0.11 \\ 0 & 0 & 0 & 0 & -0.11 & 0.93 \end{bmatrix} \begin{Bmatrix} -ik \\ 0 \\ 0 \\ 0 \\ 0 \\ 0 \end{Bmatrix} {}^{\circ}u_x e^{ik(vt-x)}$$

$$= 10^{11}(-ik) \begin{Bmatrix} 2.33 \\ 0.47 \\ 0.80 \\ -0.11 \\ 0 \\ 0 \end{Bmatrix} {}^{\circ}u_x e^{ik(vt-x)}$$

After further multiplication with the $[\nabla_s]$ matrix to the left side, we obtain

$$[\nabla_s][c][\nabla_s]^T\{u\} = 10^{11}(-ik) \begin{bmatrix} \frac{\partial}{\partial x} & 0 & 0 & 0 & \frac{\partial}{\partial z} & \frac{\partial}{\partial y} \\ 0 & \frac{\partial}{\partial y} & 0 & \frac{\partial}{\partial z} & 0 & \frac{\partial}{\partial x} \\ 0 & 0 & \frac{\partial}{\partial z} & \frac{\partial}{\partial y} & \frac{\partial}{\partial x} & 0 \end{bmatrix} \begin{Bmatrix} 2.33 \\ 0.47 \\ 0.80 \\ -0.11 \\ 0 \\ 0 \end{Bmatrix} {}^{\circ}u_x e^{ik(vt-x)}$$

$$= 10^{11}(-k^2) \begin{Bmatrix} 2.33 \\ 0 \\ 0 \end{Bmatrix} {}^{\circ}u_x e^{ik(vt-x)}$$

We now look at the second term in Eq. (2.18) by taking second order derivatives of the displacement field with respect to time

$$\rho \frac{\partial^2}{\partial t^2}\{u\} = 7450 \begin{Bmatrix} (-ikv)^2 \\ 0 \\ 0 \end{Bmatrix} {}^{\circ}u_x e^{ik(vt-x)} = 7450(-k^2v^2) \begin{Bmatrix} 1 \\ 0 \\ 0 \end{Bmatrix} {}^{\circ}u_x e^{ik(vt-x)}$$

By plugging these two terms into Eq. (2.18) and ignoring the $\{f\}$ term we have

$$10^{11}(-k^2)\begin{Bmatrix} 2.33 \\ 0 \\ 0 \end{Bmatrix} {}^{\circ}u_x e^{ik(vt-x)} - 7450v^2(-k^2)\begin{Bmatrix} 1 \\ 0 \\ 0 \end{Bmatrix} {}^{\circ}u_x e^{ik(vt-x)} = 0$$

Through simplification by eliminating the common factors (i.e., k^2 and ${}^{\circ}u_x e^{ik(vt-x)}$), we arrive at the following governing equation for the L wave traveling in the x direction in $LiTaO_3$ crystal:

$$2.33 \times 10^{11} - 7450v^2 = 0$$

Apparently, the positive root for this equation is

$$v = 5592.4m/s$$

This is the wave velocity of the L wave traveling in the x direction of the $LiTaO_3$ crystal.

Example 11: For the problem given in Example 10, consider all three wave components.

Answer: In this case, we consider the following particle displacement field to capture all three waves traveling in the x direction, namely, an L wave polarizing in the x direction, a T wave polarizing in the y direction and a T wave polarizing in the z direction:

$$\{u\} = \begin{Bmatrix} {}^{\circ}u_x \\ {}^{\circ}u_y \\ {}^{\circ}u_z \end{Bmatrix} e^{ik(vt-x)}$$

in which ${}^{\circ}u_x$, ${}^{\circ}u_y$ and ${}^{\circ}u_z$ are the respective magnitude of polarization of each wave.

By taking the same steps as in Example 10, we begin by determining the first term.

$$[\nabla_s]^T\{u\} = \begin{bmatrix} \frac{\partial}{\partial x} & 0 & 0 \\ 0 & \frac{\partial}{\partial y} & 0 \\ 0 & 0 & \frac{\partial}{\partial z} \\ 0 & \frac{\partial}{\partial z} & \frac{\partial}{\partial y} \\ \frac{\partial}{\partial z} & 0 & \frac{\partial}{\partial x} \\ \frac{\partial}{\partial y} & \frac{\partial}{\partial x} & 0 \end{bmatrix} \begin{Bmatrix} {}^\circ u_x \\ {}^\circ u_y \\ {}^\circ u_z \end{Bmatrix} e^{ik(vt-x)} = (-ik) \begin{Bmatrix} {}^\circ u_x \\ 0 \\ 0 \\ 0 \\ {}^\circ u_z \\ {}^\circ u_y \end{Bmatrix} e^{ik(vt-x)}$$

With multiplication of the $[c]$ matrix to the left side, we have

$$[c][\nabla_s]^T\{u\}$$

$$= 10^{11}(-ik) \begin{bmatrix} 2.33 & 0.47 & 0.80 & -0.11 & 0 & 0 \\ 0.47 & 2.33 & 0.80 & 0.11 & 0 & 0 \\ 0.80 & 0.80 & 2.75 & 0 & 0 & 0 \\ -0.11 & 0.11 & 0 & 0.94 & 0 & 0 \\ 0 & 0 & 0 & 0 & 0.94 & -0.11 \\ 0 & 0 & 0 & 0 & -0.11 & 0.93 \end{bmatrix} \begin{Bmatrix} {}^\circ u_x \\ 0 \\ 0 \\ 0 \\ {}^\circ u_z \\ {}^\circ u_y \end{Bmatrix} e^{ik(vt-x)}$$

$$= 10^{11}(-ik) \begin{Bmatrix} 2.33 \times {}^\circ u_x \\ 0.47 \times {}^\circ u_x \\ 0.80 \times {}^\circ u_x \\ -0.11 \times {}^\circ u_x \\ -0.11 \times {}^\circ u_y + 0.94 \times {}^\circ u_z \\ 0.93 \times {}^\circ u_y - 0.11 \times {}^\circ u_z \end{Bmatrix} e^{ik(vt-x)}$$

After further multiplication with the $[\nabla_s]$ matrix to the left side, we obtain

$$[\nabla_s][c][\nabla_s]^T\{u\}$$

$$= 10^{11}(-ik)\begin{bmatrix} \frac{\partial}{\partial x} & 0 & 0 & 0 & \frac{\partial}{\partial z} & \frac{\partial}{\partial y} \\ 0 & \frac{\partial}{\partial y} & 0 & \frac{\partial}{\partial z} & 0 & \frac{\partial}{\partial x} \\ 0 & 0 & \frac{\partial}{\partial z} & \frac{\partial}{\partial y} & \frac{\partial}{\partial x} & 0 \end{bmatrix}\begin{Bmatrix} 2.33 \times {}^\circ u_x \\ 0.47 \times {}^\circ u_x \\ 0.80 \times {}^\circ u_x \\ -0.11 \times {}^\circ u_x \\ -0.11 \times {}^\circ u_y + 0.94 \times {}^\circ u_z \\ 0.93 \times {}^\circ u_y - 0.11 \times {}^\circ u_z \end{Bmatrix} e^{ik(vt-x)}$$

$$= 10^{11}(-k^2)\begin{Bmatrix} 2.33 \times {}^\circ u_x \\ 0.93 \times {}^\circ u_y - 0.11 \times {}^\circ u_z \\ -0.11 \times {}^\circ u_y + 0.94 \times {}^\circ u_z \end{Bmatrix} e^{ik(vt-x)}$$

For the acceleration term, by taking second order derivatives of the displacement field with respect to time

$$\rho\frac{\partial^2}{\partial t^2}\{u\} = 7450v^2(-k^2)\begin{Bmatrix} {}^\circ u_x \\ {}^\circ u_y \\ {}^\circ u_z \end{Bmatrix} e^{ik(vt-x)}$$

By plugging these two terms into Eq. (2.18) and ignoring the $\{f\}$ term we have

$$10^{11}(-k^2)\begin{Bmatrix} 2.33 \times {}^\circ u_x \\ 0.93 \times {}^\circ u_y - 0.11 \times {}^\circ u_z \\ -0.11 \times {}^\circ u_y + 0.94 \times {}^\circ u_z \end{Bmatrix} e^{ik(vt-x)}$$

$$-7450v^2(-k^2)\begin{Bmatrix} {}^\circ u_x \\ {}^\circ u_y \\ {}^\circ u_z \end{Bmatrix} e^{ik(vt-x)} = 0$$

Through simplification we arrive at the following set of three equations for the acoustic waves traveling in the x direction in $LiTaO_3$:

$$2.33 \times 10^{11} \times {}^{\circ}u_x - 7450v^2 \times {}^{\circ}u_x = 0$$

$$0.93 \times 10^{11} \times {}^{\circ}u_y - 7450v^2 \times {}^{\circ}u_y - 0.11 \times 10^{11} \times {}^{\circ}u_z = 0$$

$$-0.11 \times 10^{11} \times {}^{\circ}u_y + 0.94 \times 10^{11} \times {}^{\circ}u_z - 7450v^2 \times {}^{\circ}u_z = 0$$

which, in a matrix form, is expressed as

$$\begin{bmatrix} 2.33 \times 10^{11} - 7450v^2 & 0 & 0 \\ 0 & 0.93 \times 10^{11} - 7450v^2 & -0.11 \times 10^{11} \\ 0 & -0.11 \times 10^{11} & 0.94 \times 10^{11} - 7450v^2 \end{bmatrix} \begin{Bmatrix} {}^{\circ}u_x \\ {}^{\circ}u_y \\ {}^{\circ}u_z \end{Bmatrix} = 0$$

This is the governing matrix equation for the acoustic waves traveling along the x direction in $LiTaO_3$ when the piezoelectric coupling effect is not considered.

The roots of this matrix equation can be found by setting

$$\det \begin{bmatrix} 2.33 \times 10^{11} - 7450v^2 & 0 & 0 \\ 0 & 0.93 \times 10^{11} - 7450v^2 & -0.11 \times 10^{11} \\ 0 & -0.11 \times 10^{11} & 0.94 \times 10^{11} - 7450v^2 \end{bmatrix} = 0$$

By solving this matrix equation, we can find three positive roots as follows:

$$v_{L(x)} = 5592.4 \text{ m/s}, \quad v_{T(y)} = 3327.5 \text{ m/s}, \quad v_{T(z)} = 3745.4 \text{ m/s}$$

representing the wave velocity of the L wave polarizing in the x direction, and T waves polarizing in the y and z directions, respectively. By comparison we note that the L wave in this case is the same as that in Example 10. This is so because the governing matrix equation can be reduced to two sub-matrices, one with the L wave as the lone element, and the other with four elements associated with the two T waves coupled together in the y and z directions.

2.6 Coupled Elastic and Piezoelectric Constitutive Relations

In this section we will discuss the effect of both the *direct piezoelectric effect* and *reverse piezoelectric effect*. Within the material's elastic deformation limit, the generated electrical field in the direct case can be regarded as linearly related to the applied strain, and when the strain is removed, the electrical field vanishes as well. In the reverse case, the strain will disappear when the electrical field is removed. Below we will examine the governing equations for piezoelectric materials by considering a 3D Cartesian Coordinates system.

2.6.1 Direct Piezoelectric Effect

Under the direct piezoelectric effect, electrical polarization will induce an electric displacement, $\{D\}_p$, by the following expression:

$$\{D\}_p = \{P\}$$

where $\{P\}$ is the electric polarization density. In a piezoelectric material, this polarization density is proportional to the external stress, or the strain, as

$$\{P\} = [d]\{T\} = [e]\{S\}$$

in which $[d]$ is the set of piezoelectric strain constants, which in a matrix form can be expressed in a 3×6 matrix as

$$[d] = \begin{bmatrix} d_{x1} & d_{x2} & d_{x3} & d_{x4} & d_{x5} & d_{x6} \\ d_{y1} & d_{y2} & d_{y3} & d_{y4} & d_{y5} & d_{y6} \\ d_{z1} & d_{z2} & d_{z3} & d_{z4} & d_{z5} & d_{z6} \end{bmatrix} \tag{2.19}$$

$[e]$ is the set of piezoelectric stress constants, which in a matrix form can be expressed in a 3×6 matrix as

$$[e] = \begin{bmatrix} e_{x1} & e_{x2} & e_{x3} & e_{x4} & e_{x5} & e_{x6} \\ e_{y1} & e_{y2} & e_{y3} & e_{y4} & e_{y5} & e_{y6} \\ e_{z1} & e_{z2} & e_{z3} & e_{z4} & e_{z5} & e_{z6} \end{bmatrix} \tag{2.20}$$

and $\{T\}$ and $\{S\}$ are the 6×1 stress and strain vectors, respectively, as given in Eq. (2.5). Clearly the matrix algebra works out nicely because in a 3D Cartesian Coordinates system $\{D\}_p$ and $\{P\}$ are 3×1 vectors:

$$\{D\}_p = \begin{Bmatrix} D_x \\ D_y \\ D_z \end{Bmatrix} \quad \text{and} \quad \{P\} = \begin{Bmatrix} P_x \\ P_y \\ P_z \end{Bmatrix}$$

In addition to this polarization induced electric displacement, the presence of electric field, $\{E\}$, will also generate an electric displacement according to the electric constitutive relation

$$\{D\}_0 = [\epsilon]\{E\}$$

where $[\epsilon]$ is the permittivity constants of the solid material, which in a matrix form is expressed in a 3×3 matrix as

$$[\epsilon] = \begin{bmatrix} \epsilon_{xx} & \epsilon_{xy} & \epsilon_{xz} \\ \epsilon_{xy} & \epsilon_{yy} & \epsilon_{yz} \\ \epsilon_{xz} & \epsilon_{yz} & \epsilon_{zz} \end{bmatrix} \tag{2.21}$$

and $\{E\}$ as

$$\{E\} = \begin{Bmatrix} E_x \\ E_y \\ E_z \end{Bmatrix}$$

By summing the two electric displacement components together, we write the total induced electric displacement vector as

$$\{D\} = \{D\}_0 + \{D\}_p$$

which, under the condition of constant stress, can be expressed as

$$\{D\} = [\epsilon]\{E\} + [d]\{T\}$$

and under the condition of constant strain, as

$$\{D\} = [\epsilon]\{E\} + [e]\{S\}$$

Note that the constants in the permittivity matrix, $[\epsilon]$, may be different under the conditions of constant stress and constant strain.

2.6.2 Reverse Piezoelectric Effect

Under the reverse piezoelectric effect, the presence of an electric field will produce a stress, or a strain, proportional to the strength of the electric field. The difference between the induced stress and strain, however, is that the stress carries a negative sign, meaning that it opposes the applied external stress. Mathematically, this statement can be expressed by the following equations:

$$\{T\} = -[e]^T\{E\} \quad \text{and} \quad \{S\} = [d]^T\{E\}$$

in which the $[e]^T$ and $[d]^T$ matrices are the transposes of these given in Eqs. (2.20) and (2.19), respectively.

By adding the mechanically induced stresses, or strains, as defined by the elastic constitutive equations in (2.6) and (2.7), we express the total induced stress as a function of strain and electric field

$$\{T\} = [c]\{S\} - [e]^T\{E\}$$

or strain as a function of stress and electric field, as

$$\{S\} = [s]\{T\} + [d]^T\{E\}$$

where $[c]$ and $[s]$ are the stiffness and compliance matrices discussed in Section 2.3.

2.6.3 Piezoelectric Coupling Equations

With the above discussions can now couple together the mechanically and electrically induced stresses and electric displacements due to the direct and reverse piezoelectric effects. One way to do this is to couple them together in terms of strain vector $\{S\}$, electric field vector $\{E\}$, and the piezoelectric stress matrix $[e]$, along with the stiffness matrix $[c]$ and the permittivity matrix $[\epsilon]$, as

$$\begin{aligned} \{T\} &= [c]\{S\} - [e]^T\{E\} \\ \{D\} &= [\epsilon]\{E\} + [e]\{S\} \end{aligned} \tag{2.22}$$

Note that in using Eq. (2.22) it is understood that the constants in the stiffness matrix $[c]$ are measured under the condition of constant electric field and the constants in the permittivity matrix $[\epsilon]$ are measured under the condition of constant strain.

In an expanded matrix form, these two equations can be expressed as

$$\begin{Bmatrix} \sigma_x \\ \sigma_y \\ \sigma_z \\ \tau_{yz} \\ \tau_{xz} \\ \tau_{xy} \\ D_x \\ D_y \\ D_z \end{Bmatrix} = \begin{bmatrix} c_{11} & c_{12} & c_{13} & c_{14} & c_{15} & c_{16} & -e_{x1} & -e_{y1} & -e_{z1} \\ c_{12} & c_{22} & c_{23} & c_{24} & c_{25} & c_{26} & -e_{x2} & -e_{y2} & -e_{z2} \\ c_{13} & c_{23} & c_{33} & c_{34} & c_{35} & c_{36} & -e_{x3} & -e_{y3} & -e_{z3} \\ c_{14} & c_{24} & c_{34} & c_{44} & c_{45} & c_{46} & -e_{x4} & -e_{y4} & -e_{z4} \\ c_{15} & c_{25} & c_{35} & c_{45} & c_{55} & c_{56} & -e_{x5} & -e_{y5} & -e_{z5} \\ c_{16} & c_{26} & c_{36} & c_{46} & c_{56} & c_{66} & -e_{x6} & -e_{y6} & -e_{z6} \\ e_{x1} & e_{x2} & e_{x3} & e_{x4} & e_{x5} & e_{x6} & \epsilon_{xx} & \epsilon_{xy} & \epsilon_{xz} \\ e_{y1} & e_{y2} & e_{y3} & e_{y4} & e_{y5} & e_{y6} & \epsilon_{xy} & \epsilon_{yy} & \epsilon_{yz} \\ e_{z1} & e_{z2} & e_{z3} & e_{z4} & e_{z5} & e_{z6} & \epsilon_{xz} & \epsilon_{yz} & \epsilon_{zz} \end{bmatrix} \begin{Bmatrix} \epsilon_x \\ \epsilon_y \\ \epsilon_z \\ \gamma_{yz} \\ \gamma_{xz} \\ \gamma_{xy} \\ E_x \\ E_y \\ E_z \end{Bmatrix}$$

Alternatively, we can couple them together in terms of stress $\{T\}$ vector, electric field $\{E\}$ vector, and the piezoelectric strain matrix $[d]$, along with the compliance matrix $[s]$ and permittivity matrix $[\epsilon]$, as

$$\begin{aligned} \{S\} &= [s]\{T\} + [d]^T\{E\} \\ \{D\} &= [\epsilon]\{E\} + [d]\{T\} \end{aligned} \tag{2.23}$$

Again, in using Eq. (2.23) it is understood that the constants in the compliance matrix $[s]$ are measured under the condition of constant electric field, and those in the permittivity matrix $[\epsilon]$ measured under the condition of constant stress. Similarly, in an expanded matrix form, they can be expressed as

$$\begin{Bmatrix} \epsilon_x \\ \epsilon_y \\ \epsilon_z \\ \gamma_{yz} \\ \gamma_{xz} \\ \gamma_{xy} \\ D_x \\ D_y \\ D_z \end{Bmatrix} = \begin{bmatrix} s_{11} & s_{12} & s_{13} & s_{14} & s_{15} & s_{16} & d_{x1} & d_{y1} & d_{z1} \\ s_{12} & s_{22} & s_{23} & s_{24} & s_{25} & s_{26} & d_{x2} & d_{y2} & d_{z2} \\ s_{13} & s_{23} & s_{33} & s_{34} & s_{35} & s_{36} & d_{x3} & d_{y3} & d_{z3} \\ s_{14} & s_{24} & s_{34} & s_{44} & s_{45} & s_{46} & d_{x4} & d_{y4} & d_{z4} \\ s_{15} & s_{25} & s_{35} & s_{45} & s_{55} & s_{56} & d_{x5} & d_{y5} & d_{z5} \\ s_{16} & s_{26} & s_{36} & s_{46} & s_{56} & s_{66} & d_{x6} & d_{y6} & d_{z6} \\ d_{x1} & d_{x2} & d_{x3} & d_{x4} & d_{x5} & d_{x6} & \epsilon_{xx} & \epsilon_{xy} & \epsilon_{xz} \\ d_{y1} & d_{y2} & d_{y3} & d_{y4} & d_{y5} & d_{y6} & \epsilon_{xy} & \epsilon_{yy} & \epsilon_{yz} \\ d_{z1} & d_{z2} & d_{z3} & d_{z4} & d_{z5} & d_{z6} & \epsilon_{xz} & \epsilon_{yz} & \epsilon_{zz} \end{bmatrix} \begin{Bmatrix} \sigma_x \\ \sigma_y \\ \sigma_z \\ \tau_{yz} \\ \tau_{xz} \\ \tau_{xy} \\ E_x \\ E_y \\ E_z \end{Bmatrix}$$

2.7 Governing Equation for Acoustic Waves in Piezoelectric Solids

By using the differential operators $[\nabla]$ and $[\nabla_s]$, given in Eqs. (2.13) and (2.14) respectively, we can express the electric field ($\{E\}$) as the negative gradient of electrical potential (ϕ) as

$$\{E\} = -[\nabla]^T\{\phi\}$$

then we rewrite the piezoelectric constitutive relations, given in Eq. (2.22), as

$$\begin{aligned} \{T\} &= [c][\nabla_s]^T\{u\} + [e]^T[\nabla]^T\{\phi\} \\ \{D\} &= [e][\nabla_s]^T\{u\} - [\epsilon][\nabla]^T\{\phi\} \end{aligned} \tag{2.24}$$

Substituting the first expression into Eq. (2.17) we obtain the following after neglecting the $\{f\}$ term:

$$[\nabla_s]\left([c][\nabla_s]^T\{u\} + [e]^T[\nabla]^T\{\phi\}\right) = \rho\frac{\partial^2}{\partial t^2}\{u\}$$

Applying Gauss's law to the second expression, we have

$$[\nabla]\{D\} = 0$$

which leads to

$$[\nabla]\left([e][\nabla_s]^T\{u\} - [\epsilon][\nabla]^T\{\phi\}\right) = 0$$

Taken together, the set of governing equations for acoustic waves in piezoelectric solids can be expressed as

$$\begin{aligned} [\nabla_s]\left([c][\nabla_s]^T\{u\} + [e]^T[\nabla]^T\{\phi\}\right) - \rho\frac{\partial^2}{\partial t^2}\{u\} &= 0 \\ [\nabla]\left([e][\nabla_s]^T\{u\} - [\epsilon][\nabla]^T\{\phi\}\right) &= 0 \end{aligned} \tag{2.25}$$

Example 12: Using the same crystal ($LiTaO_3$) with properties given in Example 10, along with the piezoelectric stress constants and permittivity given below, apply Eq. (2.25) to express the governing equation for the acoustic waves and determine the wave velocity of the three sub waves.

$$[e] = \begin{bmatrix} 0 & 0 & 0 & 0 & 2.6 & -1.6 \\ -1.6 & 1.6 & 0 & 2.6 & 0 & 0 \\ 0 & 0 & 1.9 & 0 & 0 & 0 \end{bmatrix} \text{C/m}^2; \quad [\epsilon] = 10^{-11} \times \begin{bmatrix} 36.3 & 0 & 0 \\ 0 & 36.3 & 0 \\ 0 & 0 & 38.1 \end{bmatrix} \text{F/m}$$

Answer: For this problem, we use the same particle displacement field as in Example 11 for the three waves traveling in the x direction. To capture the electric field, we write

$$\{\phi\} = {}^0\phi e^{ik(vt-x)}$$

in which ${}^0\phi$ is the magnitude of the electric field.

We first examine the second equation in (2.25), which we rewrite as

$$[\nabla][e][\nabla_s]^T\{u\} - [\nabla][\epsilon][\nabla]^T\{\phi\} = 0$$

From Example 11, we have

$$[\nabla_s]^T\{u\} = (-ik)\begin{Bmatrix} {}^\circ u_x \\ 0 \\ 0 \\ 0 \\ {}^\circ u_z \\ {}^\circ u_y \end{Bmatrix} e^{ik(vt-x)}$$

By multiplying the $[e]$ matrix to the left side, we have

$$[e][\nabla_s]^T\{u\} = (-ik)\begin{bmatrix} 0 & 0 & 0 & 0 & 2.6 & -1.6 \\ -1.6 & 1.6 & 0 & 2.6 & 0 & 0 \\ 0 & 0 & 1.9 & 0 & 0 & 0 \end{bmatrix}\begin{Bmatrix} {}^\circ u_x \\ 0 \\ 0 \\ 0 \\ {}^\circ u_z \\ {}^\circ u_y \end{Bmatrix} e^{ik(vt-x)}$$

$$= (-ik)\begin{Bmatrix} 2.6 \times {}^\circ u_z - 1.6 \times {}^\circ u_y \\ -1.6 \times {}^\circ u_x \\ 0 \end{Bmatrix} e^{ik(vt-x)}$$

With further multiplication with the $[\nabla]$ matrix to the left side, we obtain

$$[\nabla][e][\nabla_s]^T\{u\} = (-ik)\begin{bmatrix} \frac{\partial}{\partial x} & \frac{\partial}{\partial y} & \frac{\partial}{\partial z} \end{bmatrix}\begin{Bmatrix} 2.6 \times {}^\circ u_z - 1.6 \times {}^\circ u_y \\ -1.6 \times {}^\circ u_x \\ 0 \end{Bmatrix} e^{ik(vt-x)}$$

$$= (-k^2)(2.6 \times {}^\circ u_z - 1.6 \times {}^\circ u_y)e^{ik(vt-x)}$$

Now we examine the second term of the second equation, step by step. First we have

$$[\nabla]^T\{\phi\} = (-ik)\begin{Bmatrix} 1 \\ 0 \\ 0 \end{Bmatrix} {}^0\phi e^{ik(vt-x)}$$

then

$$[\epsilon][\nabla]^T\{\phi\} = 10^{-11}(-ik)\begin{bmatrix}36.3 & 0 & 0\\ 0 & 36.3 & 0\\ 0 & 0 & 38.1\end{bmatrix}\begin{Bmatrix}1\\0\\0\end{Bmatrix}{}^0\phi e^{ik(vt-x)}$$
$$= 10^{-11}(-ik)\begin{Bmatrix}36.3\\0\\0\end{Bmatrix}{}^0\phi e^{ik(vt-x)}$$

and

$$[\nabla][\epsilon][\nabla]^T\{\phi\} = 10^{-11}(-ik)\left[\frac{\partial}{\partial x}\ \frac{\partial}{\partial y}\ \frac{\partial}{\partial z}\right]\begin{Bmatrix}36.3\\0\\0\end{Bmatrix}{}^0\phi e^{ik(vt-x)}$$
$$= 36.3 \times 10^{-11}(-k^2)\,{}^0\phi e^{ik(vt-x)}$$

Putting the two terms back in the second equation, with $\{\phi\} = {}^0\phi e^{ik(vt-x)}$ in mind, we write

$$(-k^2)(2.6 \times {}^\circ u_z - 1.6 \times {}^\circ u_y)e^{ik(vt-x)}$$
$$-36.3 \times 10^{-11}(-k^2)\,{}^0\phi e^{ik(vt-x)} = 0$$

which leads us to

$$\{\phi\} = 2.75 \times 10^9(2.6 \times {}^\circ u_z - 1.6 \times {}^\circ u_y)e^{ik(vt-x)}$$

Now we go back to the first equation in (2.25). From Example 11, we already know that

$$[\nabla_s][c][\nabla_s]^T\{u\} = 10^{11}(-k^2)\begin{Bmatrix}2.33 \times {}^\circ u_x\\ 0.93 \times {}^\circ u_y - 0.11 \times {}^\circ u_z\\ -0.11 \times {}^\circ u_y + 0.94 \times {}^\circ u_z\end{Bmatrix}e^{ik(vt-x)}$$

and that

$$\rho\frac{\partial^2}{\partial t^2}\{u\} = 7450v^2(-k^2)\begin{Bmatrix}{}^\circ u_x\\ {}^\circ u_y\\ {}^\circ u_z\end{Bmatrix}e^{ik(vt-x)}$$

so we only need to look at the second term, that is $[\nabla_s][e]^T[\nabla]^T\{\phi\}$. From the above derivations, first we express

$$[\nabla]^T\{\phi\} = (-ik)\begin{Bmatrix}1\\0\\0\end{Bmatrix}{}^0\phi e^{ik(vt-x)} = (-ik)\begin{Bmatrix}1\\0\\0\end{Bmatrix}\{\phi\}$$

$$= 2.75\times 10^9(-ik)\begin{Bmatrix}1\\0\\0\end{Bmatrix}(2.6\times {}^\circ u_z - 1.6\times {}^\circ u_y)e^{ik(vt-x)}$$

then calculate

$$[e]^T[\nabla]^T\{\phi\}$$

$$= 2.75\times 10^9(-ik)\begin{bmatrix}0 & -1.6 & 0\\0 & 1.6 & 0\\0 & 0 & 1.9\\0 & 2.6 & 0\\2.6 & 0 & 0\\-1.6 & 0 & 0\end{bmatrix}\begin{Bmatrix}1\\0\\0\end{Bmatrix}(2.6\times {}^\circ u_z - 1.6\times {}^\circ u_y)e^{ik(vt-x)}$$

$$= 10^9(-ik)\begin{Bmatrix}0\\0\\0\\0\\7.15\\-4.4\end{Bmatrix}(2.6\times {}^\circ u_z - 1.6\times {}^\circ u_y)e^{ik(vt-x)}$$

Furthermore, we determine

$$[\nabla_s][e]^T[\nabla]^T\{\phi\}$$

$$= 10^9(-ik)\begin{bmatrix}\frac{\partial}{\partial x} & 0 & 0 & 0 & \frac{\partial}{\partial z} & \frac{\partial}{\partial y}\\ 0 & \frac{\partial}{\partial y} & 0 & \frac{\partial}{\partial z} & 0 & \frac{\partial}{\partial x}\\ 0 & 0 & \frac{\partial}{\partial z} & \frac{\partial}{\partial y} & \frac{\partial}{\partial x} & 0\end{bmatrix}\begin{Bmatrix}0\\0\\0\\0\\7.15\\-4.4\end{Bmatrix}(2.6\times {}^\circ u_z - 1.6\times {}^\circ u_y)e^{ik(vt-x)}$$

$$= \times 10^9(-k^2)\begin{Bmatrix}0\\-4.4\\7.15\end{Bmatrix}(2.6\times {}^\circ u_z - 1.6\times {}^\circ u_y)e^{ik(vt-x)}$$

By putting all the terms back in the first equation and eliminating the common factors, we have

$$10^{11}\left\{\begin{array}{c}2.33\times{}^{\circ}u_x\\ 0.93\times{}^{\circ}u_y-0.11\times{}^{\circ}u_z\\ -0.11\times{}^{\circ}u_y+0.94\times{}^{\circ}u_z\end{array}\right\}+10^{9}\left\{\begin{array}{c}0\\ -4.4\\ 7.15\end{array}\right\}(2.6\times{}^{\circ}u_z-1.6\times{}^{\circ}u_y)-7450v^2\left\{\begin{array}{c}{}^{\circ}u_x\\ {}^{\circ}u_y\\ {}^{\circ}u_z\end{array}\right\}=0$$

Through simplification we arrive at the following set of three equations for the acoustic waves traveling in the x direction in $LiTaO_3$ with consideration of the piezoelectric coupling effect.

$$\begin{aligned}2.33\times10^{11\circ}u_x-7450v^{2\circ}u_x&=0\\ 1.01\times10^{11\circ}u_y-7450v^{2\circ}u_y-0.22\times10^{11\circ}u_z&=0\\ -0.22\times10^{11\circ}u_y+1.12\times10^{11\circ}u_z-7450v^{2\circ}u_z&=0\end{aligned}$$

which, in a matrix form, is expressed as

$$\begin{bmatrix}2.33\times10^{11}-7450v^2 & 0 & 0\\ 0 & 1.01\times10^{11}-7450v^2 & -0.22\times10^{11}\\ 0 & -0.22\times10^{11} & 1.12\times10^{11}-7450v^2\end{bmatrix}\left\{\begin{array}{c}{}^{\circ}u_x\\ {}^{\circ}u_y\\ {}^{\circ}u_z\end{array}\right\}=0$$

This is the governing matrix equation for the acoustic waves traveling along the x direction in $LiTaO_3$ when the piezoelectric coupling effect is considered.

The roots of this matrix equation can be found by setting

$$\det\begin{bmatrix}2.33\times10^{11}-7450v^2 & 0 & 0\\ 0 & 1.01\times10^{11}-7450v^2 & -0.22\times10^{11}\\ 0 & -0.22\times10^{11} & 1.12\times10^{11}-7450v^2\end{bmatrix}=0$$

By solving this matrix equation, we can find three positive roots as follows:

$$v_{L(x)}=5592.4\ \text{m/s},\quad v_{T(y)}=3338.1\ \text{m/s},\quad v_{T(z)}=4171.8\ \text{m/s}$$

representing the wave velocity of the L wave polarizing in the x direction, and T waves polarizing in the y and z directions, respectively. In a comparison with the results obtained in Example 11, we note that when the piezoelectric coupling effect is considered the L wave appears to be not affected at all, but the T waves polarizing in both the y and z directions are stiffened with faster wave velocity in both cases.

2.8 Unit Analysis

For complex governing equations such as (2.18), (2.22) and (2.23), it would be helpful to attach units to all the variables and constants in the piezoelectric property matrices. In this section, we will do so through a unit analysis for these equations.

Table 2.1 lists the units associated with all the variable and constant terms in these equations, where F represents farads, N newtons, C coulombs, V volts, m meters, Kg kilograms, and s seconds.

Knowing the conversion relationships between N and Kg: $N = Kg \cdot m/s^2$; between F, C and V: $F = C/V$, or alternatively, $C = F \cdot V$; and between C, V, N and m: $C \cdot V = N \cdot m$ or alternatively, $N/C = V/m$ and $N = C \cdot V/m$, we examine the governing equation for acoustic waves in elastic solids by rewriting Eq. (2.18) in a manner that the units for the corresponding terms are given below the equation:

$$[\nabla_s]^T[c][\nabla_s]\{u\} - \rho\frac{\partial^2}{\partial t^2}\{u\} + \{f\} = 0:$$

$$\begin{array}{ccccccc} [\nabla_s]^T[c][\nabla_s]\{u\} & - & \rho\frac{\partial^2}{\partial t^2}\{u\} & + & \{f\} & = 0 \\ \downarrow & & \downarrow & & \downarrow & \\ 1/m \times N/m^2 \times 1/m \times m & & Kg/m^3 \times 1/s^2 \times m & & N/m^3 & \end{array} \tag{2.26}$$

Because $1/m \times N/m^2 \times 1/m \times m = N/m^3$ and $Kg/m^3 \times 1/s^2 \times m = Kg \cdot m/s^2/m^3 = N/m^3$ (due to $N = Kg \cdot m/s^2$), it is clear that the units for all the terms in the governing equation for acoustic waves in elastic solids match perfectly.

Example 13: Examine the units in the governing equations for piezoelectric solids.

Table 2.1 Units for various elastic and piezoelectric constants

Variable/Constant Matrices	Units
$[c]$	N/m^2
$[s]$	m^2/N
$[e]$	C/m^2
$[d]$	C/N
$[\epsilon]$	F/m
$\{D\}$	C/m^2
$\{E\}$	V/m
$\{T\}$	N/m^2
$\{S\}$	1
$\{f\}$	N/m^3
$\{u\}$	m
ρ	Kg/m^3
$[\nabla_s]$ or $[\nabla_s]^T$	$1/m$
$\partial^2/\partial t^2$	$1/s^2$

Answer: By applying the same unit analysis steps discussed earlier, we rewrite Eq. (2.22) as

$$\{T\} = [c]\{S\} - [e]^T\{E\}:$$

$$\begin{array}{ccccc} \{T\} & = & [c]\{S\} & - & [e]^T\{E\} \\ \downarrow & & \downarrow & & \downarrow \\ N/m^2 & & N/m^2 \times 1 & & C/m^2 \times V/m \rightarrow N/m^2 \end{array} \tag{2.27}$$

and

$$\{D\} = [\epsilon]\{E\} + [e]\{S\}:$$

$$\begin{array}{ccccc} \{D\} & = & [e]\{S\} & + & [\epsilon]\{E\} \\ || & & || & & || \\ C/m^2 & & C/m^2 \times 1 & & F/m \times V/m \rightarrow C/m^2 \end{array} \tag{2.28}$$

we can see that because $C/m^2 \times V/m = N/m^2$ (due to $N = C \cdot V/m$), and $F/m \times V/m = C/m^2$ (due to $C = F \cdot V$), the units for the terms on both sides of the equations match nicely.

2.9 Study Problems

1. For the stiffness matrix given below, find its corresponding compliance matrix.

$$[c] = 10^{10} \times \begin{bmatrix} 20.3 & 5.3 & 7.5 & 0.9 & 0 & 0 \\ 5.3 & 20.3 & 7.5 & -0.9 & 0 & 0 \\ 7.5 & 7.5 & 24.5 & 0 & 0 & 0 \\ 0.9 & -0.9 & 0 & 6.0 & 0 & 0 \\ 0 & 0 & 0 & 0 & 6.0 & 0.9 \\ 0 & 0 & 0 & 0 & 0.9 & 7.5 \end{bmatrix}$$

2. Find the compliance matrix for $LiTaO_3$ discussed in Example 10.
3. How many different types of anisotropic materials are discussed in this chapter?
4. Find the stiffness matrix $[c]$ for a 2D plane-stress situation and plane-strain situation for an isotropic material with E = 85 GPa and ν = 0.42.
5. What will happen if someone has mistaken a plane-stress problem as a plane-strain problem in the situation as the previous problem?
6. Following Example 10, find the governing equation and wave velocity for the L wave using $LiNbO_3$ crystal.
7. Repeat Example 11 for $LiNbO_3$ crystal.
8. Repeat Example 12 for $LiNbO_3$ crystal.
9. Perform a unit analysis for Eq. (2.23):

$$\{S\} = [s]\{T\} + [d]^T\{E\}, \{D\} = [d]\{T\} + [\epsilon]\{E\}$$

Chapter 3

Coordinate Rotation and Matrix Transformation

Since piezoelectric crystals are anisotropic materials, their properties are direction dependent, unlike isotropic materials. This is the reason we often speak of cut angles when we describe devices made of piezoelectric materials. Because of this anisotropy, different cuts (small pieces in the shapes of slabs or plates cut from a bulk piece of a piezoelectric crystal) will possess different mechanical and piezoelectric properties. In this chapter, we will discuss how such a change in material properties is determined by using the concept of coordinate rotation and matrix transformation.

3.1 Coordinate Rotation and Transformation Matrix

A practical way to deal with changes in mechanical properties in anisotropic materials of a different cut is to first define a fixed XYZ Cartesian Coordinate system in which all material constants are known, and then assign a rotated *xyz* Cartesian Coordinate system according to the known set of cut angles (more discussions

Bulk and Surface Acoustic Waves: Fundamentals, Devices, and Applications
Guigen Zhang

ISBN 978-981-4877-73-2 (Hardcover), 978-1-003-25662-5 (eBook)
www.jennystanford.com

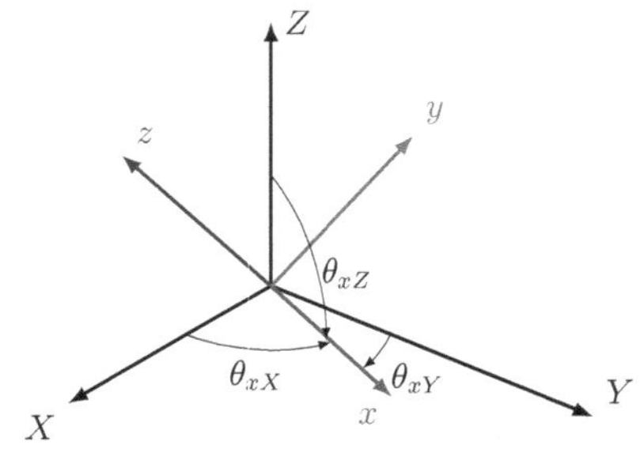

Figure 3.1 Rotation of coordinates.

in Section 3.4). From a linear algebra standpoint, we have now a coordinate rotation problem: the rotation from the XYZ to *xyz* coordinates systems as illustrated in Figure 3.1. Mathematically, this is just a simple operation of applying a transformation matrix, $[a]$, given below:

$$[a] = \begin{bmatrix} a_{xX} \ a_{xY} \ a_{xZ} \\ a_{yX} \ a_{yY} \ a_{yZ} \\ a_{zX} \ a_{zY} \ a_{zZ} \end{bmatrix} \tag{3.1}$$

where its nine elements are cosines of the angles between any pairs of two axes

$$a_{xX} = \cos(\theta_{xX}), a_{xY} = \cos(\theta_{xY}), a_{xZ} = \cos(\theta_{xZ})$$

$$a_{yX} = \cos(\theta_{yX}), a_{yY} = \cos(\theta_{yY}), a_{yZ} = \cos(\theta_{yZ})$$

$$a_{zX} = \cos(\theta_{zX}), a_{zY} = \cos(\theta_{zY}), a_{zZ} = \cos(\theta_{zZ})$$

Example 1: In a rotation situation, knowing the following angles, determine the transformation matrix $[a]$ using Eq. (3.1).

$$\theta_{xX} = 45^\circ, \ \theta_{xY} = 45^\circ, \ \theta_{xZ} = 90^\circ$$

$$\theta_{yX} = 135^\circ, \ \theta_{yY} = 45^\circ, \ \theta_{yZ} = 90^\circ$$

$$\theta_{zX} = 90^\circ, \ \theta_{zY} = 90^\circ, \ \theta_{zZ} = 0^\circ$$

Answer: With these known angles, we first calculate the cosines as follows:

$$cos(\theta_{xX}) = \frac{\sqrt{2}}{2},\ cos(\theta_{xY}) = \frac{\sqrt{2}}{2},\ cos(\theta_{xZ}) = 0$$

$$cos(\theta_{yX}) = -\frac{\sqrt{2}}{2},\ cos(\theta_{yY}) = \frac{\sqrt{2}}{2},\ cos(\theta_{yZ}) = 0$$

$$cos(\theta_{zX}) = 0,\ cos(\theta_{zY}) = 0,\ cos(\theta_{zZ}) = 1$$

By plugging these terms into the transformation matrix, we have

$$[a] = \begin{bmatrix} \frac{\sqrt{2}}{2} & \frac{\sqrt{2}}{2} & 0 \\ -\frac{\sqrt{2}}{2} & \frac{\sqrt{2}}{2} & 0 \\ 0 & 0 & 1 \end{bmatrix}$$

3.2 Transformation of Stress and Strain Matrices

3.2.1 Transformation of Matrices of Primary and Secondary Variables

With this transformation matrix, $[a]$, changes in material properties can be determined by multiplying the original properties (expressed in a matrix form) by the transformation matrix, $[a]$. One cautious note is that the multiplication is not a simple two factor multiplication, but instead, it depends on the type (or rank) of the variables in the property matrix.

When a property matrix contains elements of primary variables such as locations and displacements, we refer to it as a matrix of the primary variables. For this type of property matrices, transformation can be done by a simple two-factor multiplication as follows:

$$\text{Matrix of primary variables:} \quad [r'] = [a][r]$$

where $[r]$ and $[r']$ are matrices of the primary variables in the original and rotated coordinates systems, respectively.

When a property matrix contains elements of the derivatives of primary variables including stresses and strains (hence secondary variables), we regard it as a matrix of secondary variables. For this type of property matrices, transformation is done by a three-factor multiplication as follows:

Matrix of secondary variables: $[R'] = [a][R][a]^T$

where $[R]$ and $[R']$ are matrices of the secondary variables in the original and rotated coordinates systems, respectively, and $[a]^T$ is the transpose matrix of $[a]$.

3.2.2 Transformation of the Stress Matrix

Since the stress ($[\mathbf{T}]$) matrix given in Eq. (2.3) is a secondary-variable matrix, its transformation from the original coordinates system ($[\mathbf{T}]$) to a rotated coordinate system ($[\mathbf{T}']$) is performed as

$$[\mathbf{T}'] = [a][\mathbf{T}][a]^T \tag{3.2}$$

With Eqs. (2.3), (3.1), and (3.2) we express

$$[\mathbf{T}'] = \begin{bmatrix} \sigma'_x & \tau'_{xy} & \tau'_{xz} \\ \tau'_{xy} & \sigma'_y & \tau'_{yz} \\ \tau'_{xz} & \tau'_{yz} & \sigma'_z \end{bmatrix} = [a][\mathbf{T}][a]^T = \begin{bmatrix} a_{xx} & a_{xy} & a_{xz} \\ a_{yx} & a_{yy} & a_{yz} \\ a_{zx} & a_{zy} & a_{zz} \end{bmatrix} \begin{bmatrix} \sigma_x & \tau_{xy} & \tau_{xz} \\ \tau_{xy} & \sigma_y & \tau_{yz} \\ \tau_{xz} & \tau_{yz} & \sigma_z \end{bmatrix} \begin{bmatrix} a_{xx} & a_{yx} & a_{zx} \\ a_{xy} & a_{yy} & a_{zy} \\ a_{xz} & a_{yz} & a_{zz} \end{bmatrix}$$

$$= \left[\begin{array}{lll}
a_{xx}^2\sigma_x + a_{xx}a_{xy}\tau_{xy} + a_{xx}a_{xz}\tau_{xz} & a_{yx}a_{xx}\sigma_x + a_{yx}a_{xy}\tau_{xy} + a_{yx}a_{xz}\tau_{xz} & a_{zx}a_{xx}\sigma_x + a_{zx}a_{xy}\tau_{xy} + a_{zx}a_{xz}\tau_{xz} \\
+ a_{xy}a_{xx}\tau_{xy} + a_{xy}^2\sigma_y + a_{xy}a_{xz}\tau_{yz} & + a_{yy}a_{xx}\tau_{xy} + a_{yy}a_{xy}\sigma_y + a_{yy}a_{xz}\tau_{yz} & + a_{zy}a_{xx}\tau_{xy} + a_{zy}a_{xy}\sigma_y + a_{zy}a_{xz}\tau_{yz} \\
+ a_{xz}a_{xx}\tau_{xz} + a_{xz}a_{xy}\tau_{yz} + a_{xz}^2\sigma_z & + a_{yz}a_{xx}\tau_{xz} + a_{yz}a_{xy}\tau_{yz} + a_{yz}a_{xz}\sigma_z & + a_{zz}a_{xx}\tau_{xz} + a_{zz}a_{xy}\tau_{yz} + a_{zz}a_{xz}\sigma_z \\
 & & \\
a_{xx}a_{yx}\sigma_x + a_{xx}a_{yy}\tau_{xy} + a_{xx}a_{yz}\tau_{xz} & a_{yx}^2\sigma_x + a_{yx}a_{yy}\tau_{xy} + a_{yx}a_{yz}\tau_{xz} & a_{zx}a_{yx}\sigma_x + a_{zx}a_{yy}\tau_{xy} + a_{zx}a_{yz}\tau_{xz} \\
+ a_{xy}a_{yx}\tau_{xy} + a_{xy}a_{yy}\sigma_y + a_{xy}a_{yz}\tau_{yz} & + a_{yy}a_{yx}\tau_{xy} + a_{yy}^2\sigma_y + a_{yy}a_{yz}\tau_{yz} & + a_{zy}a_{yx}\tau_{xy} + a_{zy}a_{yy}\sigma_y + a_{zy}a_{yz}\tau_{yz} \\
+ a_{xz}a_{yx}\tau_{xz} + a_{xz}a_{yy}\tau_{yz} + a_{xz}a_{yz}\sigma_z & + a_{yz}a_{yx}\tau_{xz} + a_{yz}a_{yy}\tau_{yz} + a_{yz}^2\sigma_z & + a_{zz}a_{yx}\tau_{xz} + a_{zz}a_{yy}\tau_{yz} + a_{zz}a_{yz}\sigma_z \\
 & & \\
a_{xx}a_{zx}\sigma_x + a_{xx}a_{zy}\tau_{xy} + a_{xx}a_{zz}\tau_{xz} & a_{yx}a_{zx}\sigma_x + a_{yx}a_{zy}\tau_{xy} + a_{yx}a_{zz}\tau_{xz} & a_{zx}^2\sigma_x + a_{zx}a_{zy}\tau_{xy} + a_{zx}a_{zz}\tau_{xz} \\
+ a_{xy}a_{zx}\tau_{xy} + a_{xy}a_{zy}\sigma_y + a_{xy}a_{zz}\tau_{yz} & + a_{yy}a_{zx}\tau_{xy} + a_{yy}a_{zy}\sigma_y + a_{yy}a_{zz}\tau_{yz} & + a_{zy}a_{zx}\tau_{xy} + a_{zy}^2\sigma_y + a_{zy}a_{zz}\tau_{xz} \\
+ a_{xz}a_{zx}\tau_{xz} + a_{xz}a_{zy}\tau_{yz} + a_{xz}a_{zz}\sigma_z & + a_{yz}a_{zx}\tau_{xz} + a_{yz}a_{zy}\tau_{yz} + a_{yz}a_{zz}\sigma_z & + a_{zz}a_{zx}\tau_{xz} + a_{zz}a_{zy}\tau_{yz} + a_{zz}^2\sigma_z
\end{array}\right]$$

Like $[\mathbf{T}]$, $[\mathbf{T}']$ is also a square and symmetric matrix.

With this $[\mathbf{T}']$ matrix, we can express $\sigma_x', \sigma_y', \sigma_z', \tau_{yz}', \tau_{xz}'$ and τ_{xy}' in terms of $\sigma_x, \sigma_y, \sigma_z, \tau_{yz}, \tau_{xz}$ and τ_{xy} as follows:

$$\sigma_x' = a_{xx}^2\sigma_x + a_{xy}^2\sigma_y + a_{xz}^2\sigma_z + 2a_{xy}a_{xz}\tau_{yz} + 2a_{xx}a_{xz}\tau_{xz} + 2a_{xx}a_{xy}\tau_{xy}$$

$$\sigma_y' = a_{yx}^2\sigma_x + a_{yy}^2\sigma_y + a_{yz}^2\sigma_z + 2a_{yy}a_{yz}\tau_{yz} + 2a_{yz}a_{yx}\tau_{xz} + 2a_{yx}a_{yy}\tau_{xy}$$

$$\sigma_z' = a_{zx}^2\sigma_x + a_{zy}^2\sigma_y + a_{zz}^2\sigma_z + 2a_{zy}a_{zz}\tau_{yz} + 2a_{zz}a_{zx}\tau_{xz} + 2a_{zx}a_{zy}\tau_{xy}$$

$$\tau_{yz}' = \begin{array}{l} a_{yx}a_{zx}\sigma_x + a_{yy}a_{zy}\sigma_y + a_{yz}a_{zz}\sigma_z + (a_{yy}a_{zz} + a_{yz}a_{zy})\tau_{yz} \\ + (a_{yx}a_{zz} + a_{yz}a_{zx})\tau_{xz} + (a_{yy}a_{zx} + a_{yx}a_{zy})\tau_{xy} \end{array}$$

$$\tau_{xz}' = \begin{array}{l} a_{zx}a_{xx}\sigma_x + a_{zy}a_{xy}\sigma_y + a_{zz}a_{xz}\sigma_z + (a_{xy}a_{zz} + a_{xz}a_{zy})\tau_{yz} \\ + (a_{xz}a_{zx} + a_{xx}a_{zz})\tau_{xz} + (a_{xx}a_{zy} + a_{xy}a_{zx})\tau_{xy} \end{array}$$

$$\tau_{xy}' = \begin{array}{l} a_{xx}a_{yx}\sigma_x + a_{xy}a_{yy}\sigma_y + a_{xz}a_{yz}\sigma_z + (a_{xy}a_{yz} + a_{xz}a_{yy})\tau_{yz} \\ + (a_{xz}a_{yx} + a_{xx}a_{yz})\tau_{xz} + (a_{xx}a_{yy} + a_{xy}a_{yx})\tau_{xy} \end{array}$$

Since the six stress components on both sides are the elements of the stress vector (see Eq. (2.5)), it becomes clear that the transformation of the stress vector from $\{T\}$ to $\{T'\}$ can be expressed by a two-factor matrix multiplication as

$$\{T'\} = [M]\{T\} \tag{3.3}$$

in which $[M]$ is a 6×6 matrix whose elements can be determined from those of the transformation matrix $[a]$

$$[M] = \begin{bmatrix} a_{xx}^2 & a_{xy}^2 & a_{xz}^2 & 2a_{xy}a_{xz} & 2a_{xz}a_{xx} & 2a_{xx}a_{xy} \\ a_{yx}^2 & a_{yy}^2 & a_{yz}^2 & 2a_{yy}a_{yz} & 2a_{yz}a_{yx} & 2a_{yx}a_{yy} \\ a_{zx}^2 & a_{zy}^2 & a_{zz}^2 & 2a_{zy}a_{zz} & 2a_{zz}a_{zx} & 2a_{zx}a_{zy} \\ a_{yx}a_{zx} & a_{yy}a_{zy} & a_{yz}a_{zz} & a_{yy}a_{zz} + a_{yz}a_{zy} & a_{yx}a_{zz} + a_{yz}a_{zx} & a_{yy}a_{zx} + a_{yx}a_{zy} \\ a_{zx}a_{xx} & a_{zy}a_{xy} & a_{zz}a_{xz} & a_{xy}a_{zz} + a_{xz}a_{zy} & a_{xz}a_{zx} + a_{xx}a_{zz} & a_{xx}a_{zy} + a_{xy}a_{zx} \\ a_{xx}a_{yx} & a_{xy}a_{yy} & a_{xz}a_{yz} & a_{xy}a_{yz} + a_{xz}a_{yy} & a_{xz}a_{yx} + a_{xx}a_{yz} & a_{xx}a_{yy} + a_{xy}a_{yx} \end{bmatrix} \tag{3.4}$$

A cautious note for using Eqs. 3.3 and 3.4 is that the arrangement of the stress components in the stress vector must follow the convention and be placed in the exact order as listed in Eq. (2.5), that is, $\sigma_x, \sigma_y, \sigma_z, \tau_{yz}, \tau_{xz}$ and τ_{xy} as the first through sixth elements.

Example 2: For the stress matrix given below, first write its equivalent stress vector, then using the transformation matrix found in Example 1 to determine the transformed stress matrix and stress vector.

$$[\mathbf{T}] = \begin{bmatrix} 25 & 15 & 9 \\ 15 & 21 & 11 \\ 9 & 11 & 19 \end{bmatrix}$$

Answer: Referring to the order of the stress elements in the stress vector given in Eq. (2.5), we write

$$\{T\} = \{25\ 21\ 19\ 11\ 9\ 15\}^T$$

With

$$[a] = \begin{bmatrix} \frac{\sqrt{2}}{2} & \frac{\sqrt{2}}{2} & 0 \\ -\frac{\sqrt{2}}{2} & \frac{\sqrt{2}}{2} & 0 \\ 0 & 0 & 1 \end{bmatrix}$$

We find

$$[M] = \begin{bmatrix} 0.5 & 0.5 & 0 & 0 & 0 & 1 \\ 0.5 & 0.5 & 0 & 0 & 0 & -1 \\ 0 & 0 & 1 & 0 & 0 & 0 \\ 0 & 0 & 0 & 0.7071 & -0.7071 & 0 \\ 0 & 0 & 0 & 0.7071 & 0.7071 & 0 \\ -0.5 & 0.5 & 0 & & 00 & 0 \end{bmatrix}$$

For the transformed stress matrix, we apply the formula, $[\mathbf{T}'] = [a][\mathbf{T}][a]^T$:

$$[\mathbf{T}'] = \begin{bmatrix} 38 & -2 & 14.14 \\ -2 & 8 & 1.414 \\ 14.142 & 1.414 & 19 \end{bmatrix}$$

and for the transformed stress vector, we apply {T'} = [M]{T}:

$$\{T'\} = \begin{Bmatrix} 38 \\ 8 \\ 19 \\ 1.414 \\ 14.142 \\ -2 \end{Bmatrix}$$

A close examination of the obtained transformed stress matrix and stress vector confirms that they represent the same stress state.

3.2.3 Transformation of the Strain Matrix

Similarly, for the strain matrix, $[\mathbf{S}]$, given in Eq. (2.4), its transformation from the original coordinates system to a rotated coordinate system can be expressed as

$$[\mathbf{S}'] = [a][\mathbf{S}][a]^T \tag{3.5}$$

By taking the same steps as to the stress matrix, we find the transformation expression for the strain matrix, starting from

$$\begin{bmatrix} \epsilon'_x & \frac{\gamma'_{xy}}{2} & \frac{\gamma'_{xz}}{2} \\ \frac{\gamma'_{xy}}{2} & \epsilon'_y & \frac{\gamma'_{yz}}{2} \\ \frac{\gamma'_{xz}}{2} & \frac{\gamma'_{yz}}{2} & \epsilon'_z \end{bmatrix} = \begin{bmatrix} a_{xx} & a_{xy} & a_{xz} \\ a_{yx} & a_{yy} & a_{yz} \\ a_{zx} & a_{zy} & a_{zz} \end{bmatrix} \begin{bmatrix} \epsilon_x & \frac{\gamma_{xy}}{2} & \frac{\gamma_{xz}}{2} \\ \frac{\gamma_{xy}}{2} & \epsilon_y & \frac{\gamma_{yz}}{2} \\ \frac{\gamma_{xz}}{2} & \frac{\gamma_{yz}}{2} & \epsilon_z \end{bmatrix} \begin{bmatrix} a_{xx} & a_{yx} & a_{zx} \\ a_{xy} & a_{yy} & a_{zy} \\ a_{xz} & a_{yz} & a_{zz} \end{bmatrix}$$

and ending with

$$\{S'\} = [N]\{S\} \tag{3.6}$$

in which

$$[N] = \begin{bmatrix} a_{xx}^2 & a_{xy}^2 & a_{xz}^2 & a_{xy}a_{xz} & a_{xz}a_{xx} & a_{xx}a_{xy} \\ a_{yx}^2 & a_{yy}^2 & a_{yz}^2 & a_{yy}a_{yz} & a_{yz}a_{yx} & a_{yx}a_{yy} \\ a_{zx}^2 & a_{zy}^2 & a_{zz}^2 & a_{zy}a_{zz} & a_{zz}a_{zx} & a_{zx}a_{zy} \\ 2a_{yx}a_{zx} & 2a_{yy}a_{zy} & 2a_{yz}a_{zz} & a_{yy}a_{zz}+a_{yz}a_{zy} & a_{yx}a_{zz}+a_{yz}a_{zx} & a_{yy}a_{zx}+a_{yx}a_{zy} \\ 2a_{zx}a_{xx} & 2a_{zy}a_{xy} & 2a_{zz}a_{xz} & a_{xy}a_{zz}+a_{xz}a_{zy} & a_{xz}a_{zx}+a_{xx}a_{zz} & a_{xx}a_{zy}+a_{xy}a_{zx} \\ 2a_{xx}a_{yx} & 2a_{xy}a_{yy} & 2a_{xz}a_{yz} & a_{xy}a_{yz}+a_{xz}a_{yy} & a_{xz}a_{yx}+a_{xx}a_{yz} & a_{xx}a_{yy}+a_{xy}a_{yx} \end{bmatrix} \tag{3.7}$$

A cautious note here for using Eqs. 3.6 and 3.7 is that not only the arrangement of the strain components in the stress vector must follow exactly the same order, namely, from the first to sixth, $\epsilon_x, \epsilon_y, \epsilon_z, \gamma_{yz}, \gamma_{xz}$ and γ_{xy}, but also the shear strain components $(\gamma_{yz}, \gamma_{xz}, \gamma_{xy})$ in the strain vector are twice that of the corresponding terms in the strain matrix.

Example 3: For the given strain matrix given below, first write its equivalent strain vector, then using the transformation matrix found in Example 1 to determine the transformed strain matrix and strain vector.

$$[\mathbf{S}] = \begin{bmatrix} 12.5 & 5.2 & 6.8 \\ 5.2 & 15.9 & 12.6 \\ 6.8 & 12.6 & 1.9 \end{bmatrix}$$

Answer: Referring to the order of the stress elements in the strain vector given in Eq. (2.5) along with considering a factor of 2 for all the shear strains in the stress vector, we write

$$\{S\} = \{12.5\ 15.9\ 1.9\ 25.2\ 13.6\ 10.4\}^T$$

With

$$[a] = \begin{bmatrix} \frac{\sqrt{2}}{2} & \frac{\sqrt{2}}{2} & 0 \\ -\frac{\sqrt{2}}{2} & \frac{\sqrt{2}}{2} & 0 \\ 0 & 0 & 1 \end{bmatrix}$$

We find

$$[N] = \begin{bmatrix} 0.5 & 0.5 & 0 & 0 & 0 & 0.5 \\ 0.5 & 0.5 & 0 & 0 & 0 & -0.5 \\ 0 & 0 & 1 & 0 & 0 & 0 \\ 0 & 0 & 0 & 0.7071 & -0.7071 & 0 \\ 0 & 0 & 0 & 0.7071 & 0.7071 & 0 \\ -1 & 1 & 0 & 0 & 0 & 0 \end{bmatrix}$$

For the transformed strain matrix, we apply the formula, $[\mathbf{S}'] = [a][\mathbf{S}][a]^T$:

$$[\mathbf{S}'] = \begin{bmatrix} 19.4 & 1.7 & 13.718 \\ 1.7 & 9 & 4.101 \\ 13.718 & 4.101 & 1.9 \end{bmatrix}$$

and for the transformed stress vector we apply {S'}=[N]{S}:

$$\{S'\} = \begin{Bmatrix} 19.4 \\ 9 \\ 1.9 \\ 8.202 \\ 27.436 \\ 3.4 \end{Bmatrix}$$

A close examination of the obtained transformed strain matrix and stress vector confirms that the respective normal strains are identical and that the shear strains in the vector are two times those in the matrix, respectively, thus confirming that they represent the same strain state.

3.3 Transformation of Piezoelectric Properties

With the obtained $[M]$ and $[N]$ matrices (note that $[M]^{-1} = [N]^T$, or $[N]^{-1} = [M]^T$), we can now conveniently perform transformation for the various property matrices for piezoelectric crystals including the stiffness matrix $[c]$, compliance matrix $[s]$, piezoelectric matrices $[e]$ and $[d]$, and permittivity matrix $[\epsilon]$.

3.3.1 Transformation of $[c]$ and $[s]$ Matrices

3.3.1.1 Transformation of $[c]$ Matrix

For the transformation of stiffness matrix $[c]$, by using the generalized Hooke's law relationships, i.e., $\{T\} = [c]\{S\}$ and $\{T'\} = [c']\{S'\}$, we write

$$\{T'\} = [M]\{T\} = [M][c]\{S\}$$

By taking the inverse of the following relationship:

$$\{S'\} = [N]\{S\}$$

we have

$$\{S\} = [N]^{-1}\{S'\}$$

We can then express

$$\{T'\} = [M][c][N]^{-1}\{S'\} = [c']\{S'\}$$

with $[N]^{-1} = [M]^T$, we find

$$[M][c][M]^T\{S'\} = [c']\{S'\}$$

which leads us to

$$[c'] = [M][c][M]^T \tag{3.8}$$

Eq. (3.8) gives the mathematical expression for the transformation of the stiffness matrix $[c]$ from the fixed XYZ coordinates system to the rotated *xyz* coordinates system.

3.3.1.2 Transformation of $[s]$ Matrix

For the transformation of compliance matrix $[s]$, with $\{S\} = [s]\{T\}$ and $\{S'\} = [s']\{T'\}$, we derive

$$\{S'\} = [N]\{S\} = [N][s][T] = [N][s][M]^{-1}\{T'\} = [N][s][N]^T\{T'\} = [s']\{T'\}$$

The above expression leads us to

$$[s'] = [N][s][N]^T \tag{3.9}$$

Eq. (3.9) is the mathematical expression for the transformation of the compliance matrix $[s]$ from the fixed XYZ coordinates system to the rotated *xyz* coordinates system.

Example 4: For the stiffness and compliance matrices listed below for orthorhombic crystal Barium Sodium Niobate ($Ba_2NaNb_5O_{15}$), determine the transformed matrices using the transformation matrix given.

$$[c] = 10^{10} \times \begin{bmatrix} 23.9 & 10.4 & 5.0 & 0 & 0 & 0 \\ 10.4 & 24.7 & 5.2 & 0 & 0 & 0 \\ 5.0 & 5.2 & 13.5 & 0 & 0 & 0 \\ 0 & 0 & 0 & 6.5 & 0 & 0 \\ 0 & 0 & 0 & 0 & 6.6 & 0 \\ 0 & 0 & 0 & 0 & 0 & 7.6 \end{bmatrix} (\text{N/m}^2)$$

$$[s] = 10^{-12} \times \begin{bmatrix} 5.30 & -1.98 & -1.20 & 0 & 0 & 0 \\ -1.98 & 5.14 & -1.25 & 0 & 0 & 0 \\ -1.20 & -1.25 & 8.33 & 0 & 0 & 0 \\ 0 & 0 & 0 & 15.38 & 0 & 0 \\ 0 & 0 & 0 & 0 & 15.15 & 0 \\ 0 & 0 & 0 & 0 & 0 & 13.16 \end{bmatrix} (\text{m}^2/N)$$

$$[a] = \begin{bmatrix} 0.50 & 0.866 & 0 \\ -0.750 & 0.433 & 0.50 \\ 0.433 & -0.250 & 0.866 \end{bmatrix}$$

Answer: Using the $[a]$ matrix given, we first determine

$$[M] = \begin{bmatrix} 0.25 & 0.75 & 0 & 0 & 0 & 0.87 \\ 0.56 & 0.19 & 0.25 & 0.43 & -0.75 & -0.65 \\ 0.19 & 0.06 & 0.75 & -0.43 & 0.75 & -0.22 \\ -0.32 & -0.12 & 0.43 & 0.25 & -0.43 & 0.37 \\ 0.22 & -0.22 & 0 & 0.75 & 0.43 & 0.25 \\ -0.37 & 0.37 & 0 & 0.43 & 0.25 & -0.43 \end{bmatrix}$$

$$[N] = \begin{bmatrix} 0.25 & 0.75 & 0 & 0 & 0 & 0.43 \\ 0.56 & 0.19 & 0.25 & 0.22 & -0.37 & -0.32 \\ 0.19 & 0.06 & 0.75 & -0.22 & 0.37 & -0.11 \\ -0.65 & -0.22 & 0.87 & 0.25 & -0.43 & 0.37 \\ 0.43 & -0.43 & 0 & 0.75 & 0.43 & 0.25 \\ -0.75 & 0.75 & 0 & 0.43 & 0.25 & -0.43 \end{bmatrix}$$

Then by $[c'] = [M][c][M]^T$ we obtain

$$[c'] = 10^{10} \times \begin{bmatrix} 24.99 & 8.72 & 6.34 & -2.06 & 0.05 & -0.09 \\ 8.72 & 21.50 & 5.37 & -2.58 & -0.21 & 0.29 \\ 6.34 & 5.37 & 15.96 & -0.22 & -0.06 & 0.17 \\ -2.06 & -2.58 & -2.22 & 6.89 & 0.06 & -0.15 \\ 0.05 & -0.21 & -0.06 & 0.06 & 6.67 & -0.25 \\ -0.09 & 0.29 & 0.17 & -0.15 & -0.25 & 6.97 \end{bmatrix} (\mathrm{N/m^2})$$

with $[s'] = [N][s][N]^T$ we obtain

$$[s'] = 10^{-12} \times \begin{bmatrix} 4.95 & -1.62 & -1.36 & 0.44 & -0.10 & 0.17 \\ -1.62 & 5.75 & -1.10 & 1.31 & 0.17 & -0.20 \\ -1.36 & -1.10 & 7.40 & 1.56 & 0.02 & -0.12 \\ 0.44 & 1.31 & 1.56 & 15.64 & -0.08 & 0.24 \\ -0.10 & 0.17 & 0.02 & -0.08 & 15.02 & 0.54 \\ 0.17 & -0.20 & -0.12 & 0.24 & 0.54 & 14.40 \end{bmatrix} (\mathrm{m^2}/N)$$

The reader is encouraged to show that $[c'] = [s']^{-1}$.

3.3.2 Transformation of $[e]$, $[d]$ and $[\epsilon]$ Matrices

3.3.2.1 Transformation of $[e]$ Matrix

For the transformation of the piezoelectric stress constants matrix $[e]$, we start with the following expressions:

$$\{D\} = [e]\{S\}, \quad \{D'\} = [e']\{S'\}$$

Since the electric displacement vector $\{D\}$ is of primary variables, we write

$$\{D'\} = [a]\{D\} = [a][e]\{S\} = [a][e][N]^{-1}\{S'\} = [a][e][M]^T\{S'\} = [e']\{S'\}$$

which leads us to

$$[e'] = [a][e][M]^T \tag{3.10}$$

Eq. (3.10) is the mathematical expression for the transformation of the piezoelectric stress constant matrix $[e]$ from the fixed XYZ to the rotated *xyz* coordinates systems.

3.3.2.2 Transformation of $[d]$ Matrix

For the transformation of the piezoelectric strain constants matrix $[d]$, we start with

$$\{D\} = [d]\{T\}, \quad \{D'\} = [d']\{T'\}$$

and write

$$\{D'\} = [a]\{D\} = [a][d]\{T\} = [a][d][M]^{-1}\{T'\} = [a][d][N]^T\{T'\} = [d']\{T'\}$$

which yields

$$[d'] = [a][d][N]^T \tag{3.11}$$

Eq. (3.11) is the mathematical expression for the transformation of the piezoelectric strain constant matrix $[d]$ from the fixed XYZ to the rotated *xyz* coordinates systems.

3.3.2.3 Transformation of $[\epsilon]$ Matrix

As we have seen in Section 2.6, the permittivity matrix, $[\epsilon]$, relates the electric displacement vector, $\{D\}$, to the electric field vector, $\{E\}$, so it is a matrix of secondary variables. Its transformation from the original coordinates system to a rotated coordinates system is in the same way as stress and strain matrices, that is

$$[\epsilon'] = [a][\epsilon][a]^T \tag{3.12}$$

Eq. (3.12) gives the mathematical expression for the transformation of the permittivity constant matrix $[\epsilon]$ from the fixed XYZ to the rotated *xyz* coordinates systems.

3.3.2.4 Relationship between $[e]$ and $[d]$ Matrices

From the constitutive and generalized Hooke's law relationships, we know that

$$\{D\} = [e]\{S\} = [d]\{T\}, \ \{T\} = [c]\{S\}, \ \{S\} = [s]\{T\}$$

With these relationships, we can derive the following relationships:

$$[e] = [d][c] \text{ and } [d] = [e][s] \tag{3.13}$$

These expressions show that the matrices of piezoelectric stress constants ($[e]$) and piezoelectric strain constants ($[d]$) are not independent, and they are related through either the stiffness matrix $[c]$ or the compliance matrix $[s]$.

Example 5: For the piezoelectric stress constant and permittivity matrices listed below, use the transformation matrix given in Example 4 to find the transformed matrices.

$$[e] = \begin{bmatrix} 0 & 0 & 0 & 0 & 2.8 & 0 \\ 0 & 0 & 0 & 3.4 & 0 & 0 \\ -0.4 & -0.3 & 4.3 & 0 & 0 & 0 \end{bmatrix} (\text{C/m}^2);$$

$$[\epsilon] = 10^{-11} \times \begin{bmatrix} 222 & 0 & 0 \\ 0 & 227 & 0 \\ 0 & 0 & 32 \end{bmatrix} (\text{F/m})$$

Answer: Referring to Example 4 for the $[a]$ and $[M]$ matrices, we determine

$$[e'] = [a][e][M]^T$$
$$= \begin{bmatrix} 0 & 0.2250 & -0.2250 & 0.1299 & 2.8146 & 1.6250 \\ -0.1625 & 2.6094 & -0.6469 & 2.2896 & 0.1840 & 0.1313 \\ -0.2815 & -0.5900 & 3.9891 & 1.0156 & -0.1312 & -0.0325 \end{bmatrix} (\mathrm{C/m^2})$$

$$[\epsilon] = 10^{-11} \times \begin{bmatrix} 225.7 & 1.9 & -1.1 \\ 1.9 & 175.4 & -82.8 \\ -1.1 & -82.8 & 79.8 \end{bmatrix} (\mathrm{F/m})$$

3.4 Transformation by Euler-Angles

Typically we use a set of three Euler angles (ϕ, θ, ψ), as illustrated in Figure 3.2, to define the selection of a crystal cut orientation. The first angle, ϕ, defines a rotation about the Z axis, the second angle, θ, defines a rotation about the x axis after the first rotation, that is, the x' axis in the figure, and the third angle, ψ, defines a rotation about the z axis after the second rotation. Very often, the direction defined by the x axis after these rotations is used as the propagation direction for an acoustic wave. In this book, we will consider that when a set of Euler angles is selected, not only the orientations of a crystal cut are defined, but the propagation direction of acoustic waves is determined as well.

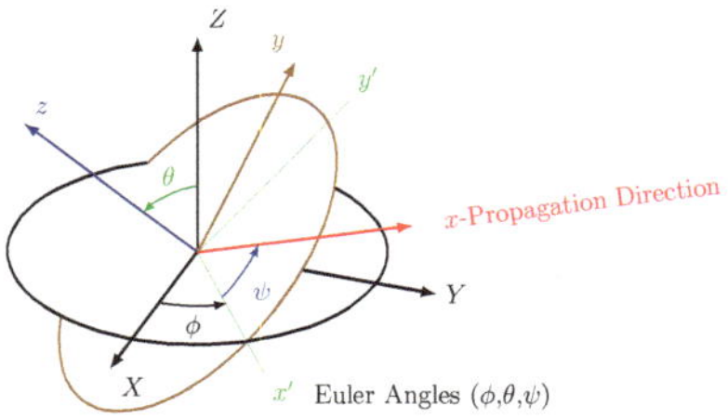

Figure 3.2 Definition of a set of Euler angles.

3.4.1 Transformation by the First Euler Angle Alone

By determining the cosines of the angles between all nine pairs of axes shown in Figure 3.2, we obtain the transformation matrix for the first rotation by ϕ alone as follows:

$$[a_\phi] = \begin{bmatrix} \cos\phi & \sin\phi & 0 \\ -\sin\phi & \cos\phi & 0 \\ 0 & 0 & 1 \end{bmatrix}$$

Using equations (3.4) and (3.7), we obtain

$$[M] = \begin{bmatrix} \cos^2\phi & \sin^2\phi & 0 & 0 & 0 & \sin 2\phi \\ \sin^2\phi & \cos^2\phi & 0 & 0 & 0 & -\sin 2\phi \\ 0 & 0 & 1 & 0 & 0 & 0 \\ 0 & 0 & 0 & \cos\phi & -\sin\phi & 0 \\ 0 & 0 & 0 & \sin\phi & \cos\phi & 0 \\ -\frac{\sin 2\phi}{2} & \frac{\sin 2\phi}{2} & 0 & 0 & 0 & \cos 2\phi \end{bmatrix}$$

$$[N] = \begin{bmatrix} \cos^2\phi & \sin^2\phi & 0 & 0 & 0 & \frac{\sin 2\phi}{2} \\ \sin^2\phi & \cos^2\phi & 0 & 0 & 0 & -\frac{\sin 2\phi}{2} \\ 0 & 0 & 1 & 0 & 0 & 0 \\ 0 & 0 & 0 & \cos\phi & -\sin\phi & 0 \\ 0 & 0 & 0 & \sin\phi & \cos\phi & 0 \\ -\sin 2\phi & \sin 2\phi & 0 & 0 & 0 & \cos 2\phi \end{bmatrix}$$

3.4.2 Transformation by the Second Euler Angle Alone

For rotation of θ about the x' axis alone, the transformation matrix is calculated as

$$[a_\theta] = \begin{bmatrix} 1 & 0 & 0 \\ 0 & \cos\theta & \sin\theta \\ 0 & -\sin\theta & \cos\theta \end{bmatrix}$$

then, using Eqs. (3.4) and (3.7), we obtain

$$[M] = \begin{bmatrix} 1 & 0 & 0 & 0 & 0 & 0 \\ 0 & \cos^2\theta & \sin^2\theta & \sin 2\theta & 0 & 0 \\ 0 & \sin^2\theta & \cos^2\theta & -\sin 2\theta & 0 & 0 \\ 0 & -\dfrac{\sin 2\theta}{2} & \dfrac{\sin 2\theta}{2} & \cos 2\theta & 0 & 0 \\ 0 & 0 & 0 & 0 & \cos\theta & -\sin\theta \\ 0 & 0 & 0 & 0 & \sin\theta & \cos\theta \end{bmatrix}$$

$$[N] = \begin{bmatrix} 1 & 0 & 0 & 0 & 0 & 0 \\ 0 & \cos^2\theta & \sin^2\theta & \dfrac{\sin 2\theta}{2} & 0 & 0 \\ 0 & \sin^2\theta & \cos^2\theta & -\dfrac{\sin 2\theta}{2} & 0 & 0 \\ 0 & -\sin 2\theta & \sin 2\theta & \cos 2\theta & 0 & 0 \\ 0 & 0 & 0 & 0 & \cos\theta & -\sin\theta \\ 0 & 0 & 0 & 0 & \sin\theta & \cos\theta \end{bmatrix}$$

3.4.3 Transformation by the Third Euler Angle Alone

For rotation of ψ about the z axis alone, the transformation matrix is

$$[a_\psi] = \begin{bmatrix} \cos\psi & \sin\psi & 0 \\ -\sin\psi & \cos\psi & 0 \\ 0 & 0 & 1 \end{bmatrix}$$

and with Eqs. (3.4) and (3.7), we have

$$[M] = \begin{bmatrix} \cos^2\psi & \sin^2\psi & 0 & 0 & 0 & \sin 2\psi \\ \sin^2\psi & \cos^2\psi & 0 & 0 & 0 & -\sin 2\psi \\ 0 & 0 & 1 & 0 & 0 & 0 \\ 0 & 0 & 0 & \cos\psi & -\sin\psi & 0 \\ 0 & 0 & 0 & \sin\psi & \cos\psi & 0 \\ -\frac{\sin 2\psi}{2} & \frac{\sin 2\psi}{2} & 0 & 0 & 0 & \cos 2\psi \end{bmatrix}$$

$$[N] = \begin{bmatrix} \cos^2\psi & \sin^2\psi & 0 & 0 & 0 & \frac{\sin 2\psi}{2} \\ \sin^2\psi & \cos^2\psi & 0 & 0 & 0 & -\frac{\sin 2\psi}{2} \\ 0 & 0 & 1 & 0 & 0 & 0 \\ 0 & 0 & 0 & \cos\psi & -\sin\psi & 0 \\ 0 & 0 & 0 & \sin\psi & \cos\psi & 0 \\ -\sin 2\psi & \sin 2\psi & 0 & 0 & 0 & \cos 2\psi \end{bmatrix}$$

3.4.4 Transformation by a Set of Euler Angles Sequentially

The transformation matrix after the rotation by all three Euler angles sequentially is determined by the following multiplication:

$$\begin{aligned} [a] &= [a_\psi] \times [a_\theta] \times [a_\phi] \\ &= \begin{bmatrix} \cos\psi & \sin\psi & 0 \\ -\sin\psi & \cos\psi & 0 \\ 0 & 0 & 1 \end{bmatrix} \times \begin{bmatrix} 1 & 0 & 0 \\ 0 & \cos\theta & \sin\theta \\ 0 & -\sin\theta & \cos\theta \end{bmatrix} \times \begin{bmatrix} \cos\phi & \sin\phi & 0 \\ -\sin\phi & \cos\phi & 0 \\ 0 & 0 & 1 \end{bmatrix} \end{aligned} \tag{3.14}$$

Note that the order of the individual rotations in the sequence matters: the transformation matrix of a later rotation is positioned to the left of that of an earlier rotation. With the nine elements known for the final transformation matrix given in Eq. (3.14), we can determine the corresponding $[M]$ and $[N]$ matrices fairly easily by using Eqs. (3.4) and (3.7).

Example 6: For a given set of Euler angles, $(45°, 0°, 0°)$ find the transformation matrix.

Answer: By using Eq. (3.14) with the substitution of given angles, we obtain

$$[a] = [a_\psi] \times [a_\theta] \times [a_\phi] = \begin{bmatrix} \frac{\sqrt{2}}{2} & \frac{\sqrt{2}}{2} & 0 \\ -\frac{\sqrt{2}}{2} & \frac{\sqrt{2}}{2} & 0 \\ 0 & 0 & 1 \end{bmatrix} \times \begin{bmatrix} 1 & 0 & 0 \\ 0 & 1 & 0 \\ 0 & 0 & 1 \end{bmatrix} \times \begin{bmatrix} 1 & 0 & 0 \\ 0 & 1 & 0 \\ 0 & 0 & 1 \end{bmatrix}$$

$$= \begin{bmatrix} \frac{\sqrt{2}}{2} & \frac{\sqrt{2}}{2} & 0 \\ -\frac{\sqrt{2}}{2} & \frac{\sqrt{2}}{2} & 0 \\ 0 & 0 & 1 \end{bmatrix}$$

This is the same matrix as the one shown in Example 1 at the beginning of this chapter. Obviously, it is much easier to find the transformation matrix by using Euler angles.

Example 7: For a given set of Euler angles, $(0°, 0°, 45°)$ find the transformation matrix.

Answer: By using Eq. (3.14) with the substitution of given angles, we obtain

$$[a] = [a_\psi] \times [a_\theta] \times [a_\phi] = \begin{bmatrix} 1 & 0 & 0 \\ 0 & 1 & 0 \\ 0 & 0 & 1 \end{bmatrix} \times \begin{bmatrix} 1 & 0 & 0 \\ 0 & 1 & 0 \\ 0 & 0 & 1 \end{bmatrix} \times \begin{bmatrix} \frac{\sqrt{2}}{2} & \frac{\sqrt{2}}{2} & 0 \\ -\frac{\sqrt{2}}{2} & \frac{\sqrt{2}}{2} & 0 \\ 0 & 0 & 1 \end{bmatrix}$$

$$= \begin{bmatrix} \frac{\sqrt{2}}{2} & \frac{\sqrt{2}}{2} & 0 \\ -\frac{\sqrt{2}}{2} & \frac{\sqrt{2}}{2} & 0 \\ 0 & 0 & 1 \end{bmatrix}$$

This is the same matrix as the one shown in Example 6. Apparently, the same axis rotation may be expressed in different, but equivalent, Euler angles.

Example 8: For the stiffness, piezoelectric stress constant and permittivity matrices given below

$$[c] = 10^{10} \times \begin{bmatrix} 20.900 & 12.110 & 10.510 & 0 & 0 & 0 \\ 12.110 & 20.900 & 10.510 & 0 & 0 & 0 \\ 10.510 & 10.510 & 21.090 & 0 & 0 & 0 \\ 0 & 0 & 0 & 4.250 & 0 & 0 \\ 0 & 0 & 0 & 0 & 4.250 & 0 \\ 0 & 0 & 0 & 0 & 0 & 4.395 \end{bmatrix} (\mathrm{N/m^2})$$

$$[e] = \begin{bmatrix} 0 & 0 & 0 & 0 & -0.480 & 0 \\ 0 & 0 & 0 & -0.480 & 0 & 0 \\ -0.573 & -0.573 & 1.320 & 0 & 0 & 0 \end{bmatrix} (\mathrm{C/m^2})$$

$$[\epsilon] = 10^{-11} \times \begin{bmatrix} 8.550 & 0 & 0 \\ 0 & 8.550 & 0 \\ 0 & 0 & 10.200 \end{bmatrix} (\mathrm{F/m})$$

determine the transformed material properties in the cuts defined by Euler angles $(0°, -90°, -90°)$ and $(0°, -90°, 90°)$.

Answer: For these cuts, by using Eq. (3.14) with the substitution of given angles, we obtain

$$[a] = \begin{bmatrix} 0 & 0 & 1 \\ 1 & 0 & 0 \\ 0 & 1 & 0 \end{bmatrix}, \quad [M] = \begin{bmatrix} 0 & 0 & 1 & 0 & 0 & 0 \\ 1 & 0 & 0 & 0 & 0 & 0 \\ 0 & 1 & 0 & 0 & 0 & 0 \\ 0 & 0 & 0 & 0 & 0 & 1 \\ 0 & 0 & 0 & 1 & 0 & 0 \\ 0 & 0 & 0 & 0 & 1 & 0 \end{bmatrix}$$

for the $(0°, -90°, -90°)$ cut, and

$$[a] = \begin{bmatrix} 0 & 0 & -1 \\ -1 & 0 & 0 \\ 0 & 1 & 0 \end{bmatrix}, [M] = \begin{bmatrix} 0 & 0 & 1 & 0 & 0 & 0 \\ 1 & 0 & 0 & 0 & 0 & 0 \\ 0 & 1 & 0 & 0 & 0 & 0 \\ 0 & 0 & 0 & 0 & 0 & -1 \\ 0 & 0 & 0 & -1 & 0 & 0 \\ 0 & 0 & 0 & 0 & 1 & 0 \end{bmatrix}$$

for the $(0°, -90°, 90°)$ cut. By substituting these matrices along with the given material properties into Eqs. (3.8), (3.10) and (3.12), respectively, we obtain

$$[c] = 10^{10} \times \begin{bmatrix} 21.090 & 10.510 & 10.510 & 0 & 0 & 0 \\ 10.510 & 20.900 & 12.110 & 0 & 0 & 0 \\ 10.510 & 12.110 & 21.090 & 0 & 0 & 0 \\ 0 & 0 & 0 & 4.395 & 0 & 0 \\ 0 & 0 & 0 & 0 & 4.250 & 0 \\ 0 & 0 & 0 & 0 & 0 & 4.250 \end{bmatrix} (\mathrm{N/m^2})$$

$$[e] = \begin{bmatrix} -1.320 & 0.573 & 0.573 & 0 & 0 & 0 \\ 0 & 0 & 0 & 0 & 0 & 0.480 \\ 0 & 0 & 0 & 0 & 0.480 & 0 \end{bmatrix} (\mathrm{C/m^2})$$

$$[\epsilon] = 10^{-11} \times \begin{bmatrix} 10.200 & 0 & 0 \\ 0 & 8.550 & 0 \\ 0 & 0 & 8.550 \end{bmatrix} (\mathrm{F/m})$$

for the $(0°, -90°, -90°)$ cut, and

$$[c] = 10^{10} \times \begin{bmatrix} 21.090 & 10.510 & 10.510 & 0 & 0 & 0 \\ 10.510 & 20.900 & 12.110 & 0 & 0 & 0 \\ 10.510 & 12.110 & 21.090 & 0 & 0 & 0 \\ 0 & 0 & 0 & 4.395 & 0 & 0 \\ 0 & 0 & 0 & 0 & 4.250 & 0 \\ 0 & 0 & 0 & 0 & 0 & 4.250 \end{bmatrix} (\mathrm{N/m^2})$$

$$[e] = \begin{bmatrix} 1.320 & -0.573 & -0.573 & 0 & 0 & 0 \\ 0 & 0 & 0 & 0 & 0 & -0.480 \\ 0 & 0 & 0 & 0 & -0.480 & 0 \end{bmatrix} (\mathrm{C/m^2})$$

$$[\epsilon] = 10^{-11} \times \begin{bmatrix} 10.200 & 0 & 0 \\ 0 & 8.550 & 0 \\ 0 & 0 & 8.550 \end{bmatrix} \text{ (F/m)}$$

for the $(0°, -90°, 90°)$ cut. Obviously, the transformed stiffness and permittivity matrices are the same for these two cuts, but the piezoelectric stress constant matrix differs by a negative sign.

Example 9: Determine the transformed material properties for the material given in Example 8 in the cuts defined by Euler angles $(0°, -45°, -90°)$ and $(0°, -45°, 90°)$.

Answer: By the same exercises, we have

$$[a] = \begin{bmatrix} 0 & -0.707 & 0.707 \\ 1 & 0 & 0 \\ 0 & 0.707 & 0.707 \end{bmatrix}, \quad [M] = \begin{bmatrix} 0 & 0.5 & 0.5 & -1 & 0 & 0 \\ 1 & 0 & 0 & 0 & 0 & 0 \\ 0 & 0.5 & 0.5 & 1 & 0 & 0 \\ 0 & 0 & 0 & 0 & 0.707 & 0.707 \\ 0 & -0.5 & 0.5 & 0 & 0 & 0 \\ 0 & 0 & 0 & 0 & 0.707 & -0.707 \end{bmatrix}$$

for the $(0°, -45°, -90°)$ cut, and

$$[a] = \begin{bmatrix} 0 & 0.707 & -0.707 \\ -1 & 0 & 0 \\ 0 & 0.707 & 0.707 \end{bmatrix}, \quad [M] = \begin{bmatrix} 0 & 0.5 & 0.5 & -1 & 0 & 0 \\ 1 & 0 & 0 & 0 & 0 & 0 \\ 0 & 0.5 & 0.5 & 1 & 0 & 0 \\ 0 & 0 & 0 & 0 & -0.707 & -0.707 \\ 0 & 0.5 & -0.5 & 0 & 0 & 0 \\ 0 & 0 & 0 & 0 & 0.707 & -0.707 \end{bmatrix}$$

for the $(0°, -45°, 90°)$ cut. By substituting these matrices along with the given material properties into Eqs. (3.8), (3.10) and (3.12), respectively, we obtain

$$[c] = 10^{10} \times \begin{bmatrix} 20.002 & 11.310 & 11.502 & 0 & 0.048 & 0 \\ 11.310 & 20.900 & 11.31 & 0 & -0.800 & 0 \\ 11.502 & 11.310 & 20.002 & 0 & 0.048 & 0 \\ 0 & 0 & 0 & 4.323 & 0 & -0.073 \\ 0.048 & -0.800 & 0.048 & 0 & 5.242 & 0 \\ 0 & 0 & 0 & -0.073 & 0 & 4.323 \end{bmatrix} \text{ (N/m}^2\text{)}$$

$$[e] = \begin{bmatrix} -0.0753 & -0.405 & 0.604 & 0 & 0.669 & 0 \\ 0 & 0 & 0 & -0.339 & 0 & -0.339 \\ 0.604 & -0.405 & -0.0753 & 0 & 0.669 & 0 \end{bmatrix} (\mathrm{C/m^2})$$

$$[\epsilon] = 10^{-11} \times \begin{bmatrix} 9.375 & 0 & 0.825 \\ 0 & 8.550 & 0 \\ 0.825 & 0 & 9.375 \end{bmatrix} (\mathrm{F/m})$$

for the $(0°, -45°, -90°)$ cut, and

$$[c] = 10^{10} \times \begin{bmatrix} 20.002 & 11.310 & 11.502 & 0 & -0.048 & 0 \\ 11.310 & 20.900 & 11.310 & 0 & 0.800 & 0 \\ 11.502 & 11.310 & 20.002 & 0 & -0.048 & 0 \\ 0 & 0 & 0 & 4.323 & 0 & 0.073 \\ -0.048 & 0.800 & -0.048 & 0 & 5.242 & 0 \\ 0 & 0 & 0 & 0.073 & 0 & 4.323 \end{bmatrix} (\mathrm{N/m^2})$$

$$[e] = \begin{bmatrix} 0.0753 & 0.405 & -0.604 & 0 & 0.669 & 0 \\ 0 & 0 & 0 & -0.339 & 0 & 0.339 \\ 0.604 & -0.405 & -0.0753 & 0 & -0.669 & 0 \end{bmatrix} (\mathrm{C/m^2})$$

$$[\epsilon] = 10^{-11} \times \begin{bmatrix} 9.375 & 0 & -0.825 \\ 0 & 8.550 & 0 \\ -0.825 & 0 & 9.375 \end{bmatrix} (\mathrm{F/m})$$

for the $(0°, -45°, 90°)$ cut. Clearly, the transformed property matrices are not exactly the same, although the difference appears to be only in the sign and not in the digits of some of the values in these matrices.

3.5 Study Problems

1. For the given set of Euler angles $(0°, 38°, 0°)$ find the corresponding transformation matrix.
2. Use the transformation matrix determined by Euler angles $(60°, 110°, 25°)$ to repeat Example 2.

3. Use the transformation matrix determined by Euler angles $(90°, 25°, 45°)$ to repeat Example 3.
4. Use the transformation matrix determined by Euler angles $(60°, 110°, 25°)$ along with the stiffness and compliance matrices given in Example 4 in to show that the transformed stiffness and compliance matrices has the inverse relationship.
5. Use the transformation matrix determined by Euler angles $(90°, 25°, 45°)$ along with the stiffness and compliance matrices given in Example 4 to show that the transformed stiffness and compliance matrices has the inverse relationship.
6. Use the transformation matrix determined by Euler angles $(0°, 65°, 25°)$ to show the follow relationships between $[M]$ and $[N]$ matrices: $[M]^{-1} = [N]^T$ and $[N]^{-1} = [M]^T$.

Chapter 4

Classes of Crystal Symmetry

Crystal materials possess many different types of structural symmetry and they are often described by the classes of symmetry. To date, 32 classes of symmetry have been identified for all the crystals known to us. Of the 32 classes, they can be categorized into seven groups, namely, the **triclinic, monoclinic, orthorhombic, tetragonal, trigonal, hexagonal**, and **cubic** groups. In this chapter, we will discuss all the 32 classes of piezoelectric crystals in these seven groups and we will do so by taking advantage of the knowledge on matrix transformation we learned in Chapter 3.

4.1 The Concept of Crystal Symmetry

By the very meaning of symmetry, a crystal should possess the same material properties when any symmetric orientation is used as a frame of reference. To understand this statement better, let us use a unit lattice of cubic crystal, gallium arsenide (GaAs), to provide some visual illustrations as depicted in Figure 4.1. Here, Figure 4.1(a) shows the unit crystal in a fixed coordinates system XYZ with the larger dots representing the arsenic (As) atoms and smaller dots the gallium (Ga) atoms, and Figure 4.1(b) shows the

Bulk and Surface Acoustic Waves: Fundamentals, Devices, and Applications
Guigen Zhang

ISBN 978-981-4877-73-2 (Hardcover), 978-1-003-25662-5 (eBook)
www.jennystanford.com

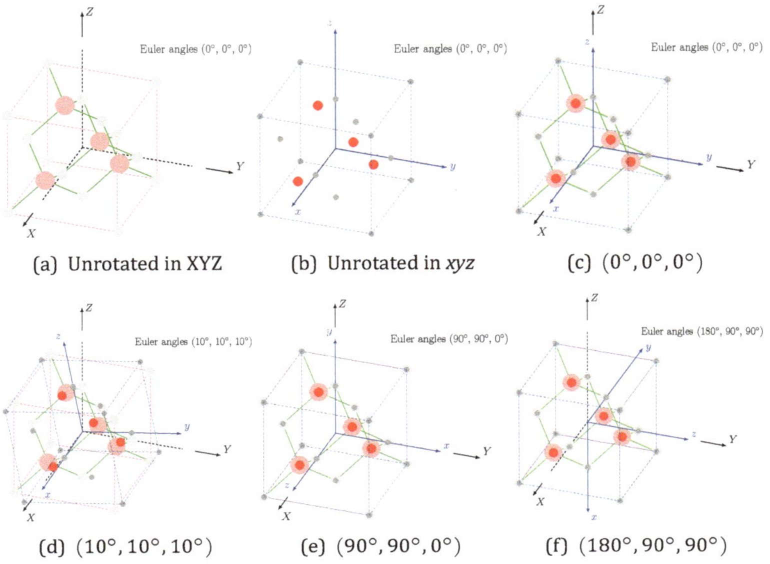

Figure 4.1 Illustration of crystal symmetry with a unit cubic lattice of gallium arsenide.

unit crystal in a rotating coordinates system *xyz*. In Figure 4.1(c), the two coordinates systems are placed at the same position in space with the same orientation, where we can see that the positions of the Ga and As atoms in both the fixed and rotating coordinates systems overlap with each other correspondingly. When the *xyz* coordinates system is rotated, the unit cubic in it will rotate accordingly. As shown in Figure 4.1(d), for a rotation by a randomly assigned set of Euler angles, e.g., $(10°, 10°, 10°)$, the atoms in the rotated cube will take different positions as those in the fixed cube. In this case, the orientation defined by the assigned Euler angles is not considered as a symmetry orientation. However, when a symmetric orientation is selected, say, a rotation by $(90°, 90°, 0°)$ as shown in Figure 4.1(e), the positions of the corresponding atoms in the rotated and fixed coordinates overlap precisely as if no rotation is made. In this case, the only visible difference in Figures 4.1(c) and 4.1(e) is in the orientations of the axes in the *xyz* coordinates. Another symmetric orientation is given by $(180°, 90°, 90°)$ as shown in Figure 4.1(f), where we can see again that aside from the different orientations of the *xyz* axes, nothing seems changed. So a symmetric orientation in a crystal means that the crystal possess the same atomic structure after changing the frame of reference (or the viewing angle) through a rotation of the coordinates system defined by the Euler angles.

4.2 Triclinic Crystals

In the triclinic group, there are two classes of triclinic crystals, namely, the **Triclinic 1** and **Triclinic** $\bar{1}$. Of the two, the former is piezoelectric and the latter is not.

4.2.1 Class 1: Triclinic 1

As illustrated in Figure 4.2, the **Triclinic 1** class does not have any symmetry; hence, we use a 3×3 identity matrix as its transformation matrix.

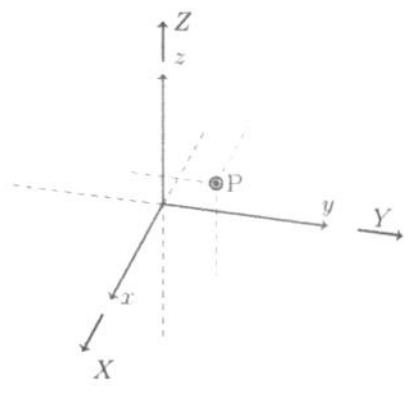

$$[a] = \begin{bmatrix} 1 & 0 & 0 \\ 0 & 1 & 0 \\ 0 & 0 & 1 \end{bmatrix}_1$$

Figure 4.2 Triclinic 1.

Using Eq. (3.4), we obtain a 6×6 identity $[M]$ matrix:

$$[M] = \begin{bmatrix} 1 & 0 & 0 & 0 & 0 & 0 \\ 0 & 1 & 0 & 0 & 0 & 0 \\ 0 & 0 & 1 & 0 & 0 & 0 \\ 0 & 0 & 0 & 1 & 0 & 0 \\ 0 & 0 & 0 & 0 & 1 & 0 \\ 0 & 0 & 0 & 0 & 0 & 1 \end{bmatrix}$$

Referring to Eqs. (3.8), (3.10), and (3.12), it becomes clear that the matrices of various material properties of this group of triclinic crystals are fully filled with non-zero elements as

$$[c] = \begin{bmatrix} c_{11} & c_{12} & c_{13} & c_{14} & c_{15} & c_{16} \\ c_{12} & c_{22} & c_{23} & c_{24} & c_{25} & c_{26} \\ c_{13} & c_{23} & c_{33} & c_{34} & c_{35} & c_{36} \\ c_{14} & c_{24} & c_{34} & c_{44} & c_{45} & c_{46} \\ c_{15} & c_{25} & c_{35} & c_{45} & c_{55} & c_{56} \\ c_{16} & c_{26} & c_{36} & c_{46} & c_{56} & c_{66} \end{bmatrix},$$

$$[e] = \begin{bmatrix} e_{x1} & e_{x2} & e_{x3} & e_{x4} & e_{x5} & e_{x6} \\ e_{y1} & e_{y2} & e_{y3} & e_{y4} & e_{y5} & e_{y6} \\ e_{z1} & e_{z2} & e_{z3} & e_{z4} & e_{z5} & e_{z6} \end{bmatrix},$$

$$[\epsilon] = \begin{bmatrix} \epsilon_{xx} & \epsilon_{xy} & \epsilon_{xz} \\ \epsilon_{xy} & \epsilon_{yy} & \epsilon_{yz} \\ \epsilon_{xz} & \epsilon_{yz} & \epsilon_{zz} \end{bmatrix}$$

Because of the symmetric nature of the $[c]$ and $[\epsilon]$ matrices with respect to their principal diagonal, these matrices of the **Triclinic 1** group have 21 and 6 constants, respectively. The piezoelectric matrix, $[e]$, has 18 constants.

4.2.2 Class 2: Triclinic $\bar{1}$

The **Triclinic $\bar{1}$** class has inverted symmetry with respect to the origin of the coordinates system (i.e., symmetric after multiplying all three coordinates by -1). Figure 4.3 illustrates this type of symmetry and gives the corresponding transformation matrix.

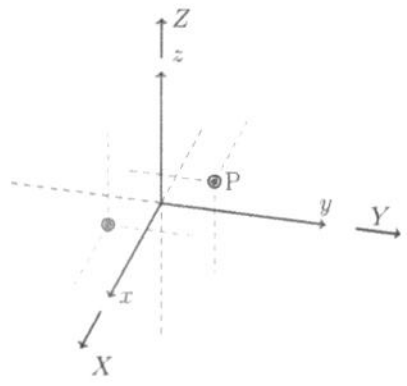

$$[a] = \begin{bmatrix} -1 & 0 & 0 \\ 0 & -1 & 0 \\ 0 & 0 & -1 \end{bmatrix}_{\bar{1}}$$

Figure 4.3 Triclinic $\bar{1}$.

By Eq. (3.4), this transformation matrix also leads to an identity $[M]$ matrix. With Eqs. (3.8) and (3.12), we find

$$[c] = \begin{bmatrix} c_{11} & c_{12} & c_{13} & c_{14} & c_{15} & c_{16} \\ c_{12} & c_{22} & c_{23} & c_{24} & c_{25} & c_{26} \\ c_{13} & c_{23} & c_{33} & c_{34} & c_{35} & c_{36} \\ c_{14} & c_{24} & c_{34} & c_{44} & c_{45} & c_{46} \\ c_{15} & c_{25} & c_{35} & c_{45} & c_{55} & c_{56} \\ c_{16} & c_{26} & c_{36} & c_{46} & c_{56} & c_{66} \end{bmatrix}, \quad [\epsilon] = \begin{bmatrix} \epsilon_{xx} & \epsilon_{xy} & \epsilon_{xz} \\ \epsilon_{xy} & \epsilon_{yy} & \epsilon_{yz} \\ \epsilon_{xz} & \epsilon_{yz} & \epsilon_{zz} \end{bmatrix}$$

However, for the $[e]$ matrix, referring to Eq. (3.10), since $[e'] = [a][e][M]^T = -[e]$, we conclude that only the follow outcome can satisfy the condition of $[e] = -[e]$:

$$[e] = \begin{bmatrix} 0 & 0 & 0 & 0 & 0 & 0 \\ 0 & 0 & 0 & 0 & 0 & 0 \\ 0 & 0 & 0 & 0 & 0 & 0 \end{bmatrix}$$

This means that the **Triclinic $\bar{\mathbf{1}}$** class is not piezoelectric. Although the stiffness ($[c]$) and permittivity ($[\epsilon]$) matrices of the **Triclinic $\bar{\mathbf{1}}$** class are full matrices, they have only 21 and 6 independent constants, respectively, due to the symmetric nature of these matrices.

4.3 Monoclinic Crystals

In the monoclinic group, there are three classes of monoclinic crystals, namely, **Monoclinic 2**, **Monoclinic m** and **Monoclinic 2/m**. Of these three classes, two (i.e., the **Monoclinic 2** and **Monoclinic m** groups) are piezoelectric and one (**Monoclinic 2/m**) is not.

4.3.1 Class 3: Monoclinic 2

The **Monoclinic 2** class has two-fold symmetry (e.g., symmetric after rotating about the Y axis by 180°). Figure 4.4 shows this type of symmetry along with the corresponding transformation matrix.

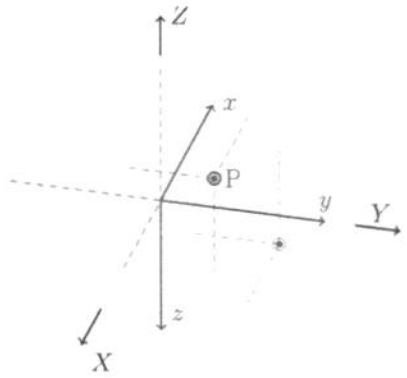

$$[a] = \begin{bmatrix} -1 & 0 & 0 \\ 0 & 1 & 0 \\ 0 & 0 & -1 \end{bmatrix}_{2Y}$$

Figure 4.4 Monoclinic 2.

Using Eq. (3.4), we obtain its corresponding $[M]$ matrix as:

$$[M] = \begin{bmatrix} 1 & 0 & 0 & 0 & 0 & 0 \\ 0 & 1 & 0 & 0 & 0 & 0 \\ 0 & 0 & 1 & 0 & 0 & 0 \\ 0 & 0 & 0 & -1 & 0 & 0 \\ 0 & 0 & 0 & 0 & 1 & 0 \\ 0 & 0 & 0 & 0 & 0 & -1 \end{bmatrix}$$

By applying Eqs. (3.8), (3.10) and (3.12), we obtain the following for the given symmetric transformation:

$$[c'] = \begin{bmatrix} c_{11} & c_{12} & c_{13} & -c_{14} & c_{15} & -c_{16} \\ c_{12} & c_{22} & c_{23} & -c_{24} & c_{25} & -c_{26} \\ c_{13} & c_{23} & c_{33} & -c_{34} & c_{35} & -c_{36} \\ -c_{14} & -c_{24} & -c_{34} & c_{44} & -c_{45} & c_{46} \\ c_{15} & c_{25} & c_{35} & -c_{45} & c_{55} & -c_{56} \\ -c_{16} & -c_{26} & -c_{36} & c_{46} & -c_{56} & c_{66} \end{bmatrix}$$

$$[e'] = \begin{bmatrix} -e_{x1} & -e_{x2} & -e_{x3} & e_{x4} & -e_{x5} & e_{x6} \\ e_{y1} & e_{y2} & e_{y3} & -e_{y4} & e_{y5} & -e_{y6} \\ -e_{z1} & -e_{z2} & -e_{z3} & e_{z4} & -e_{z5} & e_{z6} \end{bmatrix}, \; [\epsilon'] = \begin{bmatrix} \epsilon_{xx} & -\epsilon_{xy} & \epsilon_{xz} \\ -\epsilon_{xy} & \epsilon_{yy} & -\epsilon_{yz} \\ \epsilon_{xz} & -\epsilon_{yz} & \epsilon_{zz} \end{bmatrix}$$

Because of the expected symmetry, these transformations are expected to yield the same material properties as the respective original matrices. By equating the corresponding original and transformed matrices, we reach the conclusion that all these negative terms should be zero. This leads to the following property matrices for the **Monoclinic 2** class of crystals:

$$[c] = \begin{bmatrix} c_{11} & c_{12} & c_{13} & 0 & c_{15} & 0 \\ c_{12} & c_{22} & c_{23} & 0 & c_{25} & 0 \\ c_{13} & c_{23} & c_{33} & 0 & c_{35} & 0 \\ 0 & 0 & 0 & c_{44} & 0 & c_{46} \\ c_{15} & c_{25} & c_{35} & 0 & c_{55} & 0 \\ 0 & 0 & 0 & c_{46} & 0 & c_{66} \end{bmatrix}$$

$$[e] = \begin{bmatrix} 0 & 0 & 0 & e_{x4} & 0 & e_{x6} \\ e_{y1} & e_{y2} & e_{y3} & 0 & e_{y5} & 0 \\ 0 & 0 & 0 & e_{z4} & 0 & e_{z6} \end{bmatrix}, \quad [\epsilon] = \begin{bmatrix} \epsilon_{xx} & 0 & \epsilon_{xz} \\ 0 & \epsilon_{yy} & 0 \\ \epsilon_{xz} & 0 & \epsilon_{zz} \end{bmatrix}$$

4.3.2 Class 4: Monoclinic m

The **Monoclinic m** class has a mirror-plane type of symmetry about the XZ plane. Figure 4.5 shows this type of symmetry along with the corresponding transformation matrix.

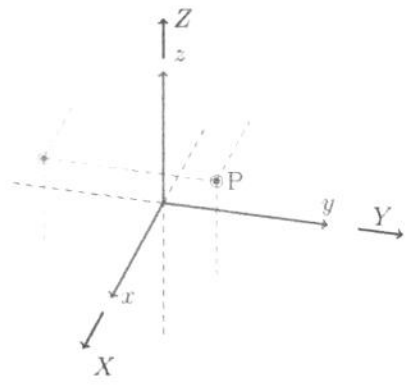

$$[a] = \begin{bmatrix} 1 & 0 & 0 \\ 0 & -1 & 0 \\ 0 & 0 & 1 \end{bmatrix}_{mXZ}$$

Figure 4.5 Monoclinic m.

Using Eq. (3.4), we obtain the same $[M]$ matrix as that for the **Monoclinic 2** class, which will yield the same $[c]$ matrix according to Eq. (3.8). By applying Eqs. (3.10) and (3.12), we obtain the following:

$$[e'] = \begin{bmatrix} e_{x1} & e_{x2} & e_{x3} & -e_{x4} & e_{x5} & -e_{x6} \\ -e_{y1} & -e_{y2} & -e_{y3} & e_{y4} & -e_{y5} & e_{y6} \\ e_{z1} & e_{z2} & e_{z3} & -e_{z4} & e_{z5} & -e_{z6} \end{bmatrix}, \quad [\epsilon'] = \begin{bmatrix} \epsilon_{xx} & -\epsilon_{xy} & \epsilon_{xz} \\ -\epsilon_{xy} & \epsilon_{yy} & -\epsilon_{yz} \\ \epsilon_{xz} & -\epsilon_{yz} & \epsilon_{zz} \end{bmatrix}$$

By setting all these negative terms to zero, we obtain the following property matrices for the **Monoclinic m** class of crystals:

$$[c] = \begin{bmatrix} c_{11} & c_{12} & c_{13} & 0 & c_{15} & 0 \\ c_{12} & c_{22} & c_{23} & 0 & c_{25} & 0 \\ c_{13} & c_{23} & c_{33} & 0 & c_{35} & 0 \\ 0 & 0 & 0 & c_{44} & 0 & c_{46} \\ c_{15} & c_{25} & c_{35} & 0 & c_{55} & 0 \\ 0 & 0 & 0 & c_{46} & 0 & c_{66} \end{bmatrix}$$

$$[e] = \begin{bmatrix} e_{x1} & e_{x2} & e_{x3} & 0 & e_{x5} & 0 \\ 0 & 0 & 0 & e_{y4} & 0 & e_{y6} \\ e_{z1} & e_{z2} & e_{z3} & 0 & e_{z5} & 0 \end{bmatrix}, \quad [\epsilon] = \begin{bmatrix} \epsilon_{xx} & 0 & \epsilon_{xz} \\ 0 & \epsilon_{yy} & 0 \\ \epsilon_{xz} & 0 & \epsilon_{zz} \end{bmatrix}$$

Note that the $[c]$ matrix of the **Monoclinic 2** class is the same as that of **Monoclinic m**. The same can be said for the $[\epsilon]$ matrix. For the $[e]$ matrix, interestingly, the non-zero terms take the positions in the matrix in a complementary way.

4.3.3 Class 5: Monoclinic 2/m

The **Monoclinic 2/m** class has both the symmetry of the **Monoclinic 2** and the **Monoclinic m** groups.

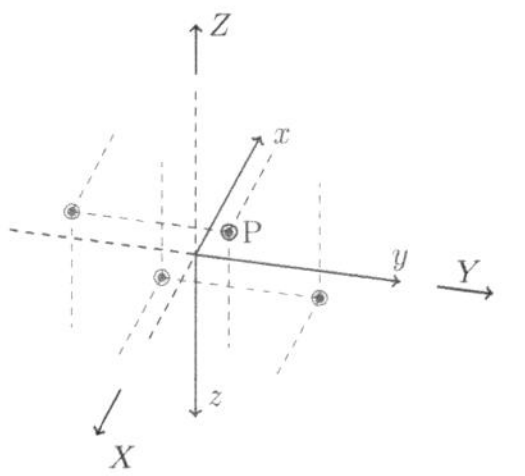

$$[a] = \begin{bmatrix} -1 & 0 & 0 \\ 0 & 1 & 0 \\ 0 & 0 & -1 \end{bmatrix}_{2Y}, \quad [a] = \begin{bmatrix} 1 & 0 & 0 \\ 0 & -1 & 0 \\ 0 & 0 & 1 \end{bmatrix}_{mXZ}$$

Figure 4.6 Monoclinic 2/m.

Figure 4.6 shows this combined symmetry and the two corresponding transformation matrices. This type of symmetry requires that symmetry hold after either and both transformation.

This means that we can simply equate the corresponding matrices for the **Monoclinic 2** and the **Monoclinic m** classes. Doing so, we find

$$[c] = \begin{bmatrix} c_{11} & c_{12} & c_{13} & 0 & c_{15} & 0 \\ c_{12} & c_{22} & c_{23} & 0 & c_{25} & 0 \\ c_{13} & c_{23} & c_{33} & 0 & c_{35} & 0 \\ 0 & 0 & 0 & c_{44} & 0 & c_{46} \\ c_{15} & c_{25} & c_{35} & 0 & c_{55} & 0 \\ 0 & 0 & 0 & c_{46} & 0 & c_{66} \end{bmatrix}, [e] = \begin{bmatrix} 0 & 0 & 0 & 0 & 0 & 0 \\ 0 & 0 & 0 & 0 & 0 & 0 \\ 0 & 0 & 0 & 0 & 0 & 0 \end{bmatrix}, [\epsilon] = \begin{bmatrix} \epsilon_{xx} & 0 & \epsilon_{xz} \\ 0 & \epsilon_{yy} & 0 \\ \epsilon_{xz} & 0 & \epsilon_{zz} \end{bmatrix}$$

Clearly, the $[c]$ and $[\epsilon]$ matrices of the **Monoclinic 2/m** class are the same as those of **Monoclinic 2** and **Monoclinic m**, respectively, but the **Monoclinic 2/m** class is non-piezoelectric because $[e] = 0$.

4.4 Orthorhombic Crystals

In the orthorhombic group, there are three classes of crystals, including **Orthorhombic 2mm**, **Orthorhombic 222**, and **Orthorhombic mmm**. Of these three classes, two (i.e., **Orthorhombic 2mm** and **Orthorhombic 222**) are piezoelectric and one (**Orthorhombic mmm**) is not.

4.4.1 Class 6: Orthorhombic 2mm

The **Orthorhombic 2mm** class has two kinds of symmetry: (i) two-fold symmetry about the Z axis and (ii) a mirror plane of XZ.

Its illustrated symmetry and transformation matrices are given in Figure 4.7. With the first transformation matrix, we obtain (using

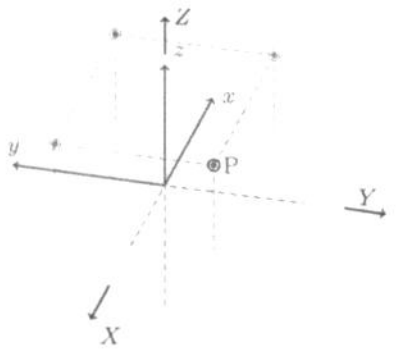

$$[a] = \begin{bmatrix} -1 & 0 & 0 \\ 0 & -1 & 0 \\ 0 & 0 & 1 \end{bmatrix}_{2Z}, \quad [a] = \begin{bmatrix} 1 & 0 & 0 \\ 0 & -1 & 0 \\ 0 & 0 & 1 \end{bmatrix}_{mXZ}$$

Figure 4.7 Orthorhombic 2mm.

Eq. (3.4))

$$[M] = \begin{bmatrix} 1 & 0 & 0 & 0 & 0 & 0 \\ 0 & 1 & 0 & 0 & 0 & 0 \\ 0 & 0 & 1 & 0 & 0 & 0 \\ 0 & 0 & 0 & -1 & 0 & 0 \\ 0 & 0 & 0 & 0 & -1 & 0 \\ 0 & 0 & 0 & 0 & 0 & 1 \end{bmatrix}$$

So this two-fold rotation will yield the following material property matrices after applying Eqs. (3.8), (3.10) and (3.12):

$$[c'] = \begin{bmatrix} c_{11} & c_{12} & c_{13} & -c_{14} & -c_{15} & c_{16} \\ c_{12} & c_{22} & c_{23} & -c_{24} & -c_{25} & c_{26} \\ c_{13} & c_{23} & c_{33} & -c_{34} & -c_{35} & c_{36} \\ -c_{14} & -c_{24} & -c_{34} & c_{44} & c_{45} & -c_{46} \\ -c_{15} & -c_{25} & -c_{35} & c_{45} & c_{55} & -c_{56} \\ c_{16} & c_{26} & c_{36} & -c_{46} & -c_{56} & c_{66} \end{bmatrix}$$

$$[e'] = \begin{bmatrix} -e_{x1} & -e_{x2} & -e_{x3} & e_{x4} & e_{x5} & -e_{x6} \\ -e_{y1} & -e_{y2} & -e_{y3} & e_{y4} & e_{y5} & -e_{y6} \\ e_{z1} & e_{z2} & e_{z3} & -e_{z4} & -e_{z5} & e_{z6} \end{bmatrix}, \ [\epsilon'] = \begin{bmatrix} \epsilon_{xx} & \epsilon_{xy} & -\epsilon_{xz} \\ \epsilon_{xy} & \epsilon_{yy} & -\epsilon_{yz} \\ -\epsilon_{xz} & -\epsilon_{yz} & \epsilon_{zz} \end{bmatrix}$$

By the nature of symmetry, all these negative terms should be zero.

With the mirroring transformation matrix, we obtain

$$[M] = \begin{bmatrix} 1 & 0 & 0 & 0 & 0 & 0 \\ 0 & 1 & 0 & 0 & 0 & 0 \\ 0 & 0 & 1 & 0 & 0 & 0 \\ 0 & 0 & 0 & -1 & 0 & 0 \\ 0 & 0 & 0 & 0 & 1 & 0 \\ 0 & 0 & 0 & 0 & 0 & -1 \end{bmatrix}$$

which leads to the following material property matrices after applying Eqs. (3.8), (3.10) and (3.12) to the corresponding ones after the two-fold rotation:

$$[c''] = \begin{bmatrix} c_{11} & c_{12} & c_{13} & 0 & 0 & -c_{16} \\ c_{12} & c_{22} & c_{23} & 0 & 0 & -c_{26} \\ c_{13} & c_{23} & c_{33} & 0 & 0 & -c_{36} \\ 0 & 0 & 0 & c_{44} & -c_{45} & 0 \\ 0 & 0 & 0 & -c_{45} & c_{55} & 0 \\ -c_{16} & -c_{26} & -c_{36} & 0 & 0 & c_{66} \end{bmatrix}$$

$$[e''] = \begin{bmatrix} 0 & 0 & 0 & -e_{x4} & e_{x5} & 0 \\ 0 & 0 & 0 & e_{y4} & -e_{y5} & 0 \\ -e_{z1} & -e_{z2} & -e_{z3} & 0 & 0 & -e_{z6} \end{bmatrix}, \quad [\epsilon''] = \begin{bmatrix} \epsilon_{xx} & -\epsilon_{xy} & 0 \\ -\epsilon_{xy} & \epsilon_{yy} & 0 \\ 0 & 0 & \epsilon_{zz} \end{bmatrix}$$

Again, by the nature of symmetry, all these negative terms should be zero. This leads to the following material property matrices for the **Orthorhombic 2mm** class of crystals:

$$[c] = \begin{bmatrix} c_{11} & c_{12} & c_{13} & 0 & 0 & 0 \\ c_{12} & c_{22} & c_{23} & 0 & 0 & 0 \\ c_{13} & c_{23} & c_{33} & 0 & 0 & 0 \\ 0 & 0 & 0 & c_{44} & 0 & 0 \\ 0 & 0 & 0 & 0 & c_{55} & 0 \\ 0 & 0 & 0 & 0 & 0 & c_{66} \end{bmatrix}, \quad [e] = \begin{bmatrix} 0 & 0 & 0 & 0 & e_{x5} & 0 \\ 0 & 0 & 0 & e_{y4} & 0 & 0 \\ e_{z1} & e_{z2} & e_{z3} & 0 & 0 & 0 \end{bmatrix},$$

$$[\epsilon] = \begin{bmatrix} \epsilon_{xx} & 0 & 0 \\ 0 & \epsilon_{yy} & 0 \\ 0 & 0 & \epsilon_{zz} \end{bmatrix}$$

4.4.2 Class 7: Orthorhombic 222

The **Orthorhombic 222** class has two kinds of two-fold symmetry: (i) symmetric after rotating about the Z axis by 180°, and (ii) symmetric after rotating about the X axis by 180°. Figure 4.8 shows these two types of rotational symmetry along with the corresponding transformation matrices. By applying the same steps as in previous cases, we obtain the material property matrices for the **Orthorhombic 222** class of crystals as given below.

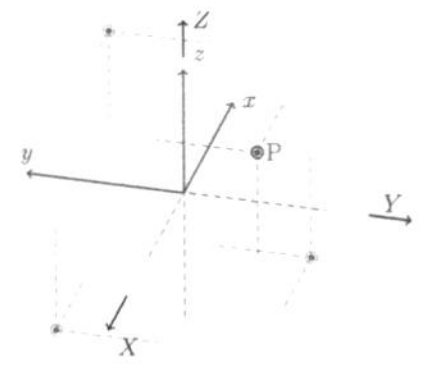

$$[a] = \begin{bmatrix} -1 & 0 & 0 \\ 0 & -1 & 0 \\ 0 & 0 & 1 \end{bmatrix}_{2Z}$$

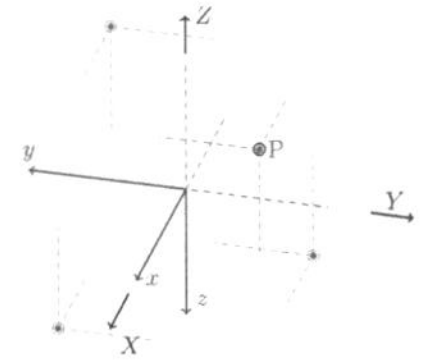

$$[a] = \begin{bmatrix} 1 & 0 & 0 \\ 0 & -1 & 0 \\ 0 & 0 & -1 \end{bmatrix}_{2X}$$

Figure 4.8 Orthorhombic 222.

$$[c] = \begin{bmatrix} c_{11} & c_{12} & c_{13} & 0 & 0 & 0 \\ c_{12} & c_{22} & c_{23} & 0 & 0 & 0 \\ c_{13} & c_{23} & c_{33} & 0 & 0 & 0 \\ 0 & 0 & 0 & c_{44} & 0 & 0 \\ 0 & 0 & 0 & 0 & c_{55} & 0 \\ 0 & 0 & 0 & 0 & 0 & c_{66} \end{bmatrix}, \; [e] = \begin{bmatrix} 0 & 0 & 0 & e_{x4} & 0 & 0 \\ 0 & 0 & 0 & 0 & e_{y5} & 0 \\ 0 & 0 & 0 & 0 & 0 & e_{z6} \end{bmatrix}. \; [\epsilon] = \begin{bmatrix} \epsilon_{xx} & 0 & 0 \\ 0 & \epsilon_{yy} & 0 \\ 0 & 0 & \epsilon_{zz} \end{bmatrix}$$

4.4.3 Class 8: Orthorhombic mmm

The **Orthorhombic mmm** class has three mirror planes, perpendicular to each of the three axes, the X, Y and Z axes, respectively. Its transformation matrices and illustrated symmetry are given in Figure 4.9.

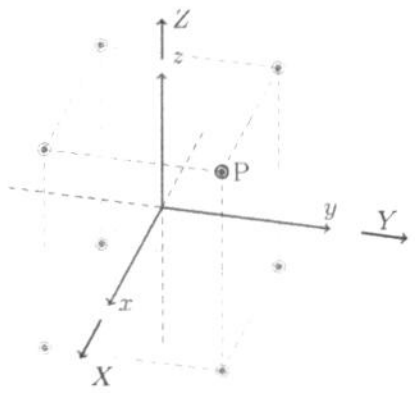

$$[a] = \begin{bmatrix} -1 & 0 & 0 \\ 0 & 1 & 0 \\ 0 & 0 & 1 \end{bmatrix}_{mYZ}, \; [a] = \begin{bmatrix} 1 & 0 & 0 \\ 0 & -1 & 0 \\ 0 & 0 & 1 \end{bmatrix}_{mXZ}, \; [a] = \begin{bmatrix} 1 & 0 & 0 \\ 0 & 1 & 0 \\ 0 & 0 & -1 \end{bmatrix}_{mXY}$$

Figure 4.9 Orthorhombic mmm.

Applying the same steps as in previous cases, we obtain the following material property matrices for the **Orthorhombic mmm** class of crystals, which is non-piezoelectric, as

$$[c] = \begin{bmatrix} c_{11} & c_{12} & c_{13} & 0 & 0 & 0 \\ c_{12} & c_{22} & c_{23} & 0 & 0 & 0 \\ c_{13} & c_{23} & c_{33} & 0 & 0 & 0 \\ 0 & 0 & 0 & c_{44} & 0 & 0 \\ 0 & 0 & 0 & 0 & c_{55} & 0 \\ 0 & 0 & 0 & 0 & 0 & c_{66} \end{bmatrix}, [e] = \begin{bmatrix} 0 & 0 & 0 & 0 & 0 & 0 \\ 0 & 0 & 0 & 0 & 0 & 0 \\ 0 & 0 & 0 & 0 & 0 & 0 \end{bmatrix}, [\epsilon] = \begin{bmatrix} \epsilon_{xx} & 0 & 0 \\ 0 & \epsilon_{yy} & 0 \\ 0 & 0 & \epsilon_{zz} \end{bmatrix}$$

Example 1: For orthorhombic 2mm crystal barium sodium niobate ($Ba_2NaNb_5O_{15}$), use the two-fold and mirror symmetry matrices listed in Case 6 to show that the stiffness matrix will not change after each rotation.

Answer: From Example 4 in Chapter 3, we copy the stiffness matrix here

$$[c] = 10^{10} \times \begin{bmatrix} 23.9 & 10.4 & 5.0 & 0 & 0 & 0 \\ 10.4 & 24.7 & 5.2 & 0 & 0 & 0 \\ 5.0 & 5.2 & 13.5 & 0 & 0 & 0 \\ 0 & 0 & 0 & 6.5 & 0 & 0 \\ 0 & 0 & 0 & 0 & 6.6 & 0 \\ 0 & 0 & 0 & 0 & 0 & 7.6 \end{bmatrix}$$

With the $[a]$ and $[M]$ for the two-fold symmetry in Case 6, by using $[c] = [M][c][M]^T$, we obtain the same stiffness matrix as listed above. Similarly with the $[a]$ and $[M]$ for the mirror symmetry in Case, we again obtain the same stiffness matrix, hence confirming the fact that the material property will not change after these transformations.

4.5 Tetragonal Crystals

In the tetragonal group, there are seven classes of crystals and they are **Tetragonal 4**, **Tetragonal $\bar{4}$**, **Tetragonal 4/m**, **Tetragonal 422**, **Tetragonal 4mm**, **Tetragonal $\bar{4}$2m**, and **Tetragonal 4/mmm**.

Of these classes, five are piezoelectric (i.e., the **Tetragonal 4**, **Tetragonal $\bar{4}$**, **Tetragonal 422**, **Tetragonal 4mm** and **Tetragonal $\bar{4}$2m**) and two (i.e., **Tetragonal 4/m** and **Tetragonal 4/mmm**) are not.

4.5.1 Class 9: Tetragonal 4

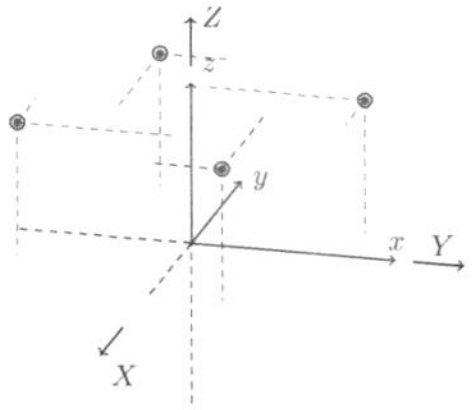

$$[a] = \begin{bmatrix} 0 & 1 & 0 \\ -1 & 0 & 0 \\ 0 & 0 & 1 \end{bmatrix}_{4Z}$$

Figure 4.10 Tetragonal 4.

The **Tetragonal 4** class has four-fold symmetry (e.g., symmetric after rotating about the Z axis by 90°). Figure 4.10 shows this type of symmetry along with its corresponding transformation matrix. With this transformation matrix, we obtain the following $[M]$ matrix by using Eq. (3.4):

$$[M] = \begin{bmatrix} 0 & 1 & 0 & 0 & 0 & 0 \\ 1 & 0 & 0 & 0 & 0 & 0 \\ 0 & 0 & 1 & 0 & 0 & 0 \\ 0 & 0 & 0 & 0 & -1 & 0 \\ 0 & 0 & 0 & 1 & 0 & 0 \\ 0 & 0 & 0 & 0 & 0 & -1 \end{bmatrix}$$

After applying Eqs. (3.8), (3.10) and (3.12), we obtain the following material property matrices after the rotation:

$$[c'] = \begin{bmatrix} c_{22} & c_{12} & c_{23} & -c_{25} & c_{24} & -c_{26} \\ c_{12} & c_{11} & c_{13} & -c_{15} & c_{14} & -c_{16} \\ c_{23} & c_{13} & c_{33} & -c_{35} & c_{34} & -c_{36} \\ -c_{25} & -c_{15} & -c_{35} & c_{55} & -c_{45} & c_{56} \\ c_{24} & c_{14} & c_{34} & -c_{45} & c_{44} & -c_{46} \\ -c_{26} & -c_{16} & -c_{36} & c_{56} & -c_{46} & c_{66} \end{bmatrix}$$

$$[e'] = \begin{bmatrix} e_{y2} & e_{y1} & e_{y3} & -e_{y5} & e_{y4} & -e_{y6} \\ -e_{x2} & -e_{x1} & -e_{x3} & e_{x5} & -e_{x4} & e_{x6} \\ e_{z2} & e_{z1} & e_{z3} & -e_{z5} & e_{z4} & -e_{z6} \end{bmatrix},$$

$$[\epsilon'] = \begin{bmatrix} \epsilon_{yy} & -\epsilon_{xy} & \epsilon_{yz} \\ -\epsilon_{xy} & \epsilon_{xx} & -\epsilon_{xz} \\ \epsilon_{yz} & -\epsilon_{xz} & \epsilon_{zz} \end{bmatrix}$$

Taking the difference of the corresponding matrices before and after the transformation, we set

$$[\Delta c] = \begin{bmatrix} c_{11}-c_{22} & 0 & c_{13}-c_{23} & c_{14}+c_{25} & c_{15}-c_{24} & c_{16}+c_{26} \\ 0 & c_{22}-c_{11} & c_{23}-c_{13} & c_{15}+c_{24} & c_{25}-c_{14} & c_{16}+c_{26} \\ c_{13}-c_{23} & c_{23}-c_{13} & 0 & c_{34}+c_{35} & c_{35}-c_{34} & 2c_{36} \\ c_{14}+c_{25} & c_{15}+c_{24} & c_{34}+c_{35} & c_{44}-c_{55} & 2c_{45} & c_{46}-c_{56} \\ c_{15}-c_{24} & c_{25}-c_{14} & c_{35}-c_{34} & 2c_{45} & c_{55}-c_{44} & c_{46}+c_{56} \\ c_{16}+c_{26} & c_{16}+c_{26} & 2c_{36} & c_{46}-c_{56} & c_{46}+c_{56} & 0 \end{bmatrix} = 0$$

$$[\Delta e] = \begin{bmatrix} e_{x1}-e_{y2} & e_{x2}-e_{y1} & e_{x3}-e_{y3} & e_{x4}+e_{y5} & e_{x5}-e_{y4} & e_{x6}+e_{y6} \\ e_{x2}+e_{y1} & e_{x1}+e_{y2} & e_{x3}+e_{y3} & e_{y4}-e_{x5} & e_{x4}+e_{y5} & e_{y6}-e_{x6} \\ e_{z1}-e_{z2} & e_{z2}-e_{z1} & 0 & e_{z4}+e_{z5} & e_{z5}-e_{z4} & 2e_{z6} \end{bmatrix} = 0$$

$$[\Delta \epsilon] = \begin{bmatrix} \epsilon_{xx}-\epsilon_{yy} & 2\epsilon_{xy} & \epsilon_{xz}-\epsilon_{yz} \\ 2\epsilon_{xy} & \epsilon_{yy}-\epsilon_{xx} & \epsilon_{xz}+\epsilon_{yz} \\ \epsilon_{xz}-\epsilon_{yz} & \epsilon_{xz}+\epsilon_{yz} & 0 \end{bmatrix} = 0$$

To meet the symmetric requirements, all elements in these three matrices have to be zero. From the $[\Delta c]$ matrix, we can see that

$$c_{22} = c_{11}, \; c_{23} = c_{13}, \; c_{26} = -c_{16}, \; c_{36} = c_{45} = 0, \; c_{55} = c_{44}$$

Also, to meet the conditions of

$$c_{14} + c_{25} = 0 \text{ and } c_{14} - c_{25} = 0;$$

$$c_{15} + c_{24} = 0 \text{ and } c_{15} - c_{24} = 0$$

$$c_{34} + c_{35} = 0 \text{ and } c_{34} - c_{35} = 0;$$

$$c_{46} + c_{56} = 0 \text{ and } c_{46} - c_{56} = 0$$

simultaneously, we arrive at

$$c_{14} = c_{15} = c_{24} = c_{25} = c_{34} = c_{35} = c_{46} = c_{56} = 0$$

Taken together, we obtain the following stiffness matrix for the **Tetragonal 4** class of crystals:

$$[c] = \begin{bmatrix} c_{11} & c_{12} & c_{13} & 0 & 0 & c_{16} \\ c_{12} & c_{11} & c_{13} & 0 & 0 & -c_{16} \\ c_{13} & c_{13} & c_{33} & 0 & 0 & 0 \\ 0 & 0 & 0 & c_{44} & 0 & 0 \\ 0 & 0 & 0 & 0 & c_{44} & 0 \\ c_{16} & -c_{16} & 0 & 0 & 0 & c_{66} \end{bmatrix}$$

By the same exercises, we find their piezoelectric and permittivity matrices as:

$$[e] = \begin{bmatrix} 0 & 0 & 0 & e_{x4} & e_{x5} & 0 \\ 0 & 0 & 0 & e_{x5} & -e_{x4} & 0 \\ e_{z1} & e_{z1} & e_{z3} & 0 & 0 & 0 \end{bmatrix}, \quad [\epsilon] = \begin{bmatrix} \epsilon_{xx} & 0 & 0 \\ 0 & \epsilon_{xx} & 0 \\ 0 & 0 & \epsilon_{zz} \end{bmatrix}$$

4.5.2 Class 10: Tetragonal $\bar{4}$

The **Tetragonal $\bar{4}$** class has four-fold inverted symmetry (e.g., symmetric after rotating about the Z axis by 90° followed by an inversion, i.e., multiplying by −1).

Its transformation matrix and illustrated symmetry are given in Figure 4.11. By taking the same steps as for the **Tetragonal 4** class,

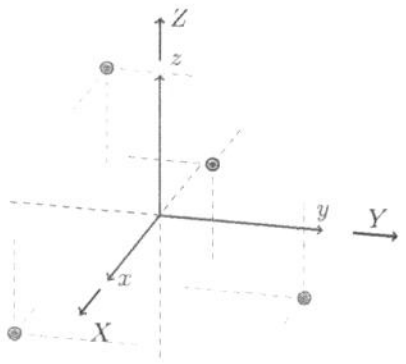

$$[a] = \begin{bmatrix} 0 & -1 & 0 \\ 1 & 0 & 0 \\ 0 & 0 & -1 \end{bmatrix}_{\bar{4}Z}$$

Figure 4.11 Tetragonal $\bar{4}$.

we obtain the following property matrices for the **Tetragonal $\bar{4}$** class of crystals:

$$[c] = \begin{bmatrix} c_{11} & c_{12} & c_{13} & 0 & 0 & c_{16} \\ c_{12} & c_{11} & c_{13} & 0 & 0 & -c_{16} \\ c_{13} & c_{13} & c_{33} & 0 & 0 & 0 \\ 0 & 0 & 0 & c_{44} & 0 & 0 \\ 0 & 0 & 0 & 0 & c_{44} & 0 \\ c_{16} & -c_{16} & 0 & 0 & 0 & c_{66} \end{bmatrix},$$

$$[e] = \begin{bmatrix} 0 & 0 & 0 & e_{x4} & e_{x5} & 0 \\ 0 & 0 & 0 & -e_{x5} & e_{x4} & 0 \\ e_{z1} & -e_{z1} & 0 & 0 & 0 & e_{z6} \end{bmatrix}, \quad [\epsilon] = \begin{bmatrix} \epsilon_{xx} & 0 & 0 \\ 0 & \epsilon_{xx} & 0 \\ 0 & 0 & \epsilon_{zz} \end{bmatrix}$$

4.5.3 Class 11: Tetragonal 4/m

The **Tetragonal 4/m** class has both four-fold symmetry about the Z axis and a mirror plane of XY .

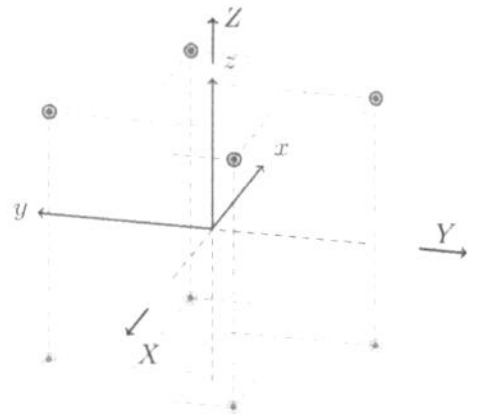

$$[a] = \begin{bmatrix} 0 & 1 & 0 \\ -1 & 0 & 0 \\ 0 & 0 & 1 \end{bmatrix}_{4Z}, \quad [a] = \begin{bmatrix} 1 & 0 & 0 \\ 0 & 1 & 0 \\ 0 & 0 & -1 \end{bmatrix}_{mXY}$$

Figure 4.12 Tetragonal 4/m.

Figure 4.12 shows the combined symmetry along with the two corresponding transformation matrices. Following the same steps twice, we can find the following property matrices for the **Tetragonal 4/m** class of crystals, which is non-piezoelectric:

$$[c] = \begin{bmatrix} c_{11} & c_{12} & c_{13} & 0 & 0 & c_{16} \\ c_{12} & c_{11} & c_{13} & 0 & 0 & -c_{16} \\ c_{13} & c_{13} & c_{33} & 0 & 0 & 0 \\ 0 & 0 & 0 & c_{44} & 0 & 0 \\ 0 & 0 & 0 & 0 & c_{44} & 0 \\ c_{16} & -c_{16} & 0 & 0 & 0 & c_{66} \end{bmatrix}, \quad [e] = \begin{bmatrix} 0 & 0 & 0 & 0 & 0 & 0 \\ 0 & 0 & 0 & 0 & 0 & 0 \\ 0 & 0 & 0 & 0 & 0 & 0 \end{bmatrix}, \quad [\epsilon] = \begin{bmatrix} \epsilon_{xx} & 0 & 0 \\ 0 & \epsilon_{xx} & 0 \\ 0 & 0 & \epsilon_{zz} \end{bmatrix}$$

4.5.4 Class 12: Tetragonal 422

The **Tetragonal 422** class has both four-fold symmetry about the Z axis and two-fold symmetry about the X axis. Figure 4.13 shows the two types of symmetry along with the corresponding transformation matrices. With these matrices, we can find the property matrices for the **Tetragonal 422** class of crystals as given below.

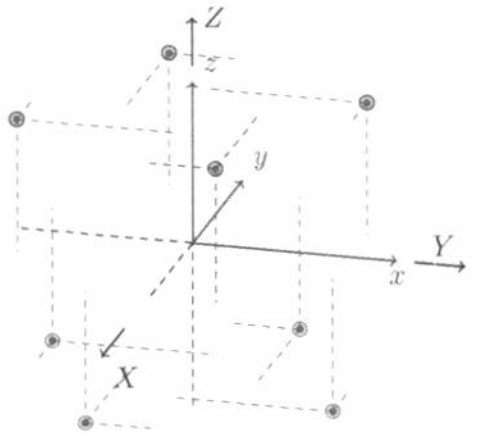

$$[a] = \begin{bmatrix} 0 & 1 & 0 \\ -1 & 0 & 0 \\ 0 & 0 & 1 \end{bmatrix}_{4Z}$$

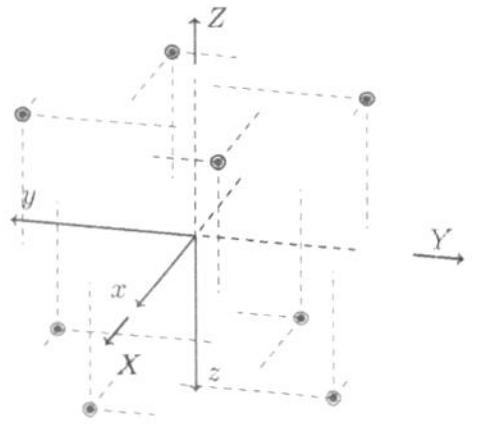

$$[a] = \begin{bmatrix} 1 & 0 & 0 \\ 0 & -1 & 0 \\ 0 & 0 & -1 \end{bmatrix}_{2X}$$

Figure 4.13 Tetragonal 422.

$$[c] = \begin{bmatrix} c_{11} & c_{12} & c_{13} & 0 & 0 & 0 \\ c_{12} & c_{11} & c_{13} & 0 & 0 & 0 \\ c_{13} & c_{13} & c_{33} & 0 & 0 & 0 \\ 0 & 0 & 0 & c_{44} & 0 & 0 \\ 0 & 0 & 0 & 0 & c_{44} & 0 \\ 0 & 0 & 0 & 0 & 0 & c_{66} \end{bmatrix}, \ [e] = \begin{bmatrix} 0 & 0 & 0 & e_{x4} & 0 & 0 \\ 0 & 0 & 0 & 0 & -e_{x4} & 0 \\ 0 & 0 & 0 & 0 & 0 & 0 \end{bmatrix}, \ [\epsilon] = \begin{bmatrix} \epsilon_{xx} & 0 & 0 \\ 0 & \epsilon_{xx} & 0 \\ 0 & 0 & \epsilon_{zz} \end{bmatrix}$$

4.5.5 Class 13: Tetragonal 4mm

The **Tetragonal 4mm** class has both four-fold symmetry and a mirror plane. Its transformation matrices and illustrated symmetry are given in Figure 4.14.

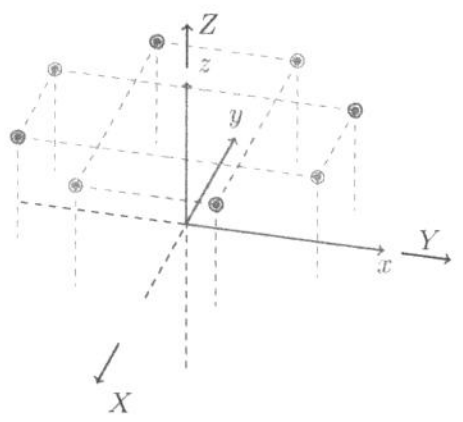

$$[a] = \begin{bmatrix} 0 & 1 & 0 \\ -1 & 0 & 0 \\ 0 & 0 & 1 \end{bmatrix}_{4Z}, \quad [a] = \begin{bmatrix} 1 & 0 & 0 \\ 0 & -1 & 0 \\ 0 & 0 & 1 \end{bmatrix}_{mXZ}$$

Figure 4.14 Tetragonal 4mm.

With these matrices, we can find the property matrices for the **Tetragonal 4mm** class of crystals as given below.

$$[c] = \begin{bmatrix} c_{11} & c_{12} & c_{13} & 0 & 0 & 0 \\ c_{12} & c_{11} & c_{13} & 0 & 0 & 0 \\ c_{13} & c_{13} & c_{33} & 0 & 0 & 0 \\ 0 & 0 & 0 & c_{44} & 0 & 0 \\ 0 & 0 & 0 & 0 & c_{44} & 0 \\ 0 & 0 & 0 & 0 & 0 & c_{66} \end{bmatrix},$$

$$[e] = \begin{bmatrix} 0 & 0 & 0 & 0 & e_{x5} & 0 \\ 0 & 0 & 0 & e_{x5} & 0 & 0 \\ e_{z1} & e_{z1} & e_{z3} & 0 & 0 & 0 \end{bmatrix}, [\epsilon] = \begin{bmatrix} \epsilon_{xx} & 0 & 0 \\ 0 & \epsilon_{xx} & 0 \\ 0 & 0 & \epsilon_{zz} \end{bmatrix}$$

4.5.6 Class 14: Tetragonal $\bar{4}$2m

The **Tetragonal $\bar{4}$2m** class has both inverted four-fold symmetry and two-fold symmetry. Figure 4.15 shows the two types of symmetry along with the corresponding transformation matrices.

With these matrices, we can find the following property matrices for the **Tetragonal $\bar{4}$2m** class of crystals:

$$[c] = \begin{bmatrix} c_{11} & c_{12} & c_{13} & 0 & 0 & 0 \\ c_{12} & c_{11} & c_{13} & 0 & 0 & 0 \\ c_{13} & c_{13} & c_{33} & 0 & 0 & 0 \\ 0 & 0 & 0 & c_{44} & 0 & 0 \\ 0 & 0 & 0 & 0 & c_{44} & 0 \\ 0 & 0 & 0 & 0 & 0 & c_{66} \end{bmatrix}, [e] = \begin{bmatrix} 0 & 0 & 0 & e_{x4} & 0 & 0 \\ 0 & 0 & 0 & 0 & e_{x4} & 0 \\ 0 & 0 & 0 & 0 & 0 & e_{z6} \end{bmatrix}, [\epsilon] = \begin{bmatrix} \epsilon_{xx} & 0 & 0 \\ 0 & \epsilon_{xx} & 0 \\ 0 & 0 & \epsilon_{zz} \end{bmatrix}$$

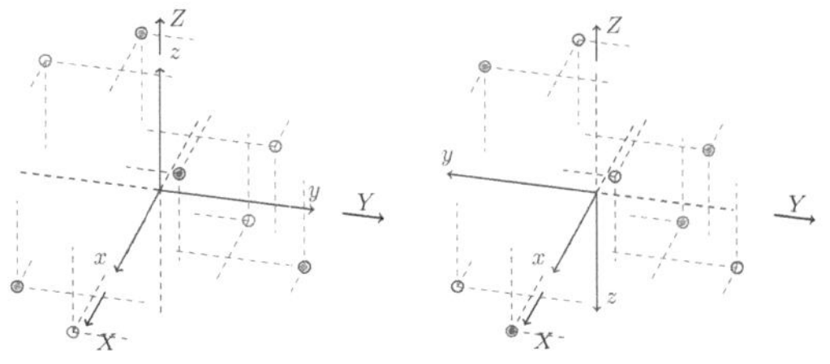

$$[a] = \begin{bmatrix} 0 & -1 & 0 \\ 1 & 0 & 0 \\ 0 & 0 & -1 \end{bmatrix}_{\bar{4}Z} \quad [a] = \begin{bmatrix} 1 & 0 & 0 \\ 0 & -1 & 0 \\ 0 & 0 & -1 \end{bmatrix}_{2X}$$

Figure 4.15 Tetragonal $\bar{4}$2m.

4.5.7 Class 15: Tetragonal 4/mmm

The **Tetragonal 4/mmm** class has four-fold symmetry and two mirror planes. Its transformation matrices and illustrated symmetry are given in Figure 4.16.

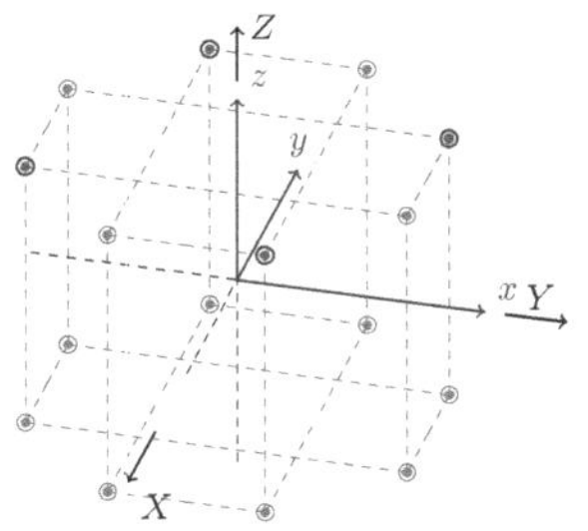

$$[a] = \begin{bmatrix} 0 & 1 & 0 \\ -1 & 0 & 0 \\ 0 & 0 & 1 \end{bmatrix}_{4}, [a] = \begin{bmatrix} 1 & 0 & 0 \\ 0 & 1 & 0 \\ 0 & 0 & -1 \end{bmatrix}_{mXY}, [a] = \begin{bmatrix} 1 & 0 & 0 \\ 0 & -1 & 0 \\ 0 & 0 & 1 \end{bmatrix}_{mXZ}$$

Figure 4.16 Tetragonal 4/mmm.

With these matrices, we can find the following property matrices for the **Tetragonal 4/mmm** class of crystals, which is non-piezoelectric, as

$$[c] = \begin{bmatrix} c_{11} & c_{12} & c_{13} & 0 & 0 & 0 \\ c_{12} & c_{11} & c_{13} & 0 & 0 & 0 \\ c_{13} & c_{13} & c_{33} & 0 & 0 & 0 \\ 0 & 0 & 0 & c_{44} & 0 & 0 \\ 0 & 0 & 0 & 0 & c_{44} & 0 \\ 0 & 0 & 0 & 0 & 0 & c_{66} \end{bmatrix}, [e] = \begin{bmatrix} 0&0&0&0&0&0 \\ 0&0&0&0&0&0 \\ 0&0&0&0&0&0 \end{bmatrix}, [\epsilon] = \begin{bmatrix} \epsilon_{xx} & 0 & 0 \\ 0 & \epsilon_{xx} & 0 \\ 0 & 0 & \epsilon_{zz} \end{bmatrix}$$

Example 2: For tetragonal 422 tellurium oxide TeO_2 crystal with its property matrices given below, show that the two symmetric transformation shown in Figure 4.13 will not change the material properties.

$$[c] = 10^{11} \begin{bmatrix} 0.56 & 0.51 & 0.22 & 0 & 0 & 0 \\ 0.51 & 0.56 & 0.22 & 0 & 0 & 0 \\ 0.22 & 0.22 & 1.06 & 0 & 0 & 0 \\ 0 & 0 & 0 & 0.25 & 0 & 0 \\ 0 & 0 & 0 & 0 & 0.26 & 0 \\ 0 & 0 & 0 & 0 & 0 & 0.66 \end{bmatrix}$$

$$[e] = \begin{bmatrix} 0 & 0 & 0 & 0.216 & 0 & 0 \\ 0 & 0 & 0 & 0 & -0.216 & 0 \\ 0 & 0 & 0 & 0 & 0 & 0 \end{bmatrix};$$

$$[\epsilon] = 10^{-9} \begin{bmatrix} 0.229 & 0 & 0 \\ 0 & 0.229 & 0 \\ 0 & 0 & 0.247 \end{bmatrix}$$

Answer: Referring to the discussion in Case 12, for the four-fold symmetry, we have

$$[a] = \begin{bmatrix} 0 & 1 & 0 \\ -1 & 0 & 0 \\ 0 & 0 & 1 \end{bmatrix}; \quad [M] = \begin{bmatrix} 0&1&0&0&0&0 \\ 1&0&0&0&0&0 \\ 0&0&1&0&0&0 \\ 0&0&0&0&-1&0 \\ 0&0&0&1&0&0 \\ 0&0&0&0&0&-1 \end{bmatrix}$$

and for the two-fold symmetry, we have

$$[a] = \begin{bmatrix} 1 & 0 & 0 \\ 0 & -1 & 0 \\ 0 & 0 & -1 \end{bmatrix}; \quad [M] = \begin{bmatrix} 1 & 0 & 0 & 0 & 0 & 0 \\ 0 & 1 & 0 & 0 & 0 & 0 \\ 0 & 0 & 1 & 0 & 0 & 0 \\ 0 & 0 & 0 & 1 & 0 & 0 \\ 0 & 0 & 0 & 0 & -1 & 0 \\ 0 & 0 & 0 & 0 & 0 & -1 \end{bmatrix}$$

By applying the respective transformation formulas, we obtain the same $[c]$, $[e]$ and $[\epsilon]$ matrices, hence confirming that the crystal will behave exactly the same after these symmetric rotations.

4.6 Trigonal Crystals

In the trigonal group of crystals, there are five classes, including **Trigonal 3**, **Trigonal $\bar{3}$**, **Trigonal 32**, **Trigonal 3m**, and **Trigonal $\bar{3}$m**. Of these classes, three (i.e., **Trigonal 3**, **Trigonal 32** and **Trigonal 3m**) are piezoelectric and two (i.e., **Trigonal $\bar{3}$** and **Trigonal $\bar{3}$m**) are not.

4.6.1 Class 16: Trigonal 3

The **Trigonal 3** class has three-fold symmetry (e.g., symmetric after rotating about the Z axis by 120°). Figure 4.17 shows this type of symmetry along with the corresponding transformation matrix.

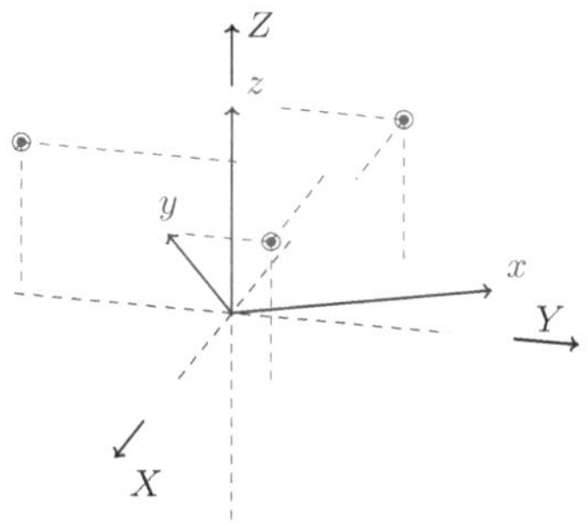

$$[a] = \begin{bmatrix} -\frac{1}{2} & \frac{\sqrt{3}}{2} & 0 \\ -\frac{\sqrt{3}}{2} & -\frac{1}{2} & 0 \\ 0 & 0 & 1 \end{bmatrix}_{3Z}$$

Figure 4.17 Trigonal 3.

With this matrix, we can determine the $[M]$ matrix. With the obtained $[a]$ and $[M]$ matrices, we can further determine the rotated property matrices. By equating the corresponding original and rotated matrices, we can find the layout of all non-zero terms in the property matrices for the **Trigonal 3** class of crystals as given below.

$$[c] = \begin{bmatrix} c_{11} & c_{12} & c_{13} & c_{14} & -c_{25} & 0 \\ c_{12} & c_{11} & c_{13} & -c_{14} & c_{25} & 0 \\ c_{13} & c_{13} & c_{33} & 0 & 0 & 0 \\ c_{14} & -c_{14} & 0 & c_{44} & 0 & c_{25} \\ -c_{25} & c_{25} & 0 & 0 & c_{44} & c_{14} \\ 0 & 0 & 0 & c_{25} & c_{14} & \dfrac{c_{11}-c_{12}}{2} \end{bmatrix}$$

$$[e] = \begin{bmatrix} e_{x1} & -e_{x1} & 0 & e_{x4} & e_{x5} & -e_{y2} \\ -e_{y2} & e_{y2} & 0 & e_{x5} & -e_{x4} & -e_{x1} \\ e_{z1} & e_{z1} & e_{z3} & 0 & 0 & 0 \end{bmatrix}, \quad [\epsilon] = \begin{bmatrix} \epsilon_{xx} & 0 & 0 \\ 0 & \epsilon_{xx} & 0 \\ 0 & 0 & \epsilon_{zz} \end{bmatrix}$$

4.6.2 Class 17: Trigonal $\bar{3}$

The **Trigonal $\bar{3}$** class has inverted three-fold symmetry (e.g., symmetric after a rotation about the Z axis by 120° followed by multiplying –1).

Its transformation matrix and illustrated symmetry are given in Figure 4.18. With this matrix, by following the same steps discussed

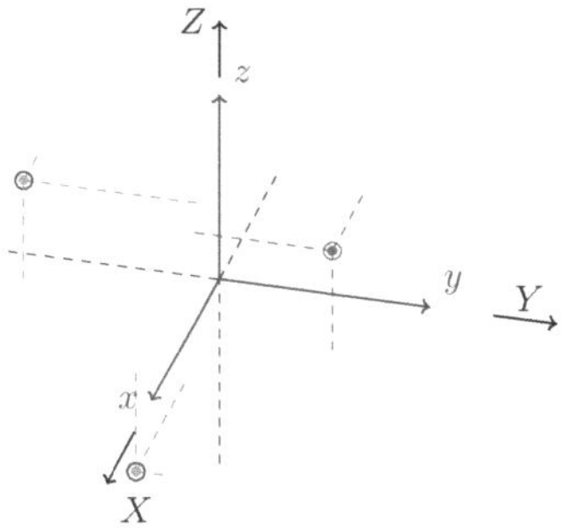

$$[a] = \begin{bmatrix} \frac{1}{2} & -\frac{\sqrt{3}}{2} & 0 \\ \frac{\sqrt{3}}{2} & \frac{1}{2} & 0 \\ 0 & 0 & -1 \end{bmatrix}_{\bar{3}Z}$$

Figure 4.18 Trigonal $\bar{3}$.

earlier, we can find the layout of the non-zero terms in the property matrices for the **Trigonal $\bar{3}$** class of crystals as given below.

$$[c] = \begin{bmatrix} c_{11} & c_{12} & c_{13} & c_{14} & -c_{25} & 0 \\ c_{12} & c_{11} & c_{13} & -c_{14} & c_{25} & 0 \\ c_{13} & c_{13} & c_{33} & 0 & 0 & 0 \\ c_{14} & -c_{14} & 0 & c_{44} & 0 & c_{25} \\ -c_{25} & c_{25} & 0 & 0 & c_{44} & c_{14} \\ 0 & 0 & 0 & c_{25} & c_{14} & \dfrac{c_{11}-c_{12}}{2} \end{bmatrix}$$

$$[e] = \begin{bmatrix} 0\,0\,0\,0\,0\,0 \\ 0\,0\,0\,0\,0\,0 \\ 0\,0\,0\,0\,0\,0 \end{bmatrix}, \quad [\epsilon] = \begin{bmatrix} \epsilon_{xx} & 0 & 0 \\ 0 & \epsilon_{xx} & 0 \\ 0 & 0 & \epsilon_{zz} \end{bmatrix}$$

This class of trigonal crystals are non-piezoelectric.

4.6.3 Class 18: Trigonal 32

The **Trigonal 32** class has both three-fold and two-fold symmetry. Figure 4.19 shows the two types of symmetry along with the corresponding transformation matrices. With these matrices, by following the same steps, we can find the layout of the non-zero terms

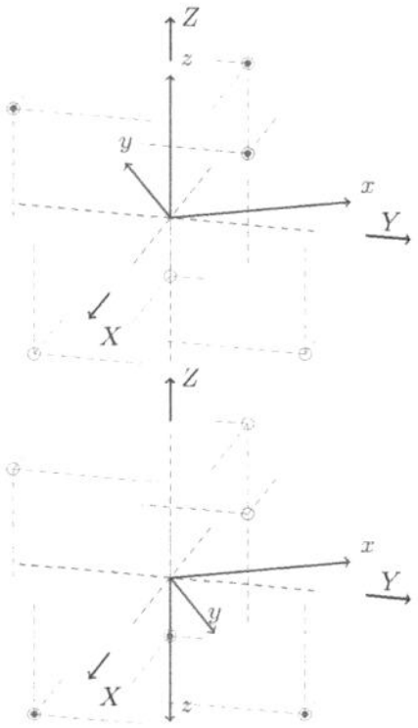

$$[a] = \begin{bmatrix} -\dfrac{1}{2} & \dfrac{\sqrt{3}}{2} & 0 \\ -\dfrac{\sqrt{3}}{2} & -\dfrac{1}{2} & 0 \\ 0 & 0 & 1 \end{bmatrix}_{3Z}$$

$$[a] = \begin{bmatrix} 1 & 0 & 0 \\ 0 & -1 & 0 \\ 0 & 0 & -1 \end{bmatrix}_{2X}$$

Figure 4.19 Trigonal 32.

in the property matrices for the **Trigonal 32** class of crystals as given below.

$$[c] = \begin{bmatrix} c_{11} & c_{12} & c_{13} & c_{14} & 0 & 0 \\ c_{12} & c_{11} & c_{13} & -c_{14} & 0 & 0 \\ c_{13} & c_{13} & c_{33} & 0 & 0 & 0 \\ c_{14} & -c_{14} & 0 & c_{44} & 0 & 0 \\ 0 & 0 & 0 & 0 & c_{44} & c_{14} \\ 0 & 0 & 0 & 0 & c_{14} & \dfrac{c_{11}-c_{12}}{2} \end{bmatrix},$$

$$[e] = \begin{bmatrix} e_{x1} & -e_{x1} & 0 & e_{x4} & 0 & 0 \\ 0 & 0 & 0 & 0 & -e_{x4} & -e_{x1} \\ 0 & 0 & 0 & 0 & 0 & 0 \end{bmatrix} \quad [\epsilon] = \begin{bmatrix} \epsilon_{xx} & 0 & 0 \\ 0 & \epsilon_{xx} & 0 \\ 0 & 0 & \epsilon_{zz} \end{bmatrix}$$

4.6.4 Class 19: Trigonal 3m

The **Trigonal 3m** class has three-fold symmetry and a mirror plane. Figure 4.20 shows the combined symmetry and the two corresponding transformation matrices.

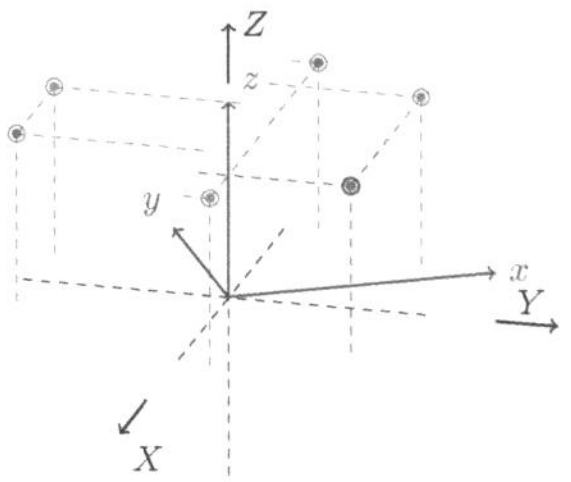

$$[a] = \begin{bmatrix} -\dfrac{1}{2} & \dfrac{\sqrt{3}}{2} & 0 \\ -\dfrac{\sqrt{3}}{2} & -\dfrac{1}{2} & 0 \\ 0 & 0 & 1 \end{bmatrix}_{3Z}, \quad [a] = \begin{bmatrix} -1 & 0 & 0 \\ 0 & 1 & 0 \\ 0 & 0 & 1 \end{bmatrix}_{mYZ}$$

Figure 4.20 Trigonal 3m.

With these matrices, by following the same steps, we can find the layout of the non-zero terms in the property matrices for the

Trigonal 3m class of crystals:

$$[c] = \begin{bmatrix} c_{11} & c_{12} & c_{13} & c_{14} & 0 & 0 \\ c_{12} & c_{11} & c_{13} & -c_{14} & 0 & 0 \\ c_{13} & c_{13} & c_{33} & 0 & 0 & 0 \\ c_{14} & -c_{14} & 0 & c_{44} & 0 & 0 \\ 0 & 0 & 0 & 0 & c_{44} & c_{14} \\ 0 & 0 & 0 & 0 & c_{14} & \dfrac{c_{11}-c_{12}}{2} \end{bmatrix}$$

$$[e] = \begin{bmatrix} 0 & 0 & 0 & 0 & e_{x5} & -e_{y2} \\ -e_{y2} & e_{y2} & 0 & e_{x5} & 0 & 0 \\ e_{z1} & e_{z1} & e_{z3} & 0 & 0 & 0 \end{bmatrix}, \quad [\epsilon] = \begin{bmatrix} \epsilon_{xx} & 0 & 0 \\ 0 & \epsilon_{xx} & 0 \\ 0 & 0 & \epsilon_{zz} \end{bmatrix}$$

4.6.5 Class 20: Trigonal $\bar{3}$m

The **Trigonal $\bar{3}$m** group has inverted three-fold symmetry and a mirror plane. Figure 4.21 shows the combined symmetry and the two corresponding transformation matrices. With these matrices, by following the same steps, we can find the layout of the non-zero terms in the property matrices for the **Trigonal $\bar{3}$m** class of crystals.

$$[c] = \begin{bmatrix} c_{11} & c_{12} & c_{13} & c_{14} & 0 & 0 \\ c_{12} & c_{11} & c_{13} & -c_{14} & 0 & 0 \\ c_{13} & c_{13} & c_{33} & 0 & 0 & 0 \\ c_{14} & -c_{14} & 0 & c_{44} & 0 & 0 \\ 0 & 0 & 0 & 0 & c_{44} & c_{14} \\ 0 & 0 & 0 & 0 & c_{14} & \dfrac{c_{11}-c_{12}}{2} \end{bmatrix}, \quad [e] = \begin{bmatrix} 0 & 0 & 0 & 0 & 0 & 0 \\ 0 & 0 & 0 & 0 & 0 & 0 \\ 0 & 0 & 0 & 0 & 0 & 0 \end{bmatrix},$$

$$[\epsilon] = \begin{bmatrix} \epsilon_{xx} & 0 & 0 \\ 0 & \epsilon_{xx} & 0 \\ 0 & 0 & \epsilon_{zz} \end{bmatrix}$$

This group of trigonal crystals are non-piezoelectric.

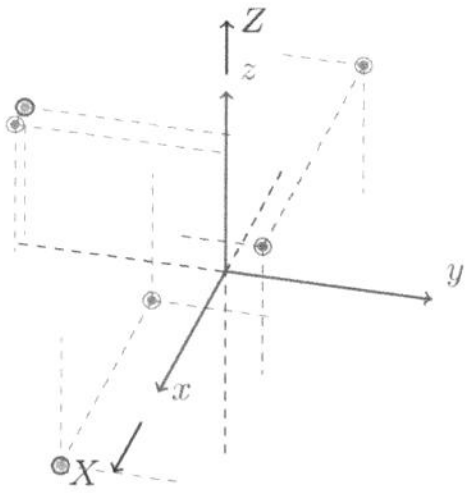

$$[a] = \begin{bmatrix} \frac{1}{2} & -\frac{\sqrt{3}}{2} & 0 \\ \frac{\sqrt{3}}{2} & \frac{1}{2} & 0 \\ 0 & 0 & -1 \end{bmatrix}_{\bar{3}Z}, \quad [a] = \begin{bmatrix} -1 & 0 & 0 \\ 0 & 1 & 0 \\ 0 & 0 & 1 \end{bmatrix}_{mYZ}$$

Figure 4.21 Trigonal $\bar{3}$m.

Example 3: For trigonal 32 langasite crystal with its property matrices given below, show that the three-fold symmetric transformation shown in Figure 4.19 will not alter the material properties.

$$[c] = 10^{11} \begin{bmatrix} 1.89 & 1.04 & 0.89 & 0.15 & 0 & 0 \\ 1.04 & 1.89 & 0.89 & -0.15 & 0 & 0 \\ 0.89 & 0.89 & 2.45 & 0 & 0 & 0 \\ 0.15 & 0.15 & 0 & 0.58 & 0 & 0 \\ 0 & 0 & 0 & 0 & 0.58 & 0 \\ 0.15 & & & & & \\ 0 & 0 & 0 & 0 & 0.150 & 0 \\ 0.425 & & & & & \end{bmatrix}$$

$$[e] = \begin{bmatrix} -0.405 & 0.405 & 0 & 0.115 & 0 & 0 \\ 0 & 0 & 0 & 0 & -0.115 & 0.405 \\ 0 & 0 & 0 & 0 & 0 & 0 \end{bmatrix};$$

$$[\epsilon] = 10^{-9} \begin{bmatrix} 0.191 & 0 & 0 \\ 0 & 0.191 & 0 \\ 0 & 0 & 0.518 \end{bmatrix}$$

Answer: Referring to the discussion in Case 18, for the three-fold symmetry, we have

$$[a] = \begin{bmatrix} -0.5 & 0.866 & 0 \\ -0.866 & -0.5 & 0 \\ 0 & 0 & 1 \end{bmatrix};$$

$$[M] = \begin{bmatrix} 0.25 & 0.75 & 0 & 0 & 0 & -0.866 \\ 0.75 & 0.25 & 0 & 0 & 0 & 0.866 \\ 0 & 0 & 1 & 0 & 0 & 0 \\ 0 & 0 & 0 & -0.5 & -0.866 & 0 \\ 0 & 0 & 0 & 0.866 & -0.5 & 0 \\ 0.433 & -0.433 & 0 & 0 & 0 & -0.5 \end{bmatrix}$$

By applying the respective transformation formulas, we obtain the same $[c]$, $[e]$ and $[\epsilon]$ matrices, hence confirming that the crystal will behave exactly the same after the three-fold symmetric rotation.

4.7 Hexagonal Crystals

In the hexagonal group of crystals, there are seven classes, including **Hexagonal 6**, **Hexagonal $\bar{6}$**, **Hexagonal 6/m**, **Hexagonal 622**, **Hexagonal 6mm**, **Hexagonal $\bar{6}$m2** and **Hexagonal 6/mmm**. Of these classes, five are piezoelectric, and they are the **Hexagonal 6**, **Hexagonal $\bar{6}$**, **Hexagonal 622**, **Hexagonal 6mm** and **Hexagonal $\bar{6}$m2**, and two (i.e., **Hexagonal 6/m** and **Hexagonal 6/mmm**) are not.

4.7.1 Class 21: Hexagonal 6

The **Hexagonal 6** class has six-fold symmetry (e.g., symmetric after rotating about the Z axis by 60°).

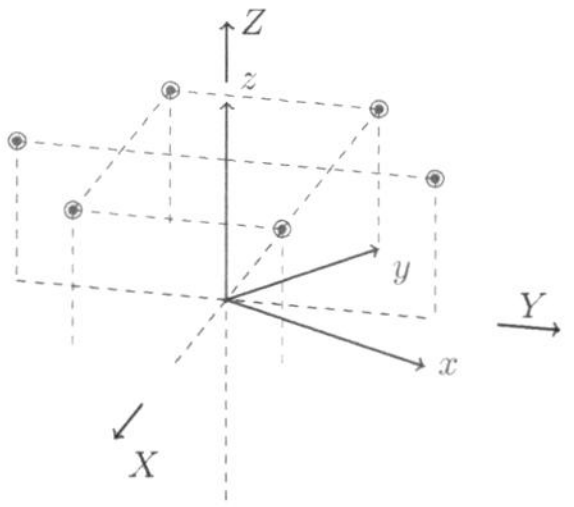

$$[a] = \begin{bmatrix} \frac{1}{2} & \frac{\sqrt{3}}{2} & 0 \\ -\frac{\sqrt{3}}{2} & \frac{1}{2} & 0 \\ 0 & 0 & 1 \end{bmatrix}_{6Z}$$

Figure 4.22 Hexagonal 6.

Figure 4.22 shows this type of symmetry along with the corresponding transformation matrix. With this matrix, by following the same steps, we can find the layout of the non-zero terms in the property matrices for the **Hexagonal 6** class of crystals as given below.

$$[c] = \begin{bmatrix} c_{11} & c_{12} & c_{13} & 0 & 0 & 0 \\ c_{12} & c_{11} & c_{13} & 0 & 0 & 0 \\ c_{13} & c_{13} & c_{33} & 0 & 0 & 0 \\ 0 & 0 & 0 & c_{44} & 0 & 0 \\ 0 & 0 & 0 & 0 & c_{44} & 0 \\ 0 & 0 & 0 & 0 & 0 & \dfrac{c_{11}-c_{12}}{2} \end{bmatrix}$$

$$[e] = \begin{bmatrix} 0 & 0 & 0 & e_{x4} & e_{x5} & 0 \\ 0 & 0 & 0 & e_{x5} & -e_{x4} & 0 \\ e_{z1} & e_{z1} & e_{z3} & 0 & 0 & 0 \end{bmatrix}, \quad [\epsilon] = \begin{bmatrix} \epsilon_{xx} & 0 & 0 \\ 0 & \epsilon_{xx} & 0 \\ 0 & 0 & \epsilon_{zz} \end{bmatrix}$$

4.7.2 Class 22: Hexagonal $\bar{6}$

The **Hexagonal $\bar{6}$** class has inverted six-fold symmetry. Its transformation matrix and illustrated symmetry are given in Figure 4.23.

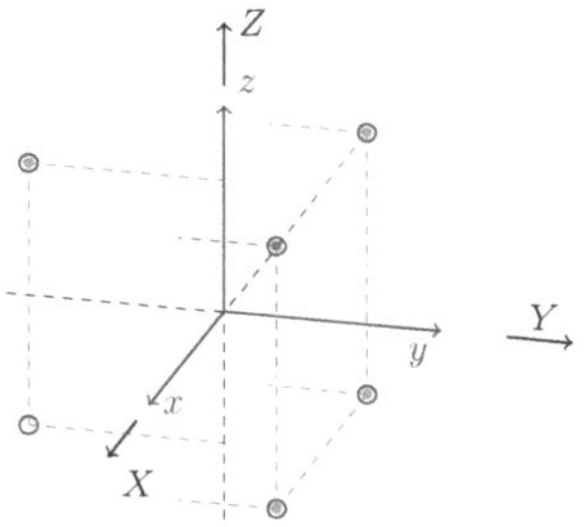

$$[a] = \begin{bmatrix} -\dfrac{1}{2} & -\dfrac{\sqrt{3}}{2} & 0 \\ \dfrac{\sqrt{3}}{2} & -\dfrac{1}{2} & 0 \\ 0 & 0 & -1 \end{bmatrix}_{\bar{6}Z}$$

Figure 4.23 Hexagonal $\bar{6}$.

With this matrix, by following the same steps, we can find the layout of the non-zero terms in the property matrices for the

Hexagonal $\bar{6}$ class of crystals:

$$[c] = \begin{bmatrix} c_{11} & c_{12} & c_{13} & 0 & 0 & 0 \\ c_{12} & c_{11} & c_{13} & 0 & 0 & 0 \\ c_{13} & c_{13} & c_{33} & 0 & 0 & 0 \\ 0 & 0 & 0 & c_{44} & 0 & 0 \\ 0 & 0 & 0 & 0 & c_{44} & 0 \\ 0 & 0 & 0 & 0 & 0 & \dfrac{c_{11}-c_{12}}{2} \end{bmatrix}$$

$$[e] = \begin{bmatrix} e_{x1} & -e_{x1} & 0 & 0 & 0 & -e_{y2} \\ -e_{y2} & e_{y2} & 0 & 0 & 0 & -e_{x1} \\ 0 & 0 & 0 & 0 & 0 & 0 \end{bmatrix}, \quad [\epsilon] = \begin{bmatrix} \epsilon_{xx} & 0 & 0 \\ 0 & \epsilon_{xx} & 0 \\ 0 & 0 & \epsilon_{zz} \end{bmatrix}$$

4.7.3 Class 23: Hexagonal 6/m

The **Hexagonal 6/m** class has both six-fold symmetry and a mirror plane. Figure 4.24 shows the combined symmetry along with the two corresponding transformation matrices.

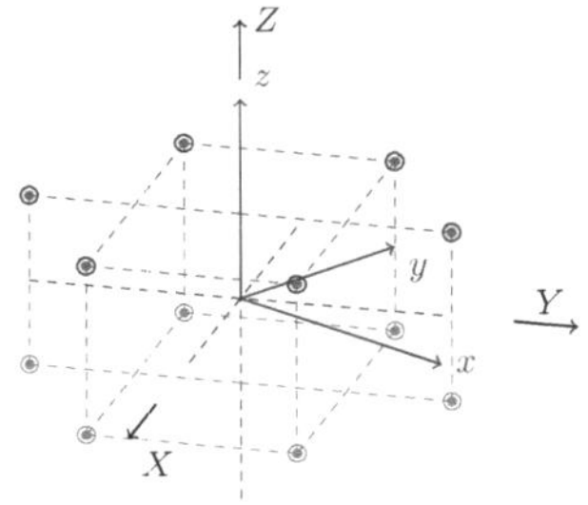

$$[a] = \begin{bmatrix} \dfrac{1}{2} & \dfrac{\sqrt{3}}{2} & 0 \\ -\dfrac{\sqrt{3}}{2} & \dfrac{1}{2} & 0 \\ 0 & 0 & 1 \end{bmatrix}_{6Z}, \quad [a] = \begin{bmatrix} 1 & 0 & 0 \\ 0 & 1 & 0 \\ 0 & 0 & -1 \end{bmatrix}_{mXY}$$

Figure 4.24 Hexagonal 6/m.

With these matrices, by following the same steps, we can find the layout of the non-zero terms in the property matrices for the

Hexagonal 6/m class of crystals:

$$[c] = \begin{bmatrix} c_{11} & c_{12} & c_{13} & 0 & 0 & 0 \\ c_{12} & c_{11} & c_{13} & 0 & 0 & 0 \\ c_{13} & c_{13} & c_{33} & 0 & 0 & 0 \\ 0 & 0 & 0 & c_{44} & 0 & 0 \\ 0 & 0 & 0 & 0 & c_{44} & 0 \\ 0 & 0 & 0 & 0 & 0 & \dfrac{c_{11}-c_{12}}{2} \end{bmatrix}, [e] = \begin{bmatrix} 0&0&0&0&0&0 \\ 0&0&0&0&0&0 \\ 0&0&0&0&0&0 \end{bmatrix}, [\epsilon] = \begin{bmatrix} \epsilon_{xx} & 0 & 0 \\ 0 & \epsilon_{xx} & 0 \\ 0 & 0 & \epsilon_{zz} \end{bmatrix}$$

This class of hexagonal crystals is non-piezoelectric.

4.7.4 Class 24: Hexagonal 622

The **Hexagonal 622** class has both six-fold symmetry and two-fold symmetry. Figure 4.25 shows the two types of symmetry and the corresponding transformation matrices.

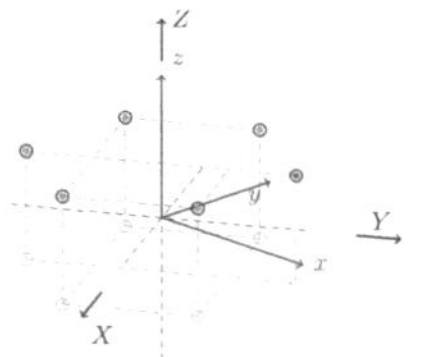

$$[a] = \begin{bmatrix} \dfrac{1}{2} & \dfrac{\sqrt{3}}{2} & 0 \\ -\dfrac{\sqrt{3}}{2} & \dfrac{1}{2} & 0 \\ 0 & 0 & 1 \end{bmatrix}_{6Z}$$

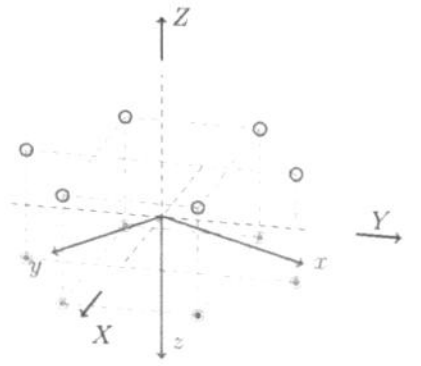

$$[a] = \begin{bmatrix} 1 & 0 & 0 \\ 0 & -1 & 0 \\ 0 & 0 & -1 \end{bmatrix}_{2X}$$

Figure 4.25 Hexagonal 622.

With these matrices, by following the same steps, we can find the layout of the non-zero terms in the property matrices for the

Hexagonal 622 class of crystals:

$$[c] = \begin{bmatrix} c_{11} & c_{12} & c_{13} & 0 & 0 & 0 \\ c_{12} & c_{11} & c_{13} & 0 & 0 & 0 \\ c_{13} & c_{13} & c_{33} & 0 & 0 & 0 \\ 0 & 0 & 0 & c_{44} & 0 & 0 \\ 0 & 0 & 0 & 0 & c_{44} & 0 \\ 0 & 0 & 0 & 0 & 0 & \dfrac{c_{11}-c_{12}}{2} \end{bmatrix}, \quad [e] = \begin{bmatrix} 0 & 0 & 0 & e_{x4} & 0 & 0 \\ 0 & 0 & 0 & 0 & -e_{x4} & 0 \\ 0 & 0 & 0 & 0 & 0 & 0 \end{bmatrix},$$

$$[\epsilon] = \begin{bmatrix} \epsilon_{xx} & 0 & 0 \\ 0 & \epsilon_{xx} & 0 \\ 0 & 0 & \epsilon_{zz} \end{bmatrix}$$

4.7.5 Class 25: Hexagonal 6mm

The **Hexagonal 6mm** class has six-fold symmetry and a mirror plane. Its transformation matrices and illustrated symmetry are given in Figure 4.26.

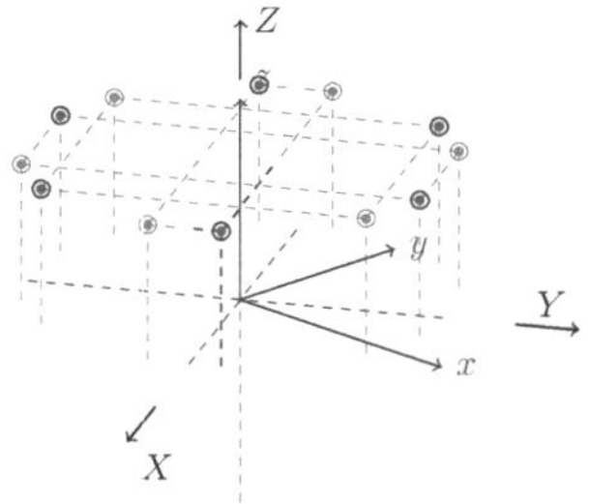

$$[a] = \begin{bmatrix} \dfrac{1}{2} & \dfrac{\sqrt{3}}{2} & 0 \\ -\dfrac{\sqrt{3}}{2} & \dfrac{1}{2} & 0 \\ 0 & 0 & 1 \end{bmatrix}_{6Z}, \quad [a] = \begin{bmatrix} 1 & 0 & 0 \\ 0 & -1 & 0 \\ 0 & 0 & 1 \end{bmatrix}_{mXZ}$$

Figure 4.26 Hexagonal 6mm.

With these matrices, by following the same steps, we can find the layout of the non-zero terms in the property matrices for the

Hexagonal 6mm class of crystals:

$$[c] = \begin{bmatrix} c_{11} & c_{12} & c_{13} & 0 & 0 & 0 \\ c_{12} & c_{11} & c_{13} & 0 & 0 & 0 \\ c_{13} & c_{13} & c_{33} & 0 & 0 & 0 \\ 0 & 0 & 0 & c_{44} & 0 & 0 \\ 0 & 0 & 0 & 0 & c_{44} & 0 \\ 0 & 0 & 0 & 0 & 0 & \dfrac{c_{11}-c_{12}}{2} \end{bmatrix}, \ [e] = \begin{bmatrix} 0 & 0 & 0 & 0 & e_{x5} & 0 \\ 0 & 0 & 0 & e_{x5} & 0 & 0 \\ e_{z1} & e_{z1} & e_{z3} & 0 & 0 & 0 \end{bmatrix},$$

$$[\epsilon] = \begin{bmatrix} \epsilon_{xx} & 0 & 0 \\ 0 & \epsilon_{xx} & 0 \\ 0 & 0 & \epsilon_{zz} \end{bmatrix}$$

4.7.6 Class 26: Hexagonal $\bar{6}$m2

The **Hexagonal $\bar{6}$m2** class has inverted six-fold symmetry and two-fold symmetry. Its transformation matrices and illustrated symmetry are given in Figure 4.27.

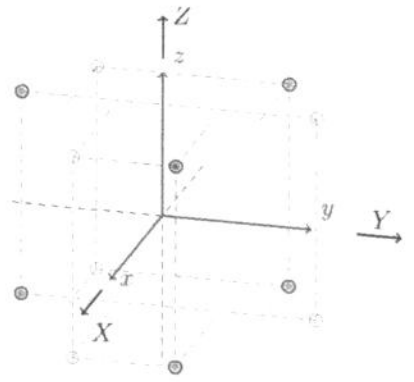

$$[a] = \begin{bmatrix} -\dfrac{1}{2} & -\dfrac{\sqrt{3}}{2} & 0 \\ \dfrac{\sqrt{3}}{2} & -\dfrac{1}{2} & 0 \\ 0 & 0 & -1 \end{bmatrix}_{\bar{6}Z}$$

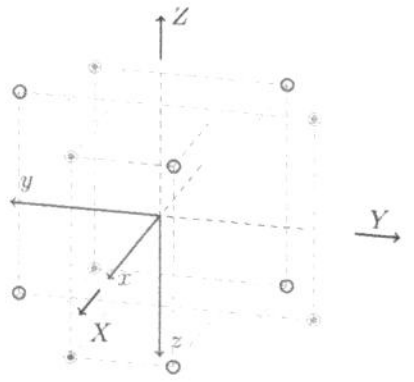

$$[a] = \begin{bmatrix} 1 & 0 & 0 \\ 0 & -1 & 0 \\ 0 & 0 & -1 \end{bmatrix}_{2X}$$

Figure 4.27 Hexagonal $\bar{6}$m2.

With these matrices, by following the same steps, we can find the layout of the non-zero terms in the property matrices for the

Hexagonal $\bar{6}$m2 class of crystals:

$$[c] = \begin{bmatrix} c_{11} & c_{12} & c_{13} & 0 & 0 & 0 \\ c_{12} & c_{11} & c_{13} & 0 & 0 & 0 \\ c_{13} & c_{13} & c_{33} & 0 & 0 & 0 \\ 0 & 0 & 0 & c_{44} & 0 & 0 \\ 0 & 0 & 0 & 0 & c_{44} & 0 \\ 0 & 0 & 0 & 0 & 0 & \dfrac{c_{11}-c_{12}}{2} \end{bmatrix}$$

$$[e] = \begin{bmatrix} e_{x1} & -e_{x1} & 0 & 0 & 0 & 0 \\ 0 & 0 & 0 & 0 & 0 & -e_{x1} \\ 0 & 0 & 0 & 0 & 0 & 0 \end{bmatrix}, \quad [\epsilon] = \begin{bmatrix} \epsilon_{xx} & 0 & 0 \\ 0 & \epsilon_{xx} & 0 \\ 0 & 0 & \epsilon_{zz} \end{bmatrix}$$

4.7.7 Class 27: Hexagonal 6/mmm

The **Hexagonal 6/mmm** class has six-fold symmetry and two mirror planes. Figure 4.28 shows the combined symmetry and the three corresponding transformation matrices. With these matrices, by following the same steps, we can find the layout of the non-zero terms

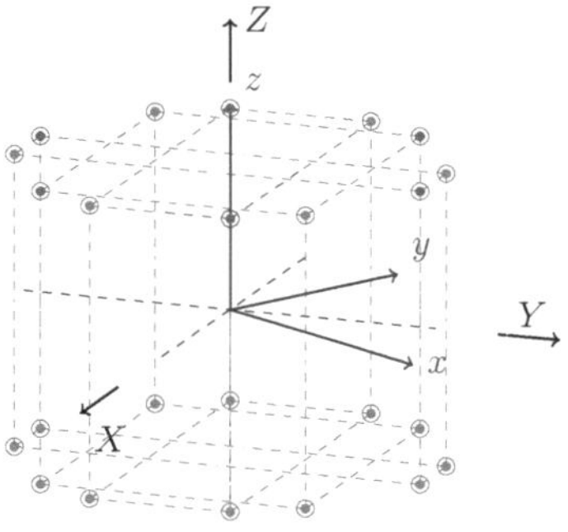

$$[a] = \begin{bmatrix} \dfrac{1}{2} & \dfrac{\sqrt{3}}{2} & 0 \\ -\dfrac{\sqrt{3}}{2} & \dfrac{1}{2} & 0 \\ 0 & 0 & 1 \end{bmatrix}_{6Z}$$

$$[a] = \begin{bmatrix} 1 & 0 & 0 \\ 0 & 1 & 0 \\ 0 & 0 & -1 \end{bmatrix}_{mXY}, \quad [a] = \begin{bmatrix} 1 & 0 & 0 \\ 0 & -1 & 0 \\ 0 & 0 & 1 \end{bmatrix}_{mXZ}$$

Figure 4.28 Hexagonal 6/mmm.

in the property matrices for the **Hexagonal 6/mmm** class of crystals.

$$[c] = \begin{bmatrix} c_{11} & c_{12} & c_{13} & 0 & 0 & 0 \\ c_{12} & c_{11} & c_{13} & 0 & 0 & 0 \\ c_{13} & c_{13} & c_{33} & 0 & 0 & 0 \\ 0 & 0 & 0 & c_{44} & 0 & 0 \\ 0 & 0 & 0 & 0 & c_{44} & 0 \\ 0 & 0 & 0 & 0 & 0 & \dfrac{c_{11}-c_{12}}{2} \end{bmatrix}, \ [e] = \begin{bmatrix} 0 & 0 & 0 & 0 & 0 & 0 \\ 0 & 0 & 0 & 0 & 0 & 0 \\ 0 & 0 & 0 & 0 & 0 & 0 \end{bmatrix}, \ [\epsilon] = \begin{bmatrix} \epsilon_{xx} & 0 & 0 \\ 0 & \epsilon_{xx} & 0 \\ 0 & 0 & \epsilon_{zz} \end{bmatrix}$$

This class of hexagonal crystals are non-piezoelectric.

Example 4: For hexagonal 6mm zinc oxide crystal with its property matrices given below, show that the six-fold symmetric transformation shown in Figure 4.26 will not alter the material properties.

$$[c] = 10^{11} \begin{bmatrix} 2.090 & 1.211 & 1.051 & 0 & 0 & \\ 1.211 & 2.090 & 1.051 & 0 & 0 & 0 \\ 1.051 & 1.051 & 2.109 & 0 & 0 & 0 \\ 0 & 0 & 0 & 0.425 & 0 & 0 \\ 0 & 0 & 0 & 0 & 0.425 & 0 \\ 0 & 0 & 0 & 0 & 0 & 0.439 \end{bmatrix}$$

$$[e] = \begin{bmatrix} 0 & 0 & 0 & 0 & -0.48 & 0 \\ 0 & 0 & 0 & -0.48 & 0 & 0 \\ -0.573 & -0.573 & 1.32 & 0 & 0 & 0 \end{bmatrix};$$

$$[\epsilon] = 10^{-9} \begin{bmatrix} 0.085 & 0 & 0 \\ 0 & 0.085 & 0 \\ 0 & 0 & 0.102 \end{bmatrix}$$

Answer: Referring to the discussion in Case 25, for the six-fold symmetry, we have

$$[a] = \begin{bmatrix} 0.5 & 0.866 & 0 \\ -0.866 & 0.5 & 0 \\ 0 & 0 & 1 \end{bmatrix}; \ [M] = \begin{bmatrix} 0.25 & 0.75 & 0 & 0 & 0 & 0.866 \\ 0.75 & 0.25 & 0 & 0 & 0 & -0.866 \\ 0 & 0 & 1 & 0 & 0 & 0 \\ 0 & 0 & 0 & 0.5 & -0.866 & 0 \\ 0 & 0 & 0 & 0.866 & 0.5 & 0 \\ -0.433 & 0.433 & 0 & 0 & 0 & -0.5 \end{bmatrix}$$

By applying the respective transformation formulas, we obtain the same $[c]$, $[e]$ and $[\epsilon]$ matrices, hence confirming that the crystal will behave exactly the same after the six-fold symmetric rotation.

4.8 Cubic Crystals

In the cubic group, there are five classes of crystals, and they are **Cubic 23**, **Cubic m3**, **Cubic 432**, **Cubic m3m** and **Cubic $\bar{4}$3m**. Of these groups, only the **Cubic 23** and **Cubic $\bar{4}$3m** classes are piezoelectric, and the rest are not.

4.8.1 Class 28: Cubic 23

The **Cubic 23** class has two-fold symmetry about the Z axis and rotation symmetry (e.g., symmetric after two rotations, one about the Z axis by 90° and the other about the rotated x axis by 90°). Figure 4.29 shows the two types of symmetry and their corresponding transformation matrices..

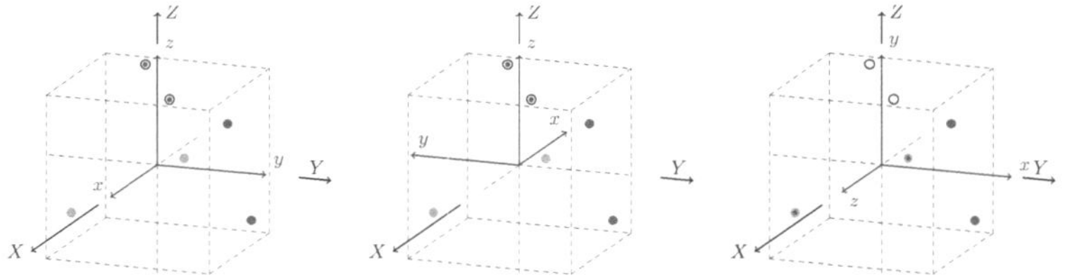

$$[a] = \begin{bmatrix} -1 & 0 & 0 \\ 0 & -1 & 0 \\ 0 & 0 & 1 \end{bmatrix}_{2Z}, \quad [a] = \begin{bmatrix} 0 & 1 & 0 \\ 0 & 0 & 1 \\ 1 & 0 & 0 \end{bmatrix}_{rZ_{90}x_{90}}$$

Figure 4.29 Cubic 23.

With these matrices, by following the same steps, we can find the layout of the non-zero terms in the property matrices for the **Cubic**

23 class of crystals:

$$[c] = \begin{bmatrix} c_{11} & c_{12} & c_{12} & 0 & 0 & 0 \\ c_{12} & c_{11} & c_{12} & 0 & 0 & 0 \\ c_{12} & c_{12} & c_{11} & 0 & 0 & 0 \\ 0 & 0 & 0 & c_{44} & 0 & 0 \\ 0 & 0 & 0 & 0 & c_{44} & 0 \\ 0 & 0 & 0 & 0 & 0 & c_{44} \end{bmatrix}, [e] = \begin{bmatrix} 0 & 0 & 0 & e_{x4} & 0 & 0 \\ 0 & 0 & 0 & 0 & e_{x4} & 0 \\ 0 & 0 & 0 & 0 & 0 & e_{x4} \end{bmatrix}, [\epsilon] = \begin{bmatrix} \epsilon_{xx} & 0 & 0 \\ 0 & \epsilon_{xx} & 0 \\ 0 & 0 & \epsilon_{xx} \end{bmatrix}$$

4.8.2 Class 29: Cubic m3

The **Cubic m3** class has two-fold symmetry, rotation symmetry and a mirror plane. Its transformation matrices and illustrated symmetry are given in Figure 4.30.

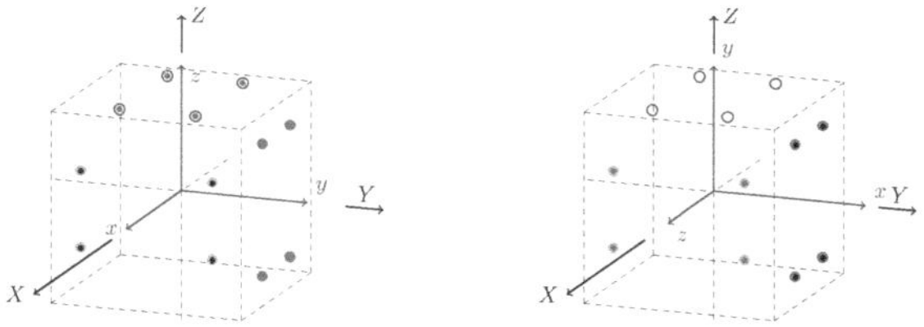

$$[a] = \begin{bmatrix} -1 & 0 & 0 \\ 0 & -1 & 0 \\ 0 & 0 & 1 \end{bmatrix}_{2Z}, \quad [a] = \begin{bmatrix} 0 & 1 & 0 \\ 0 & 0 & 1 \\ 1 & 0 & 0 \end{bmatrix}_{rZ_{90}x_{90}}, \quad [a] = \begin{bmatrix} 1 & 0 & 0 \\ 0 & 1 & 0 \\ 0 & 0 & -1 \end{bmatrix}_{mXY}$$

Figure 4.30 Cubic m3.

With these matrices, by following the same steps, we can find the layout of the non-zero terms in the property matrices for the **Cubic m3** class of crystals:

$$[c] = \begin{bmatrix} c_{11} & c_{12} & c_{12} & 0 & 0 & 0 \\ c_{12} & c_{11} & c_{12} & 0 & 0 & 0 \\ c_{12} & c_{12} & c_{11} & 0 & 0 & 0 \\ 0 & 0 & 0 & c_{44} & 0 & 0 \\ 0 & 0 & 0 & 0 & c_{44} & 0 \\ 0 & 0 & 0 & 0 & 0 & c_{44} \end{bmatrix}, [e] = \begin{bmatrix} 0 & 0 & 0 & 0 & 0 & 0 \\ 0 & 0 & 0 & 0 & 0 & 0 \\ 0 & 0 & 0 & 0 & 0 & 0 \end{bmatrix}, [\epsilon] = \begin{bmatrix} \epsilon_{xx} & 0 & 0 \\ 0 & \epsilon_{xx} & 0 \\ 0 & 0 & \epsilon_{xx} \end{bmatrix}$$

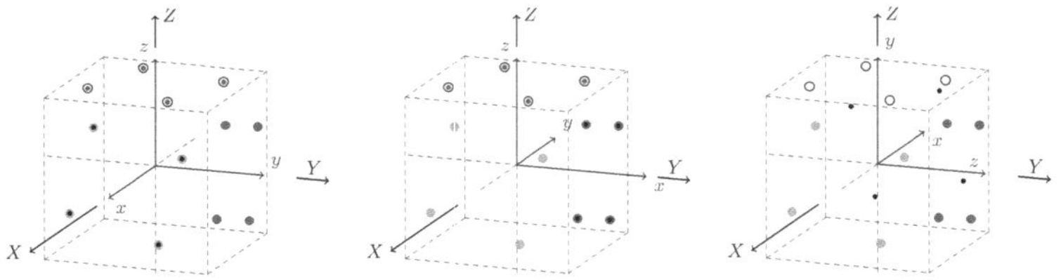

$$[a] = \begin{bmatrix} 0 & 1 & 0 \\ -1 & 0 & 0 \\ 0 & 0 & 1 \end{bmatrix}_{4Z}, \quad [a] = \begin{bmatrix} -1 & 0 & 0 \\ 0 & 0 & 1 \\ 0 & 1 & 0 \end{bmatrix}_{rZ_{180}x_{90}}$$

Figure 4.31 Cubic 432.

This class of cubic crystals is non-piezoelectric.

4.8.3 Class 30: Cubic 432

The **Cubic 432** class has four-fold symmetry and rotational symmetry of a different kind (e.g., symmetric after two rotations, one about the Z axis by 180° and the other about the rotated x axis by 90°). Figure 4.31 illustrates these types of symmetry along with its transformation matrices.

With these matrices, by following the same steps, we can find the layout of the non-zero terms in the property matrices for the **Cubic 432** class of crystals:

$$[c] = \begin{bmatrix} c_{11} & c_{12} & c_{12} & 0 & 0 & 0 \\ c_{12} & c_{11} & c_{12} & 0 & 0 & 0 \\ c_{12} & c_{12} & c_{11} & 0 & 0 & 0 \\ 0 & 0 & 0 & c_{44} & 0 & 0 \\ 0 & 0 & 0 & 0 & c_{44} & 0 \\ 0 & 0 & 0 & 0 & 0 & c_{44} \end{bmatrix}, \quad [e] = \begin{bmatrix} 0 & 0 & 0 & 0 & 0 & 0 \\ 0 & 0 & 0 & 0 & 0 & 0 \\ 0 & 0 & 0 & 0 & 0 & 0 \end{bmatrix}, \quad [\epsilon] = \begin{bmatrix} \epsilon_{xx} & 0 & 0 \\ 0 & \epsilon_{xx} & 0 \\ 0 & 0 & \epsilon_{xx} \end{bmatrix}$$

This class of cubic crystals is non-piezoelectric.

4.8.4 Class 31: Cubic m3m

The **Cubic m3m** class has four-fold symmetry, rotational symmetry and a mirror plane. Figure 4.32 shows the symmetry and the corresponding transformation matrices.

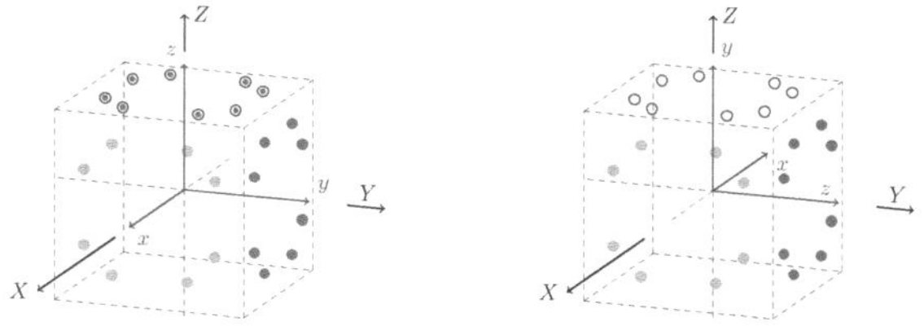

$$[a] = \begin{bmatrix} 0 & 1 & 0 \\ -1 & 0 & 0 \\ 0 & 0 & 1 \end{bmatrix}_{4Z}, \quad [a] = \begin{bmatrix} -1 & 0 & 0 \\ 0 & 0 & 1 \\ 0 & 1 & 0 \end{bmatrix}_{rZ_{180}X_{90}}, \quad \begin{bmatrix} 1 & 0 & 0 \\ 0 & 1 & 0 \\ 0 & 0 & -1 \end{bmatrix}_{mXY}$$

Figure 4.32 Cubic m3m.

With these matrices, by following the same steps, we can find the layout of the non-zero terms in the property matrices for the **Cubic m3m** class of crystals:

$$[c] = \begin{bmatrix} c_{11} & c_{12} & c_{12} & 0 & 0 & 0 \\ c_{12} & c_{11} & c_{12} & 0 & 0 & 0 \\ c_{12} & c_{12} & c_{11} & 0 & 0 & 0 \\ 0 & 0 & 0 & c_{44} & 0 & 0 \\ 0 & 0 & 0 & 0 & c_{44} & 0 \\ 0 & 0 & 0 & 0 & 0 & c_{44} \end{bmatrix}, \quad [e] = \begin{bmatrix} 0 & 0 & 0 & 0 & 0 & 0 \\ 0 & 0 & 0 & 0 & 0 & 0 \\ 0 & 0 & 0 & 0 & 0 & 0 \end{bmatrix}, \quad [\epsilon] = \begin{bmatrix} \epsilon_{xx} & 0 & 0 \\ 0 & \epsilon_{xx} & 0 \\ 0 & 0 & \epsilon_{xx} \end{bmatrix}$$

This class of cubic crystals is non-piezoelectric.

4.8.5 Class 32: Cubic $\bar{4}$3m

The **Cubic $\bar{4}$3m** class has inverted four-fold symmetry and a mirror plane about a diagonal plane (e.g., the one going through the X axis). Its transformation matrices and illustrated symmetry are given in Figure 4.33.

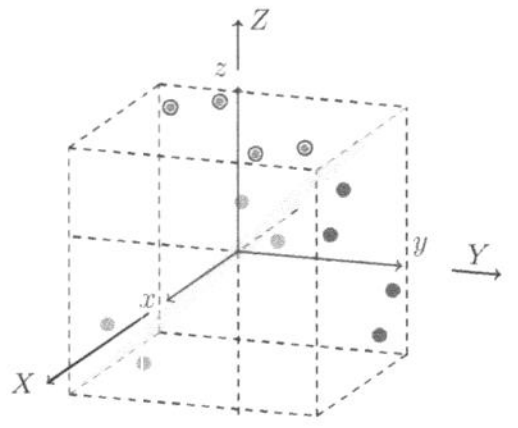

$$[a] = \begin{bmatrix} 0 & -1 & 0 \\ 1 & 0 & 0 \\ 0 & 0 & -1 \end{bmatrix}_{\bar{4}Z}, \quad [a] = \begin{bmatrix} 1 & 0 & 0 \\ 0 & 0 & 1 \\ 0 & 1 & 0 \end{bmatrix}_{mXD}$$

Figure 4.33 Cubic $\bar{4}$3m.

With these matrices, by following the same steps, we can find the layout of the non-zero terms in the property matrices for the **Cubic $\bar{4}$3m** class of crystals:

$$[c] = \begin{bmatrix} c_{11} & c_{12} & c_{12} & 0 & 0 & 0 \\ c_{12} & c_{11} & c_{12} & 0 & 0 & 0 \\ c_{12} & c_{12} & c_{11} & 0 & 0 & 0 \\ 0 & 0 & 0 & c_{44} & 0 & 0 \\ 0 & 0 & 0 & 0 & c_{44} & 0 \\ 0 & 0 & 0 & 0 & 0 & c_{44} \end{bmatrix}$$

$$[e] = \begin{bmatrix} 0 & 0 & 0 & e_{x4} & 0 & 0 \\ 0 & 0 & 0 & 0 & e_{x4} & 0 \\ 0 & 0 & 0 & 0 & 0 & e_{x4} \end{bmatrix}, \quad [\epsilon] = \begin{bmatrix} \epsilon_{xx} & 0 & 0 \\ 0 & \epsilon_{xx} & 0 \\ 0 & 0 & \epsilon_{xx} \end{bmatrix}$$

4.9 Properties of Some Common Crystal Materials

In this section, we provide the constants for the elements in the $[c]$, $[e]$ and $[\epsilon]$ matrices for a few selected piezoelectric materials, including orthorhombic, tetragonal, trigonal, hexagonal and cubic crystals. When taking these constants, it is necessary to assign proper units and scales. For the constants in the $[c]$ matrix and the $[\epsilon]$ matrix, all values are to be multiplied by the following factors with units

$$[c] \times 10^{10}\text{N/m}^2; \quad [\epsilon] \times 10^{-11}\text{F/m}$$

For the values in the $[e]$ matrix and for ρ, they are just assigned with units

$$[e] : \text{C/m}^2 \quad \text{and} \quad \rho : \text{Kg/m}^3$$

Of the crystal materials listed here, the classes of trigonal $\bar{3}$m crystals which includes Alumina (Al_2O_3), the class of cubic m3m crystals which includes silicon (Si), and the class of tetragonal 4/m and 4/mmm crystals which include lead molybdate ($PbMoO_4$), calcium molybdate ($CaMoO_4$) and titanium oxide (TiO_2) are all non-piezoelectric, thus the constants in their piezoelectric stress matrix $[e]$ are zero. Because of this, there will be no coupling of the electrical fields with mechanical strains in these materials. Moreover, referring to Eqs. (2.8) and (3.13), we know that the piezoelectric strain matrix $[d]$ is linearly related to $[e]$. Thus from $[e] = 0$ we know $[d] = 0$ for all these crystals as well.

In Chapters 5 and 6, we will use these material properties to examine the polarization and propagation of various bulk and surface acoustic waves in these piezoelectric crustal materials.

Table 4.1 Orthorhombic crystals

Material	Class	c_{11}	c_{12}	c_{13}	c_{22}	c_{23}	c_{33}	c_{44}	c_{55}	c_{66}	ρ
Barium sodium	2mm	23.9	10.4	5.0	24.7	5.2	13.5	6.5	6.6	7.6	5300
Niobate		e_{x5}	e_{y4}	e_{z1}	e_{z2}	e_{z3}	ϵ_{xx}	ϵ_{yy}	ϵ_{zz}		
($Ba_2NaNb_5O_{15}$)		2.8	3.4	−0.4	−0.3	4.3	222	227	32		

Table 4.2 Tetragonal crystals

Material	Class	c_{11}	c_{12}	c_{13}	c_{16}	c_{33}	c_{44}	c_{66}	ρ
Tellurium oxide(TeO_2)	422	5.6	5.1	2.2	0	10.6	2.65	6.6	6000
Barium titanate ($BaTiO_3$)	4mm	27.5	17.9	15.2	0	16.5	5.43	11.3	6020
Lead molybdate ($PbMoO_4$)	4/m	10.92	6.83	5.28	1.36	9.17	2.67	3.37	6950
Calcium molybdate ($CaMoO_4$)	4/m	14.5	6.6	4.46	1.3	12.65	3.69	4.5	4255
Titanium oxide (TiO_2)	4/mmm	27.3	17.6	14.9	0	48.4	12.5	19.4	4250
Material	Class	e_{x4}	e_{x5}	e_{z1}	e_{z3}	ϵ_{xx}	ϵ_{zz}		
Tellurium oxide (TeO_2)	422	0.22	0	0	0	22.9	24.7		
Barium titanate ($BaTiO_3$)	4mm	0	21.3	−2.74	3.7	1744	97		

Note: The 4/m and 4/mmm classes are non-piezoelectric.

Table 4.3 Hexagonal crystals

Material	Class	c_{11}	c_{12}	c_{13}	c_{33}	c_{44}	ρ
Zinc oxide (ZnO)	6mm	20.9	12.11	10.51	21.09	4.25	5676
Cadmium sulfide (CdS)	6mm	8.56	5.32	4.62	9.36	1.49	4824
Material	**Class**	e_{x5}	e_{z1}	e_{z3}	ϵ_{xx}	ϵ_{zz}	
Zinc oxide (ZnO)	6mm	−0.48	−0.57	1.32	8.55	10.2	
Cadmium sulfide (CdS)	6mm	−0.21	−0.24	0.44	9.02	9.53	

Table 4.4 Cubic crystals

Material	Class	c_{11}	c_{12}	c_{44}	ρ	e_{x4}	ϵ_{xx}
Bismuth germanium oxide ($Bi_{12}GeO_{20}$)	23	12.8	3.05	2.55	9230	0.99	34.2
Gallium arsenide (GaAs)	$\bar{4}3m$	11.88	5.38	5.94	5307	0.154	9.73
Silicon (Si)	m3m	16.56	6.39	7.59	2329	0	11.7

Note: The m3m class is non-piezoelectric.

Table 4.5 Trigonal crystals

Material	Class	c_{11}	c_{12}	c_{13}	c_{14}	c_{33}	c_{44}	ρ	
Lithium niobate ($LiNbO_3$)	3m	20.3	5.3	7.5	0.9	24.5	6.0	4700	
Lithium tantalate ($LiTaO_3$)	3m	22.9	4.4	8.12	−1.04	27.98	9.68	7450	
Langasite ($La_3Ga_5SiO_{14}$)	32	18.9	10.4	8.9	1.5	24.5	5.8	5739	
Quartz (SiO_2)	32	8.7	0.7	1.2	−1.8	10.72	6.0	2648	
Alumina (Al_2O_3)	$\bar{3}$m	49.7	16.3	11.1	−2.35	49.8	14.7	3986	
Material	**Class**	e_{x1}	e_{x4}	e_{x5}	e_{y2}	e_{z1}	e_{z3}	ϵ_{xx}	ϵ_{zz}
Lithium niobate ($LiNbO_3$)	3m	0	0	3.7	2.5	0.2	1.3	38.9	25.7
Lithium tantalate ($LiTaO_3$)	3m	0	0	2.72	1.67	−0.38	1.09	37.7	37.9
Langasite ($La_3Ga_5SiO_{14}$)	32	−0.405	0.115	0	0	0	0	19.1	51.8
Quartz (SiO_2)	32	0.171	−0.04	0	0	0	0	4.4	4.6

Note: The $\bar{3}$m class is non-piezoelectric.

4.10 Study Problems

1. With the material properties given in Section 4.9 for tetragonal 4mm barium titanate crystal, show that its properties will not change after any symmetric rotation shown in Figure 4.14.
2. With the material properties given in Section 4.9 for cubic m3m silicon crystal, show that its properties will not change after any symmetric rotation shown in Figure 4.32.
3. With the material properties given in Section 4.9 for trigonal 32 SiO_2 crystal, show that its properties will not change after any symmetric rotation shown in Figure 4.19.
4. For the monoclinic 2/m class of crystals, use the symmetric transformation matrices given in Figure 4.6 to show that some of the terms in the full 6×6 stiffness matrix and 3×3 permittivity matrix are zero such that this class of crystals will have the following reduced forms for the two matrices, respectively. Also, show that this class of crystals is non-piezoelectric.

$$[c] = \begin{bmatrix} c_{11} & c_{12} & c_{13} & 0 & c_{15} & 0 \\ c_{12} & c_{22} & c_{23} & 0 & c_{25} & 0 \\ c_{13} & c_{23} & c_{33} & 0 & c_{35} & 0 \\ 0 & 0 & 0 & c_{44} & 0 & c_{46} \\ c_{15} & c_{25} & c_{35} & 0 & c_{55} & 0 \\ 0 & 0 & 0 & c_{46} & 0 & c_{66} \end{bmatrix}, \quad [\epsilon] = \begin{bmatrix} \epsilon_{xx} & 0 & \epsilon_{xz} \\ 0 & \epsilon_{yy} & 0 \\ \epsilon_{xz} & 0 & \epsilon_{zz} \end{bmatrix}$$

Chapter 5

Bulk Acoustic Waves

Like seismic waves, acoustic waves also travel in various different forms, including bulk waves and surface waves. Bulk acoustic waves (BAWs) propagate through the interior of a solid material in the form of particulate vibration occurring in a so-called wave-front plane that cuts through the entire material. BAWs can occur in any directions and any materials, regardless of their isotropy (isotropic or anisotropic) or peizoelectricity (piezoelectric or non-piezoelectric). In this chapter, we will discuss the propagation characteristics of BAWs in various elastic and piezoelectric materials.

5.1 Different Types of Bulk Acoustic Waves

5.1.1 Longitudinal and Transverse Acoustic Waves

When dealing with acoustic waves, the term polarization refers to the oscillatory movements of particles in a solid material, and the term propagation refers to the transmission of the oscillatory mechanical energy in time from one point to another.

Bulk and Surface Acoustic Waves: Fundamentals, Devices, and Applications
Guigen Zhang

ISBN 978-981-4877-73-2 (Hardcover), 978-1-003-25662-5 (eBook)
www.jennystanford.com

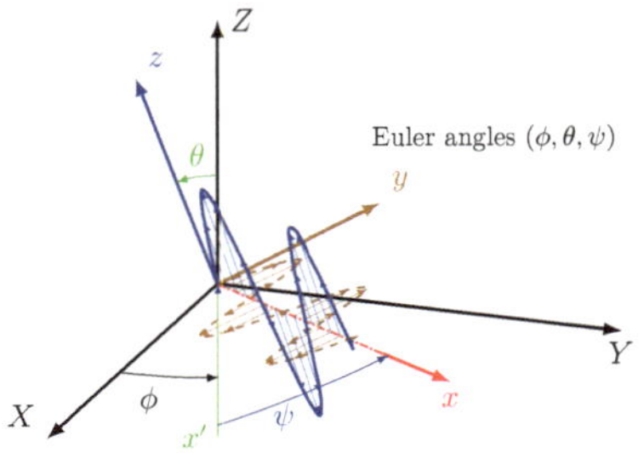

Figure 5.1 Illustration of polarization and propagation in bulk acoustic waves.

In Figure 5.1, bulk waves are depicted as traveling in a randomly selected direction, *x*, defined by a set of Euler angles. Here the *xyz*-coordinates is regarded as the rotated system and the XYZ-coordinates the fixed system. For BAWs propagating along the *x* direction, we often express the waves in three sub waves in an orthogonal manner in the *xyz* coordinates system. For instance, we consider the first sub wave with its particles vibrating in the *x* direction, the second with its particles vibrating in the *y* direction, and the third with its particles vibrating in the *z* direction. Here the particle vibrational movement is regarded as wave polarization. For the first sub wave, since its polarization direction coincides with its propagation direction, it is known as the *x*-propagating and *x*-polarizing longitudinal (L) wave. For the other two sub waves, since their particle vibration occurs in a direction perpendicular to the *x* axis, they can be depicted using sine waves. In both cases, the rise and fall of the wave patterns capture the polarization of the waves. The sub wave sketched in the *xy* plane is regarded as the *x*-propagating and *y*-polarizing wave, and the sub wave in the *xz* plane is known as the *x*-propagating and *z*-polarizing wave. Because these waves are polarizing in a direction perpendicular to their propagation direction, they are known as transverse (T) waves. In Figure 5.1 we use a dash-dotted (red) line to represent the L wave, a dashed (brown) line to mark the T wave polarizing in the *xy* plane, and a solid (blue) line to label the T wave polarizing in the *xz* plane. Note that when the polarization direction of a T wave is horizontal, the wave is commonly referred to as a traverse horizontal wave, or TH wave. Similarly, when the polarization direction of a T wave is vertical, the wave is referred to as a traverse vertical wave, or TV wave.

Figure 5.2 shows the particular displacements of an L wave having its polarization in the same direction as its propagation direction along the *x* axis (Figure 5.2(a)), and a TH wave having its polarization in the *y* direction and propagation in the *x* direction (Figure 5.2(b)), and a TV wave having its polarization in the *z* direction and propagation in the *x* direction (Figure 5.2(c)).

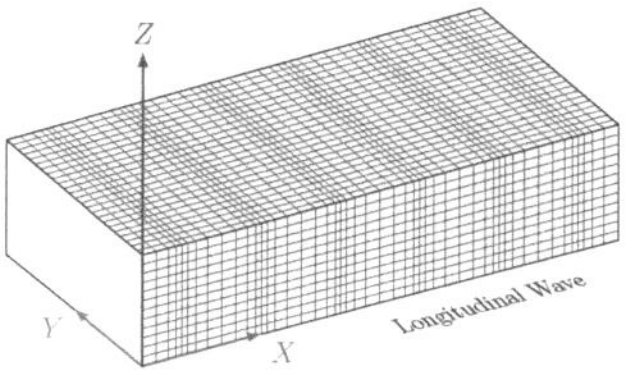

(a) L wave

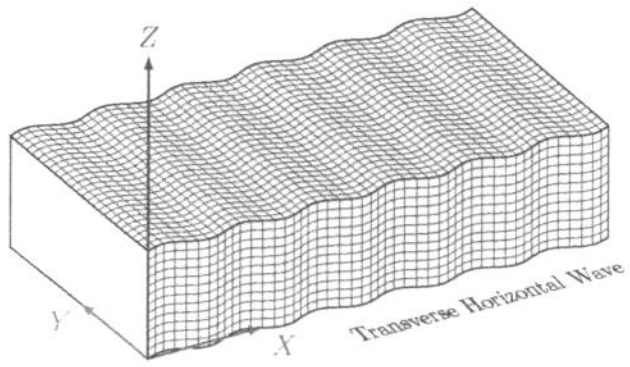

(b) TH wave

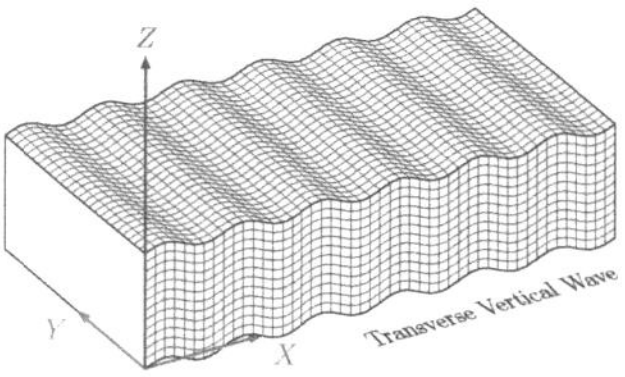

(c) TV wave

Figure 5.2 Illustration of particles displacements in L, TH and TV BAWs.

5.1.2 Polarization and Propagation of BAWs

Once the propagation direction of BAWs is selected, the wave-fronts of the L and T waves will occur in respective planes and they will travel along the direction of wave propagation. For example, when the selected propagation direction (*x*) takes an angle within the XY section plane, as depicted in Figure 5.3(a), the wave-front planes of the L and T waves will be parallel with the Z axis propagating out radially. In this case, the L wave polarizes and propagates in the same direction along the *x* axis. Of the two T waves, one polarizes in the *y* direction, which can be regarded as polarizing in a plane parallel with the section plane (denoted with a || symbol hereafter), and other polarizes in the *z* direction, perpendicular to the section plane (denoted with a $\perp$ symbol). To distinguish these waves, we use the same line designations as in Figure 5.1 to mark them: dash-dotted (red) lines for the L waves, dashed (brown) lines for the T($\|$) wave, and solid (blue) lines for the T($\perp$) wave. In an Euler angles notion, these three BAWs are said to be traveling in the direction of (ϕ_{XY}, 0°, 0°) and therefore their characteristics are often examined as a function of ϕ_{XY} in the XY section plane. Often we follow the right-hand convention for determining the sign for the rotation angle, that is, with the thumb pointing to the *z* direction, the rotation direction given by the four fingers is assigned with a positive sign, and the opposite rotation is assigned with a negative sign.

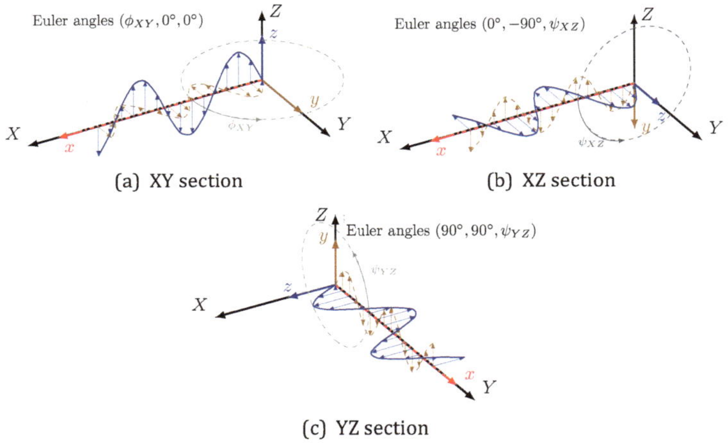

Figure 5.3 Illustration of BAWs polarization and propagation in 3D Cartesian coordinates system in three orthogonal sections.

For the cases of having the propagation direction (*x*) take a direction within the XZ section plane, as depicted in Figure 5.3(b), the wave-front planes will be parallel with the Y axis. These BAWs are regarded as traveling in the direction of (0°, −90°, ψ_{XZ}) and the characteristics of these waves are commonly examined as a function of ψ_{XZ} in the XZ section plane. Similarly, for the cases of having the propagation direction (*x*) take a direction in the YZ section plane, as depicted in Figure 5.3(c), the wave-front planes will be parallel with the X axis. These BAWs are traveling in the direction of (90°, 90°, ψ_{YZ}) and their characteristics are examined as a function of ψ_{YZ} in the YZ section plane.

5.2 Bulk Acoustic Waves in Elastic Solids

5.2.1 Christoffel Equation in a Matrix Form

By ignoring the body force term in the wave governing equation in elastic solids as given in Eq. (2.18), which is provided below for convenience

$$[\nabla_s][c][\nabla_s]^T\{u\} - \rho\frac{\partial^2}{\partial t^2}\{u\} = 0$$

and substituting the differential operator matrices using Eq. (2.14), we expand the wave governing equation in a matrix form as

$$\begin{bmatrix} \frac{\partial}{\partial x} & 0 & 0 & 0 & \frac{\partial}{\partial z} & \frac{\partial}{\partial y} \\ 0 & \frac{\partial}{\partial y} & 0 & \frac{\partial}{\partial z} & 0 & \frac{\partial}{\partial x} \\ 0 & 0 & \frac{\partial}{\partial z} & \frac{\partial}{\partial y} & \frac{\partial}{\partial x} & 0 \end{bmatrix} \begin{bmatrix} c_{11} & c_{12} & c_{13} & c_{14} & c_{15} & c_{16} \\ c_{12} & c_{22} & c_{23} & c_{24} & c_{25} & c_{26} \\ c_{13} & c_{23} & c_{33} & c_{34} & c_{35} & c_{36} \\ c_{14} & c_{24} & c_{34} & c_{44} & c_{45} & c_{46} \\ c_{15} & c_{25} & c_{35} & c_{45} & c_{55} & c_{56} \\ c_{16} & c_{26} & c_{36} & c_{46} & c_{56} & c_{66} \end{bmatrix}$$

$$\times \begin{bmatrix} \frac{\partial}{\partial x} & 0 & 0 \\ 0 & \frac{\partial}{\partial y} & 0 \\ 0 & 0 & \frac{\partial}{\partial z} \\ 0 & \frac{\partial}{\partial z} & \frac{\partial}{\partial y} \\ \frac{\partial}{\partial z} & 0 & \frac{\partial}{\partial x} \\ \frac{\partial}{\partial y} & \frac{\partial}{\partial x} & 0 \end{bmatrix} \begin{Bmatrix} u_x \\ u_y \\ u_z \end{Bmatrix} - \rho\frac{\partial^2}{\partial t^2}\begin{Bmatrix} u_x \\ u_y \\ u_z \end{Bmatrix} = 0 \tag{5.1}$$

We now consider an arbitrary point (x, y, z) in the XYZ coordinates system and use the direction formed between the origin of the coordinates and this point as the wave propagation direction.

The length of this vector can be expressed as follows:

$$r = l_x x + l_y y + l_z z$$

where l_x, l_y and l_z are the cosines of the angles between the r vector and the X, Y and Z axes, respectively. With this expression, we note

$$\frac{\partial r}{\partial x} = l_x, \quad \frac{\partial r}{\partial y} = l_y, \quad \frac{\partial r}{\partial z} = l_z$$

For waves propagating in the r direction their particle displacement fields can be expresses as

$$\{u\} = \begin{Bmatrix} u_x \\ u_y \\ u_z \end{Bmatrix} = \begin{Bmatrix} {}^\circ u_x \\ {}^\circ u_y \\ {}^\circ u_z \end{Bmatrix} e^{i(\omega t - kr)} = \begin{Bmatrix} {}^\circ u_x \\ {}^\circ u_y \\ {}^\circ u_z \end{Bmatrix} e^{ik(vt - r)} \tag{5.2}$$

where ${}^\circ u_x$, ${}^\circ u_y$ and ${}^\circ u_z$ are the respective polarization magnitude in each of the x, y and z dietections, respectively, ω and k are wave angular frequency and wavenumber, respectively, $v = \omega/k$ is the *phase velocity* of the wave propagation, and $i = \sqrt{-1}$ is the imaginary unit. Note that the wavenumber (k) is related to the wave wavelength (λ) as in $k = 2\pi/\lambda$.

Here the phase velocity of a wave is defined as the speed at which the phase of the wave propagates in space. For instance, the waves given in Eq. (5.2) are considered to propagate with a constant phase of $k(vt - r)$ in time. This statement can be expressed by the following derivative equation:

$$\frac{\partial[k(vt - r)]}{\partial t} = 0$$

By further simplifying this expression, we obtain the velocity ($\partial r/\partial t$) at which the phase propagates

$$\frac{\partial r}{\partial t} = \frac{\partial(kvt)}{k\partial t} = v$$

Apparently the waves represented by Eq. (5.2) propagate in space with a phase velocity of $v = \omega/k$.

By using Eq. (5.2) to simplify Eq. (5.1), we note

$$\begin{bmatrix} \frac{\partial}{\partial x} & 0 & 0 \\ 0 & \frac{\partial}{\partial y} & 0 \\ 0 & 0 & \frac{\partial}{\partial z} \\ 0 & \frac{\partial}{\partial z} & \frac{\partial}{\partial y} \\ \frac{\partial}{\partial z} & 0 & \frac{\partial}{\partial x} \\ \frac{\partial}{\partial y} & \frac{\partial}{\partial x} & 0 \end{bmatrix} \begin{Bmatrix} u_x \\ u_y \\ u_z \end{Bmatrix} = -ik \begin{bmatrix} l_x & 0 & 0 \\ 0 & l_y & 0 \\ 0 & 0 & l_z \\ 0 & l_z & l_y \\ l_z & 0 & l_x \\ l_y & l_x & 0 \end{bmatrix} \begin{Bmatrix} u_x \\ u_y \\ u_z \end{Bmatrix}$$

and

$$\rho \frac{\partial^2}{\partial t^2} \begin{Bmatrix} u_x \\ u_y \\ u_z \end{Bmatrix} = -\rho k^2 v^2 \begin{Bmatrix} u_x \\ u_y \\ u_z \end{Bmatrix} = -\rho k^2 v^2 \begin{bmatrix} 1 & 0 & 0 \\ 0 & 1 & 0 \\ 0 & 0 & 1 \end{bmatrix} \begin{Bmatrix} u_x \\ u_y \\ u_z \end{Bmatrix}$$

Putting these together in Eq. (5.2) we derive the following:

$$\left\{ -k^2 \begin{bmatrix} l_x & 0 & 0 & 0 & l_z & l_y \\ 0 & l_y & 0 & l_z & 0 & l_x \\ 0 & 0 & l_z & l_y & l_x & 0 \end{bmatrix} \begin{bmatrix} c_{11} & c_{12} & c_{13} & c_{14} & c_{15} & c_{16} \\ c_{12} & c_{22} & c_{23} & c_{24} & c_{25} & c_{26} \\ c_{13} & c_{23} & c_{33} & c_{34} & c_{35} & c_{36} \\ c_{14} & c_{24} & c_{34} & c_{44} & c_{45} & c_{46} \\ c_{15} & c_{25} & c_{35} & c_{45} & c_{55} & c_{56} \\ c_{16} & c_{26} & c_{36} & c_{46} & c_{56} & c_{66} \end{bmatrix} \begin{bmatrix} l_x & 0 & 0 \\ 0 & l_y & 0 \\ 0 & 0 & l_z \\ 0 & l_z & l_y \\ l_z & 0 & l_x \\ l_y & l_x & 0 \end{bmatrix} \right.$$

$$\left. + \rho k^2 v^2 \begin{bmatrix} 1 & 0 & 0 \\ 0 & 1 & 0 \\ 0 & 0 & 1 \end{bmatrix} \right\} \begin{Bmatrix} u_x \\ u_y \\ u_z \end{Bmatrix} = 0$$

For BAWs satisfying this equation, they will have to meet the following condition:

$$\begin{bmatrix} l_x & 0 & 0 & 0 & l_z & l_y \\ 0 & l_y & 0 & l_z & 0 & l_x \\ 0 & 0 & l_z & l_y & l_x & 0 \end{bmatrix} \begin{bmatrix} c_{11} & c_{12} & c_{13} & c_{14} & c_{15} & c_{16} \\ c_{12} & c_{22} & c_{23} & c_{24} & c_{25} & c_{26} \\ c_{13} & c_{23} & c_{33} & c_{34} & c_{35} & c_{36} \\ c_{14} & c_{24} & c_{34} & c_{44} & c_{45} & c_{46} \\ c_{15} & c_{25} & c_{35} & c_{45} & c_{55} & c_{56} \\ c_{16} & c_{26} & c_{36} & c_{46} & c_{56} & c_{66} \end{bmatrix} \begin{bmatrix} l_x & 0 & 0 \\ 0 & l_y & 0 \\ 0 & 0 & l_z \\ 0 & l_z & l_y \\ l_z & 0 & l_x \\ l_y & l_x & 0 \end{bmatrix}$$

$$-\rho v^2 \begin{bmatrix} 1 & 0 & 0 \\ 0 & 1 & 0 \\ 0 & 0 & 1 \end{bmatrix} = 0 \tag{5.3}$$

The first term in the above equation is the product of three matrices with dimensions of $3 \times 6, 6 \times 6$ and 6×3, respectively, in the given order, thus the outcome will be a 3×3 matrix, which can be represented by a $[\Gamma]$ matrix as follows:

$$[\Gamma] = \begin{bmatrix} \Gamma_{11} & \Gamma_{12} & \Gamma_{13} \\ \Gamma_{12} & \Gamma_{22} & \Gamma_{23} \\ \Gamma_{13} & \Gamma_{23} & \Gamma_{33} \end{bmatrix}$$

$$= \begin{bmatrix} l_x & 0 & 0 & 0 & l_z & l_y \\ 0 & l_y & 0 & l_z & 0 & l_x \\ 0 & 0 & l_z & l_y & l_x & 0 \end{bmatrix} \begin{bmatrix} c_{11} & c_{12} & c_{13} & c_{14} & c_{15} & c_{16} \\ c_{12} & c_{22} & c_{23} & c_{24} & c_{25} & c_{26} \\ c_{13} & c_{23} & c_{33} & c_{34} & c_{35} & c_{36} \\ c_{14} & c_{24} & c_{34} & c_{44} & c_{45} & c_{46} \\ c_{15} & c_{25} & c_{35} & c_{45} & c_{55} & c_{56} \\ c_{16} & c_{26} & c_{36} & c_{46} & c_{56} & c_{66} \end{bmatrix} \begin{bmatrix} l_x & 0 & 0 \\ 0 & l_y & 0 \\ 0 & 0 & l_z \\ 0 & l_z & l_y \\ l_z & 0 & l_x \\ l_y & l_x & 0 \end{bmatrix}$$

Then Eq. (5.3) can be condensed into

$$\begin{bmatrix} \Gamma_{11} & \Gamma_{12} & \Gamma_{13} \\ \Gamma_{12} & \Gamma_{22} & \Gamma_{23} \\ \Gamma_{13} & \Gamma_{23} & \Gamma_{33} \end{bmatrix} - \rho v^2 \begin{bmatrix} 1 & 0 & 0 \\ 0 & 1 & 0 \\ 0 & 0 & 1 \end{bmatrix} = 0 \tag{5.4}$$

This equation is known as the *Christoffel equation* in its matrix form for governing the BAWs traveling in elastic solids. It has three real positive roots as solutions for governing the polarization and propagation of the three waves, namely, one L wave and two T waves in elastic solids.

Referring to Figure 3.1 and Eq. (3.1), we know that if the wave propagation direction is determined by using a set of Euler angles, (ϕ, θ, ψ), we can determine the three cosines from the elements of the first row of the transformation matrix $[a] = [a_\psi] \times [a_\theta] \times [a_\phi]$ as

$$l_x = a_{11}, \quad l_y = a_{12}, \quad l_z = a_{13} \tag{5.5}$$

Clearly for a given material and propagation direction, we can determine the phase velocity of the three sub-waves polarizing longitudinally and transversely. Note that the phase velocity is also commonly referred to as wave velocity.

Example 1: Use the matrix form *Christoffel equation* presented in this section to resolve the problem given in Example 11 in Chapter 2 .

Answer: For convenience sake, we copy the stiffness matrix below along with ρ = 7450 kg/m^3.

$$[c] = 10^{11} \begin{bmatrix} 2.33 & 0.47 & 0.80 & -0.11 & 0 & 0 \\ 0.47 & 2.33 & 0.80 & 0.11 & 0 & 0 \\ 0.80 & 0.80 & 2.75 & 0 & 0 & 0 \\ -0.11 & 0.11 & 0 & 0.94 & 0 & 0 \\ 0 & 0 & 0 & 0 & 0.94 & -0.11 \\ 0 & 0 & 0 & 0 & -0.11 & 0.93 \end{bmatrix} (\text{N/m}^2)$$

Since the waves are considered traveling in the X direction, we have

$$l_x = 1, l_y = 0, l_z = 0$$

Knowing these terms we calculate

$$[\Gamma] = 10^{11} \times \begin{bmatrix} 1 & 0 & 0 & 0 & 0 & 0 \\ 0 & 0 & 0 & 0 & 0 & 1 \\ 0 & 0 & 0 & 0 & 1 & 0 \end{bmatrix} \begin{bmatrix} 2.33 & 0.47 & 0.80 & -0.11 & 0 & 0 \\ 0.47 & 2.33 & 0.80 & 0.11 & 0 & 0 \\ 0.80 & 0.80 & 2.75 & 0 & 0 & 0 \\ -0.11 & 0.11 & 0 & 0.94 & 0 & 0 \\ 0 & 0 & 0 & 0 & 0.94 & -0.11 \\ 0 & 0 & 0 & 0 & -0.11 & 0.93 \end{bmatrix} \begin{bmatrix} 1 & 0 & 0 \\ 0 & 0 & 0 \\ 0 & 0 & 0 \\ 0 & 0 & 0 \\ 0 & 0 & 1 \\ 0 & 1 & 0 \end{bmatrix}$$

$$= 10^{11} \times \begin{bmatrix} 1 & 0 & 0 & 0 & 0 & 0 \\ 0 & 0 & 0 & 0 & 0 & 1 \\ 0 & 0 & 0 & 0 & 1 & 0 \end{bmatrix} \begin{bmatrix} 2.33 & 0 & 0 \\ 0.47 & 0 & 0 \\ 0.8 & 0 & 0 \\ -0.11 & 0 & 0 \\ 0 & -0.11 & 0.94 \\ 0 & 0.93 & -0.11 \end{bmatrix} = 10^{11} \times \begin{bmatrix} 2.33 & 0 & 0 \\ 0 & 0.93 & -0.11 \\ 0 & -0.11 & 0.94 \end{bmatrix}$$

With Eq. (5.4), we express

$$10^{11} \times \begin{bmatrix} 2.33 & 0 & 0 \\ 0 & 0.93 & -0.11 \\ 0 & -0.11 & 0.94 \end{bmatrix} - 7450v^2 \begin{bmatrix} 1 & 0 & 0 \\ 0 & 1 & 0 \\ 0 & 0 & 1 \end{bmatrix} = 0$$

With some rearrangements, we have

$$\begin{bmatrix} 2.33 \times 10^{11} - 7450v^2 & 0 & 0 \\ 0 & 0.93 \times 10^{11} - 7450v^2 & -0.11 \times 10^{11} \\ 0 & -0.11 \times 10^{11} & 0.94 \times 10^{11} - 7450v^2 \end{bmatrix} = 0$$

This is exactly the same governing matrix equation obtained in Example 11 in Chapter 2 as expected. Its three positive roots are

$$v_{L(x)} = 5592.4 \text{ m/s}, \quad v_{T(y)} = 3327.5 \text{ m/s}, \quad v_{T(z)} = 3745.4 \text{ m/s}$$

representing the wave velocity of the L wave polarizing in the x direction, and T waves polarizing in the y and z directions, respectively.

5.2.2 Slowness Curves in Anisotropic Solids

Typically, when investigating the polarization and propagation of these BAWs, it is customary to examine the inverse value of the wave velocity, which is commonly known as the slowness of the waves. In this section, we will use case studies to examine slowness curves in each of the sections depicted in Figure 5.3.

Example 2: Find the slowness curves for BAWs traveling in Al_2O_3, a non-piezoelectric trigonal crystal (Class $\bar{3}$m).

Answer: Referring to Chapter 4, we know that Class $\bar{3}$m trigonal crystals have $[c]$ matrix in the following form:

$$[c] = \begin{bmatrix} c_{11} & c_{12} & c_{13} & c_{14} & 0 & 0 \\ c_{12} & c_{22} & c_{13} & -c_{14} & 0 & 0 \\ c_{13} & c_{13} & c_{33} & 0 & 0 & 0 \\ c_{14} & -c_{14} & 0 & c_{44} & 0 & 0 \\ 0 & 0 & 0 & 0 & c_{44} & -c_{14} \\ 0 & 0 & 0 & 0 & -c_{14} & \frac{c_{11}-c_{12}}{2} \end{bmatrix}$$

We now determine the slowness curves in three section planes, namely, the XY, XZ and YZ plans, section-by-section.

Slowness curves examined in the XY section plane

Referring to Figure 5.3(a), by plugging the Euler angles $(\phi_{XY}, 0°, 0°)$ for this section into Eq. (3.14), we find

$$[a_{XY}] = \begin{bmatrix} \cos(\phi_{XY}) & \sin(\phi_{XY}) & 0 \\ -\sin(\phi_{XY}) & \cos(\phi_{XY}) & 0 \\ 0 & 0 & 1 \end{bmatrix}$$

With Eq. (5.5), we have

$$l_x = \cos(\phi_{XY}),\ l_y = \sin(\phi_{XY}),\ l_z = 0$$

Referring to Eq. (5.3), we obtain the $[\Gamma]$ matrix as

$$[\Gamma_{XY}] = \begin{bmatrix} c_{11} - \dfrac{(c_{11}+c_{12})\sin^2(\phi_{XY})}{2} & \dfrac{(c_{11}+c_{12})\sin(2\phi_{XY})}{4} & c_{14}\sin(2\phi_{XY}) \\ \dfrac{(c_{11}+c_{12})\sin(2\phi_{XY})}{4} & c_{11}\sin^2(\phi_{XY}) + \dfrac{(c_{11}-c_{12})\cos^2(\phi_{XY})}{2} & c_{14}\cos(2\phi_{XY}) \\ c_{14}\sin(2\phi_{XY}) & c_{14}\cos(2\phi_{XY}) & c_{44} \end{bmatrix}$$

Using Eq. (5.4) along with the following material constants for Al_2O_3 given in Table 4.5:

$$[c] = 10^{10} \times \begin{bmatrix} 49.7 & 16.3 & 11.1 & -2.35 & 0 & 0 \\ 16.3 & 49.7 & 11.1 & 2.35 & 0 & 0 \\ 11.1 & 11.1 & 49.8 & 0 & 0 & 0 \\ -2.35 & 2.35 & 0 & 14.7 & 0 & 0 \\ 0 & 0 & 0 & 0 & 14.7 & -2.35 \\ 0 & 0 & 0 & 0 & -2.35 & 16.7 \end{bmatrix} \mathrm{N/m^2},$$

$$\rho = 3986 \ \mathrm{Kg/m^3}$$

we obtain the following matrix equation:

$$\begin{bmatrix} 49.7 - 33\sin^2(\phi_{XY}) - \rho' v^2 & 6.5\sin(2\phi_{XY}) & -2.35\sin(2\phi_{XY}) \\ 6.5\sin(2\phi_{XY}) & 49.7\sin^2(\phi_{XY}) + 16.7\cos^2(\phi_{XY}) - \rho' v^2 & -2.35\cos(2\phi_{XY}) \\ -2.35\sin(2\phi_{XY}) & -2.35\cos(2\phi_{XY}) & 14.7 - \rho' v^2 \end{bmatrix} = 0$$

as a function of the propagation direction defined by ϕ_{XY} in the XY section plane. Here $\rho' = 39.86 \times 10^{-8}$.

For a given value of ϕ_{XY}, we first solve this matrix equation (using MATLAB or other software packages, if necessary) for the velocity (v) of the three waves (i.e., one L wave and two T waves) from the three positive real solutions. We then take the inverse of the velocity value for each angle to determine the corresponding slowness ($1/v$). Table 5.1 lists some of the obtained velocity and slowness results for the angles selected.

By varying the direction angle ϕ_{XY} in a full span from $0°$ to $360°$ with finer increments, we will obtain three sets of values for wave velocity as a function of ϕ_{XY}, and then taking the inverse of the obtained velocity values, we can find three sets of slowness values as a function of ϕ_{XY}. Sketching out the three data sets in a polar-plot manner we will obtain three curves for the XY section. Figure 5.4(a) shows the obtained curves, where the L wave is shown in dash-dotted (red) line, the T($\perp$) wave is shown in solid (blue) line, and the T($\|$) wave is shown with dashed (brown) line.

Table 5.1 Velocity and slowness values for some selected angles

	Velocity (m/s $\times 10^3$)			Slowness (s/m $\times 10^{-4}$)		
Angle	L	T(⊥)	T(‖)	L	T(⊥)	T(‖)
0°	11.1663	5.7429	6.7672	0.8956	1.7413	1.4777
30°	11.1839	6.0403	6.4728	0.8941	1.6556	1.5449
60°	11.1663	5.7429	6.7672	0.8956	1.7413	1.4777
90°	11.1839	6.0403	6.4728	0.8941	1.6556	1.5449
120°	11.1663	5.7429	6.7672	0.8956	1.7413	1.4777
150°	11.1839	6.0403	6.4728	0.8941	1.6556	1.5449
180°	11.1663	5.7429	6.7672	0.8956	1.7413	1.4777
210°	11.1839	6.0403	6.4728	0.8941	1.6556	1.5449
240°	11.1663	5.7429	6.7672	0.8956	1.7413	1.4777
270°	11.1839	6.0403	6.4728	0.8941	1.6556	1.5449
300°	11.1663	5.7429	6.7672	0.8956	1.7413	1.4777
330°	11.1839	6.0403	6.4728	0.8941	1.6556	1.5449
360°	11.1663	5.7429	6.7672	0.8956	1.7413	1.4777

Slowness curves examined in the XZ and YZ section planes

By repeating the same exercises we can find the slowness curves for the L and T waves propagating in a direction within the XZ and YZ section planes, respectively.

For the XZ section, referring to Figure 5.3(b) we have the following $[a]$ and $[\Gamma]$ matrices:

$$[a_{XZ}] = \begin{bmatrix} \cos(\psi_{XZ}) & 0 & \sin(\psi_{XZ}) \\ -\sin(\psi_{XZ}) & 0 & \cos(\psi_{XZ}) \\ 0 & -1 & 0 \end{bmatrix}$$

and

$$[\Gamma_{XZ}] = \begin{bmatrix} c_{11}\cos^2(\psi_{XZ}) + c_{44}\sin(\psi_{XZ})^2 & c_{14}\sin(2\psi_{XZ}) & \dfrac{(c_{13}+c_{44})\sin(2\psi_{XZ})}{2} \\ c_{14}\sin(2\psi_{XZ}) & \Gamma_{22} & c_{14}\cos^2(\psi_{XZ}) \\ \dfrac{(c_{13}+c_{44})\sin(2\psi_{XZ})}{2} & c_{14}\cos^2(\psi_{XZ}) & c_{44}\cos^2(\psi_{XZ}) + c_{33}\sin^2(\psi_{XZ}) \end{bmatrix}$$

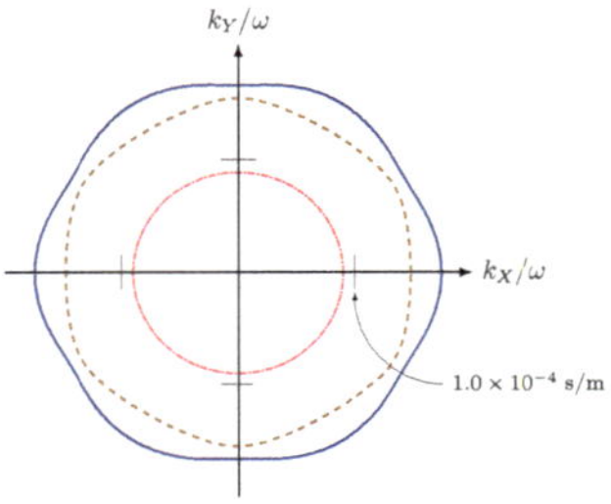

(a) XY section

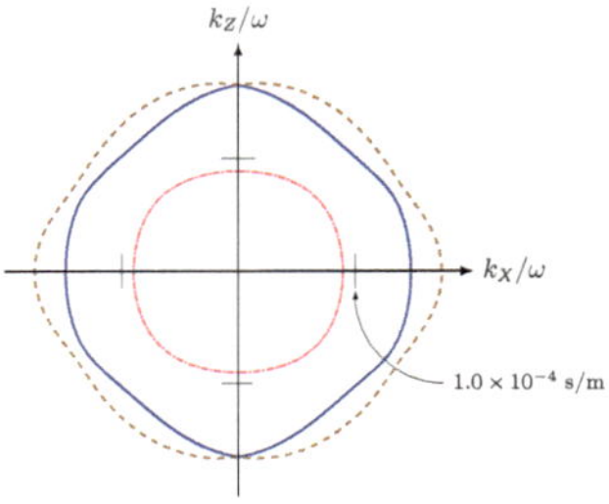

(b) XZ section

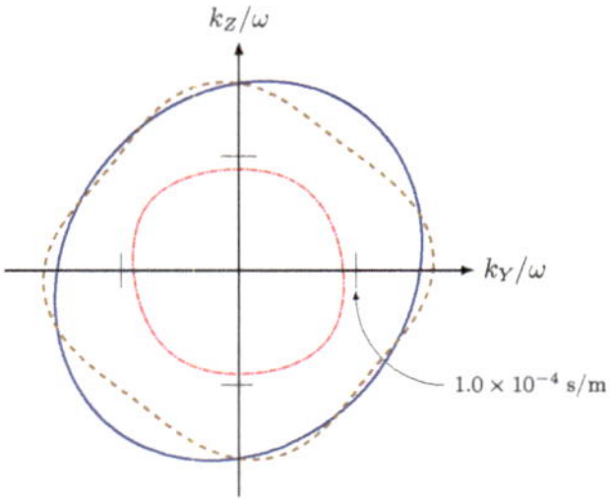

(c) YZ section

Figure 5.4 Slowness curves for BAWs in trigonal crystal (Class $\bar{3}$m) Al_2O_3 in three orthogonal sections.

in which

$$\Gamma_{22} = c_{44}\cos^2(\psi_{XZ}) + \frac{(c_{11}-c_{12})\cos^2(\psi_{XZ})}{2}$$

For the YZ section, referring to Figure 5.3(c), we have

$$[a_{YZ}] = \begin{bmatrix} 0 & \cos(\psi_{YZ}) & \sin(\psi_{YZ}) \\ 0 & -\sin(\psi_{YZ}) & \cos(\psi_{YZ}) \\ 1 & 0 & 0 \end{bmatrix}$$

and

$$[\Gamma_{YZ}] = \begin{bmatrix} \Gamma_{11} & 0 & 0 \\ 0 & \Gamma_{22} & \Gamma_{23} \\ 0 & \Gamma_{23} & \Gamma_{33} \end{bmatrix}$$

where

$$\Gamma_{11} = c_{14}\sin(2\psi_{XZ}) + c_{44}\sin^2(\psi_{XZ}) + \frac{(c_{11} - c_{12})\cos^2(\psi_{YZ})}{2}$$

$$\Gamma_{22} = c_{11}\cos^2(\psi_{YZ}) - c_{14}\sin(2\psi_{YZ})) + c_{44}\sin^2(\psi_{YZ})$$

$$\Gamma_{23} = \cos(\psi_{YZ})\left[c_{13}\sin(\psi_{YZ}) - c_{14}\cos(\psi_{YZ}) + c_{44}\sin(\psi_{YZ})\right]$$

$$\Gamma_{33} = c_{44}\cos^2(\psi_{YZ}) + c_{33}\sin^2(\psi_{YZ})$$

With these matrices along with the material constants, we can solve for the three sets of velocity values first in the XY and YZ sections, respectively, and then take the inverse of the velocity values to obtain the data sets for the slowness curvesSlowness curves: Al_2O_3. The results for the XY section are presented in Figure 5.4(b), and those for the YZ section in Figure 5.4(c). In both cases, the L waves are shown in dash-dotted (red) lines, the T($\perp$) waves are shown in solid (blue) line, and the T($\|$) waves are shown with dashed (brown) line.

5.3 Slowness Curves Interpreted

5.3.1 Information Contained in Slowness Curves

A point on a slowness curve tells not only how fast (or, precisely, how slow) a wave propagates but also the orientation of the wave-front and its propagation direction. The radial direction pointed by the line between the point and the origin of the coordinates system defines the wave propagation direction, the distance from the point to the origin gives the magnitude of the slowness (or velocity after taking its inverse) of the wave, the tangent line of the curve at the point defines the wave-front plane, and its outward normal defines the direction of wave energy which is often termed as the *Poynting vector*, to honor John Henry Poynting for his discovery of it in 1884 (see more discussion in Section 5.3.4). The angle between the radial direction and Poynting vector is commonly known as the *Poynting angle*.

Take the XY section shown in Figure 5.4(a), for example, since the slowness curves for the L wave has the smallest magnitude and it takes the shape of a circle, this means that the L wave in Al_2O_3 travels faster than other waves and its velocity remain constant independent of the orientation angle. Moreover, because the normal of the wave-front plane of any point on a circle always coincides with the radial direction, this means that the Poynting angle is zero, so the L wave in this case is pure longitudinal wave. When the Poynting angle is not zero, such an L wave is called quasi-longitudinal (quasi-L) wave. By contrast, the two T waves in this case are not pure, but quasi-transverse (quasi-T) waves. The T($\perp$) wave (marked by the solid blue line) is the outer curve, indicating that this T wave travels the slowest. The T($\|$) wave (marked by the dashed brown line) has its slowness or velocity in between those of the other two waves. With careful inspections, we can see, from the shapes of the three slowness curves obtained in this sectional view, an inverted three-fold symmetry as expected for the Class $\bar{3}$m trigonal crystals.

For the XZ section (Figure 5.4(b)), the three waves are quasi-L and quasi-T waves, respectively, with the quasi-L wave possessing the smallest slowness (or the highest velocity). Of the two quasi-T waves, the T($\|$) wave in dashed brown line has the largest slowness (or the smallest velocity), and the T($\perp$) wave in solid blue line has its slowness or velocity in between those of the other two waves. In both the XY and XZ sectional views, we can see that all slowness curves possess a YZ mirror plane as expected for the Class $\bar{3}$m trigonal crystals. For the YZ section (Figure 5.4(c)), aside from the commonality noted already, the slowness curves for the two quasi-T waves intersect and cross each other. This characteristic is not seen in the other two sections. This means that the velocity of the two quasi-T waves will vary depending on their propagation direction: in some directions, the quasi-T wave polarizing perpendicular to the YZ section plane (solid blue line) travels faster than the quasi-T wave polarizing in the YZ section plane (dashed brown line), and in other directions, it will travel slower.

The exercises in Example 2 also shed some insight into the relationship between the wave propagation angle and the shape of slowness curves. For the sake of discussion, let β be the angle of interest that defines the wave propagation direction. Of the three section cases discussed, we note that terms in the $[\Gamma]$ matrix contain the sine ($\sin(\beta)$) and cosine ($\cos(\beta)$) factors either in a squared form like $\sin^2(\beta)$ and $\cos^2(\beta)$, or in a double angle form such as $\sin(2\beta)$ and $\cos(2\beta)$. From trigonometry we know that $\sin(\beta \pm \pi) = -\sin(\beta)$, $\cos(\beta \pm \pi) = -\cos(\beta)$, $\sin(2(\beta \pm \pi)) = \sin(2\beta)$, and $\cos(2(\beta \pm \pi)) = \cos(2\beta)$. This means that the $[\Gamma]$ matrix will remain unchanged if we replace $\beta \pm \pi$ for β (note that the period of both the sine and cosine functions is 2π). This fact suggests that slowness values are the same for the direction defined either by β or by $\beta \pm \pi$. This simply says that the slowness curves will repeat after a rotation of 180°.

5.3.2 Slowness Shells or Surfaces

We now examine the relationships between these slowness curves obtained in the three different section planes. Figure 5.5 shows the slowness curves for trigonal crystal lithium niobate ($LiNbO_3$, Class 3m) in the XY, XZ and YZ sections after ignoring the piezoelectric effect.

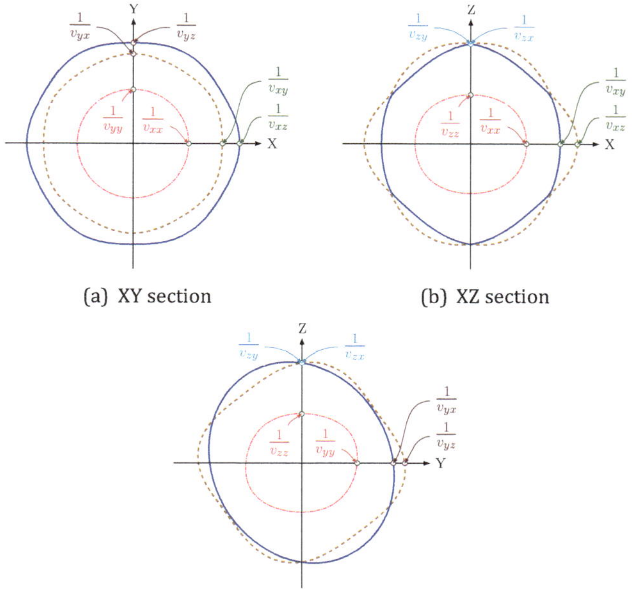

Figure 5.5 Slowness curves for BAWs in trigonal crystal (Class 3m) lithium niobate in three orthogonal sections.

Of all the slowness curves shown here, the L waves in the center are readily distinguishable due to their small magnitudes and non-crossover with other curves. They are marked with dash-dotted red lines. For the two T waves, it is not as easy to tell them apart. In this book we mark them using the line designations discussed earlier for easy identification of different waves, that is, to mark the T(⊥) waves in solid blue lines, and the T(||) wave in dashed brown lines. In this way, because the orientations of section planes are recognizable by the axis labels in the plots, we can quickly tell the two T waves apart.

One issue with this labeling method is that the curves marked by the same line type (and color) in different sections could be mistaken as belonging to the same wave. They are not. The dimensions of these slowness curves, however, will help us out in this regard. Take a look of the L wave curves (dash-dotted red curves) in all three section planes first. Letting v_{xx}, v_{yy} and v_{zz} be the wave velocity of the L waves traveling in the respective axial direction, with the first index representing the wave propagation direction and the second the polarization direction, we note that $1/v_{xx}$ in the XY section equals that in the XZ section, $1/v_{yy}$ in the YZ section equals that in the XY section, and $1/v_{zz}$ in the XZ section equals that in the YZ section. This observation makes a perfect sense because they are the respective intersects of the slowness curve of the L waves with each of the three axes, respectively. Similar observations can be made for the two T waves, that is, $1/v_{xy}$ and $1/v_{xz}$ in the *XY* section equal those in the XZ section, $1/v_{yx}$ and $1/v_{yz}$ in the YZ section equal those in the XY section, and $1/v_{zx}$ and $1/v_{zy}$ in the XZ section equal those in the YZ section. Again, of the two indexes of these slowness values, $1/v_{xy}$, $1/v_{xz}$, $1/v_{yx}$, $1/v_{yz}$, $1/v_{zx}$ and $1/v_{zy}$, the first one represents wave propagation direction and the second one polarization direction.

Based on these observations, we can conclude that the slowness curves having the same intersect value with the same axis belong to the same wave. We now gain a sense that these slowness curves in different section planes are actually sectional views of three slowness shells in the 3D Cartesian Coordinates system, representing the components of one L wave and two T waves. By putting these sectional views together, one may be able to imagine what these three 3D slowness shells may look like.

Of the three slowness shells, the smallest one inside is easily recognized as the L-wave shell, but the two T-wave shells intersect each other. Let us take a moment to think about how we can best label and separate them apart. Labeling by their size will not work when the shells possess the behavior observed in Figure 5.5(c), namely, when the slowness of one T wave becomes larger than that of the other, or vice versa, as the propagation direction changes. If we label them by the direction of polarization and propagation, it will not work either, because the same slowness shell can generate a XY T wave in one section but a XZ T wave in another section. This can be seen in Figure 5.5, where the outer curve in the *XY* section (Figure 5.5(a)) is a XY T wave and the out curve in the XZ section (Figure 5.5(b)) is a XZ T wave, but judging from their dimensions, we know that they are section curves from the same slowness shell in the respective sections. To avoid these confusions, the label method used in this book, namely by their propagation and polarization directions with respect to a selected section plan as discussed earlier, is informative.

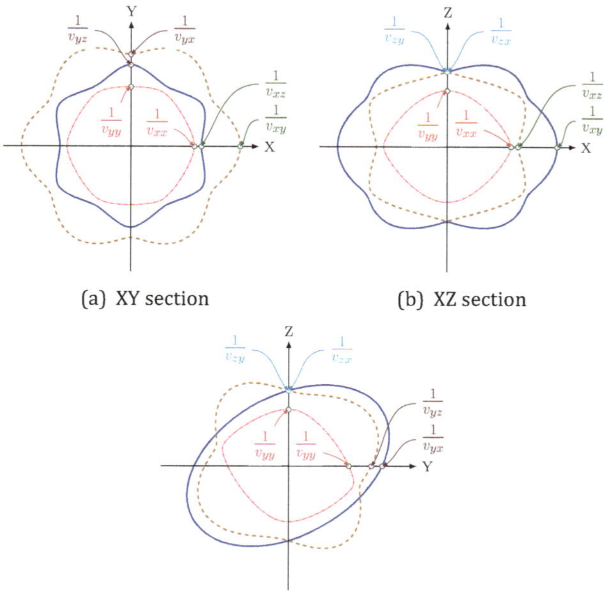

(a) XY section (b) XZ section

(c) YZ section

Figure 5.6 Slowness curves for BAWs in trigonal crystal (Class 32) quartz (SiO_2) in three orthogonal sections.

Figure 5.6 gives another example, in which three orthogonal sections of the slowness shells for trigonal crystal quartz (SiO_2, Class 32), after ignoring the piezoelectric coupling effect, are shown. Different from lithium niobate shown in Figure 5.5, in the XY section the T(||) wave travels the slowest in quartz crystal. Also, these slowness curves exhibit different shapes as compared with those for $LiNbO_3$, due to the difference in crystal symmetry in these two materials. The reader is encouraged to examine these curves in detail to gain better appreciation in terms of the observations and conclusions made earlier from the slowness shells and sectional curves for $LiNbO_3$. The example cases we have seen to this point are all for trigonal crystals. In the next section we will list a few more examples for other types of crystals, including cubic, tetragonal, hexagonal and orthorhombic crystals.

5.3.3 More Examples of Slowness Curves

5.3.3.1 Slowness Curves for Tetragonal Crystals

Figure 5.7 shows slowness curves for two piezoelectric tetragonal crystals. The curves for tellurium oxide (TeO_2), a Class 422 crystal, are shown in Figures 5.7(a) and 5.7(b), and those for barium titanate ($BaTiO_3$), a Class 4mm crystal, are given in Figures 5.7(c) and 5.7(d).

In these tetragonal crystals (see Section 4.5), four-fold symmetry (about Z) is seen in the XY section, and the curves for the Class 422 (TeO_2) has two-fold symmetry (about X), while the ones for the Class 4mm ($BaTiO_3$) showing mirror plane symmetry (about XZ). Because of that, the respective slowness curves in the XZ and YZ section planes have the same shapes and dimensions, only the curves in the XZ section for TeO_2 and the curves in the YZ section for $BaTiO_3$ are provided along with their corresponding curves in the XY section.

In Figure 5.8, slowness curves for three non-piezoelectric tetragonal crystals, including lead molybdate ($PbMoO_4$, Class 4/m), calcium molybdate ($CaMoO_4$, Class 4/m) and titanium oxide (TiO_2, Class 4/mmm). In these tetragonal crystals, aside from their four-fold symmetry, the two Class 4/m ones have a mirror plane of XY and the Class 4/mmm one has two mirror planes of XY and XZ. Due to these types of symmetry, slowness curves in the XZ and YZ section planes

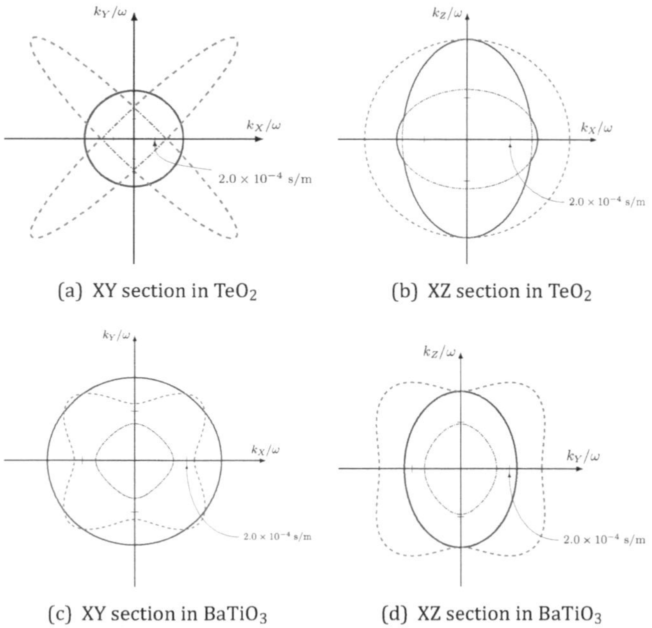

Figure 5.7 Slowness curves for BAWs in tetragonal TeO_2 and $BaTiO_3$ in two selected sections. Note that to fit side-by-side, (a) and (b) are plotted in different scales.

have the same shapes and dimensions, thus only the curves in the XZ section plane are provided along with those in the XY section.

5.3.3.2 Slowness Curves for Cubic Crystals

Figure 5.9 shows the slowness curves for three cubic crystals: those for Silicon (Si), a non-piezoelectric Class m3m crystal, are given in Figures 5.9(a) and 5.9(b)); those for gallium arsenide (GaAs), a piezoelectric Class $\bar{4}3m$) crystal, are in Figures 5.9(c) and 5.9(d); and those for bismuth germanium oxide ($Bi_{12}GeO_{20}$), a piezoelectric Class 23 crystal, are in Figures 5.9(e) and 5.9(f). Due to their unique symmetric structures (see Section 4.8), the respective slowness curves of these cubic crystals in the XY, XZ and YZ section planes have the same shapes and dimensions, thus only the curves in the XY section are plotted here. Because of that, in Figures 5.9(b), reffig:GaAsdz and 5.9(f), slowness curves in a section plane that goes

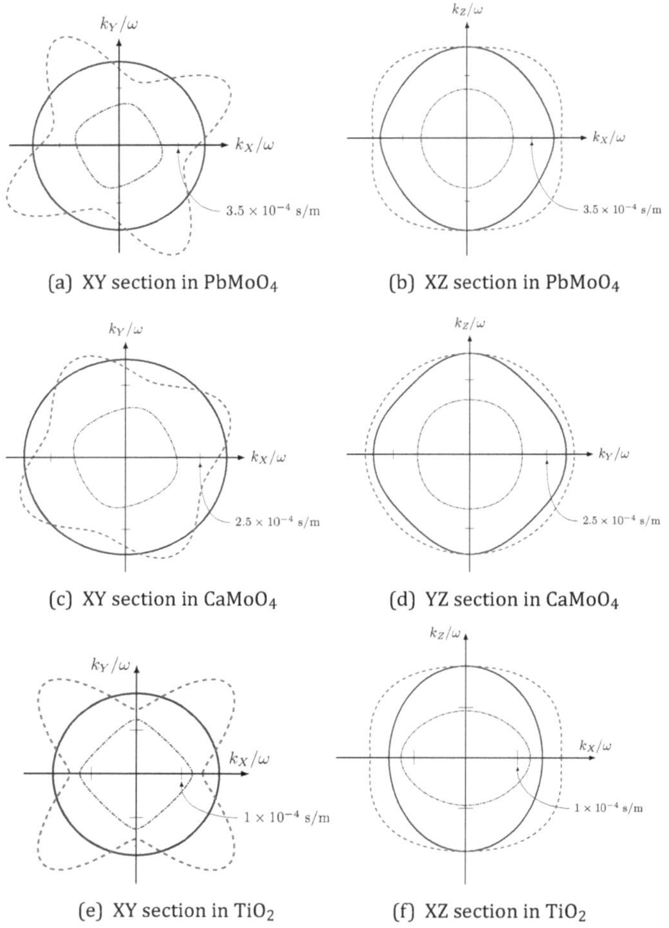

Figure 5.8 Slowness curves for BAWs in tetragonal $PbMoO_4$, $CaMoO_4$ and TiO_2 in two selected sections.

through the diagonal line in the XY plane and the Z axis are shown to provide some more insight into the 3D slowness shells.

5.3.3.3 Slowness Curves for Hexagonal Crystals

Figure 5.10 shows slowness curves for two piezoelectric tetragonal crystals of Class 6mm, namely, zinc oxide (ZnO) and cadmium sulfide (CdS). These crystals possess six-fold symmetry (about Z) and a mirror of XZ (see Section 4.7). From these lowness curves we note

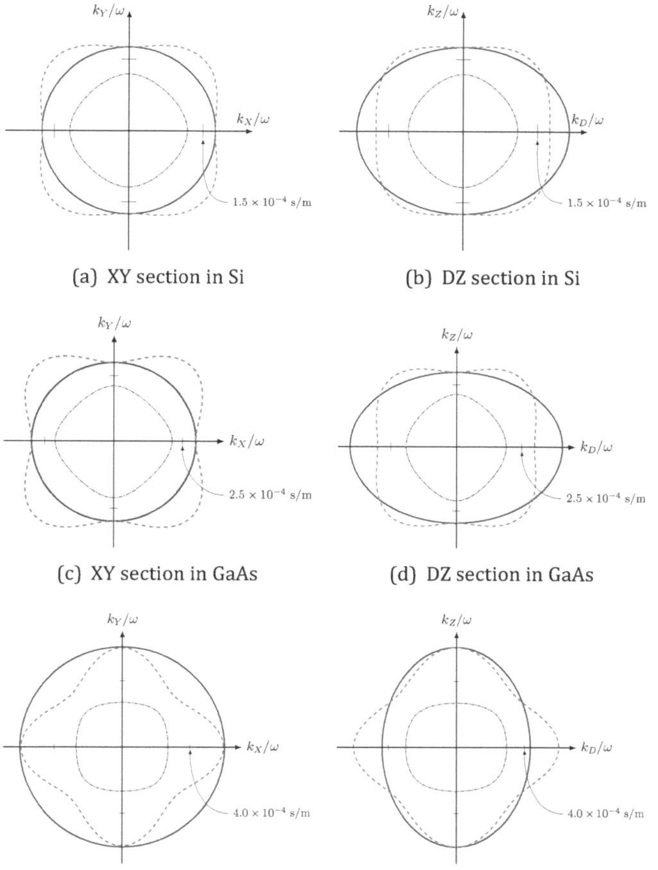

(a) XY section in Si (b) DZ section in Si

(c) XY section in GaAs (d) DZ section in GaAs

(e) XY section in $Bi_{12}GeO_{20}$ (f) DZ section in $Bi_{12}GeO_{20}$

Figure 5.9 Slowness curves for BAWs in cubic Si, GaAs and $Bi_{12}GeO_{20}$ in two selected sections, in which D represents the 45° diagonal direction in the XY plane.

that those in the XZ and YZ sections have the same shapes and dimensions; hence, only the curves in the XZ section plane are provided along with those in the XY section.

5.3.3.4 Slowness Curves for Orthorhombic Crystals

Figure 5.11 shows slowness curves for orthorhombic crystal (Class 2mm) barium sodium niobate. For this type of crystals, because of

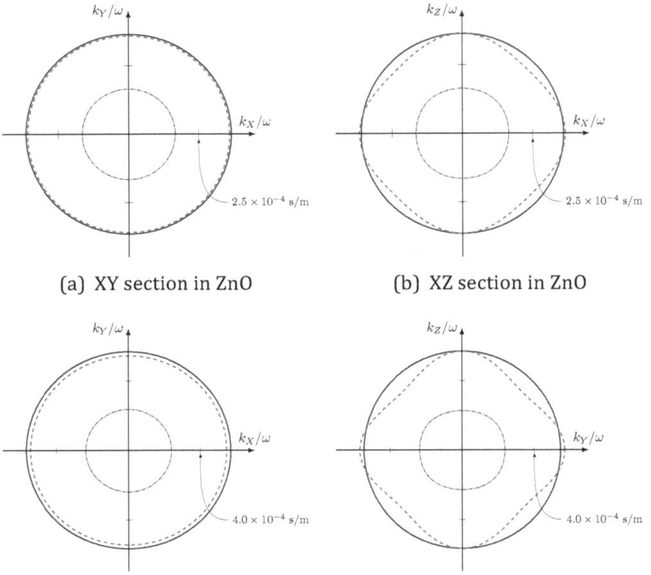

(a) XY section in ZnO (b) XZ section in ZnO

(c) XY section in CdS (d) YZ section in CdS

Figure 5.10 Slowness curves for BAWs in hexagonal ZnO and CdS.

their two-fold symmetry (about Z) and a mirror of XZ (see Section 4.4), the respective slowness curves in the XZ and YZ section planes have the same shapes and dimensions, thus only the curves in the XZ section plane are provided along with those in the XY section.

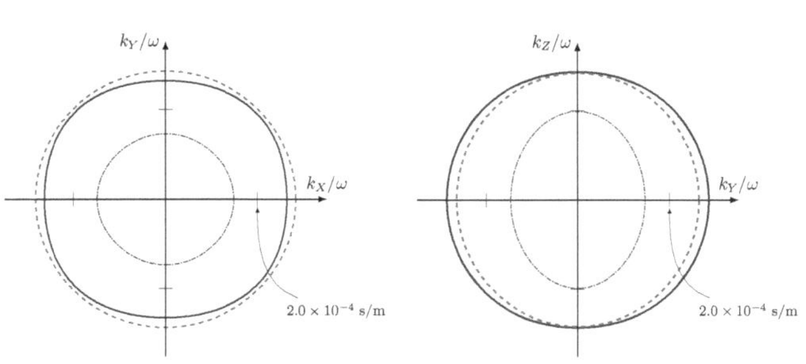

(a) XY section in $Ba_2NaNb_5O_{15}$ (b) YZ section in $Ba_2NaNb_5O_{15}$

Figure 5.11 Slowness curves for BAWs in orthorhombic crystal $Ba_2NaNb_5O_{15}$.

Of all the slowness curves shown we can clearly see that slowness values remain the same if we rotates either +180° or -180° from a preselected orientation, thus confirming that the slowness curves will repeat after a rotation of 180°.

5.3.4 Poynting Vector, Energy Velocity, Wave and Velocity Surfaces

Poynting vector captures the directional energy flux, that is the energy transfer per unit area per unit time, of an acoustic wave. It is named after John Henry Poynting who first derived it in 1884. Mathematically, Poynting vector is expressed as (referring to Eq. (2.3)):

$$\mathbf{P} = -[\mathbf{T}]\frac{\partial u}{\partial t} = -\begin{bmatrix} \sigma_x & \tau_{xy} & \tau_{xz} \\ \tau_{xy} & \sigma_y & \tau_{yz} \\ \tau_{xz} & \tau_{yz} & \sigma_z \end{bmatrix} \frac{\partial}{\partial t} \begin{Bmatrix} u_x \\ u_y \\ u_z \end{Bmatrix}$$

Energy velocity is then defined as the Poynting vector divided by the energy per unit volume. It can be expressed as

$$v_e = \frac{\mathbf{P}}{U_{\text{unit-volume}}}$$

Energy velocity (v_e) is related to the wave phase velocity (v) by the following relationship via the Poynting angle (θ_{Poynting}) as

$$v = \frac{v_e}{\cos\theta_{\text{Poynting}}} \tag{5.6}$$

where the Poynting angle (θ_{Poynting}) is the angle between the direction of the energy velocity and the direction of the wave propagation. A positive θ_{Poynting} means that the angle of energy velocity is larger than the angle of propagation, and vice versa. The above relationship indicates that energy velocity, phase velocity and slowness are all related.

Aside from slowness shells or surfaces, there are also velocity surfaces and wave surfaces. Velocity surfaces are surfaces constructed in a polar coordinate system by connecting the end point

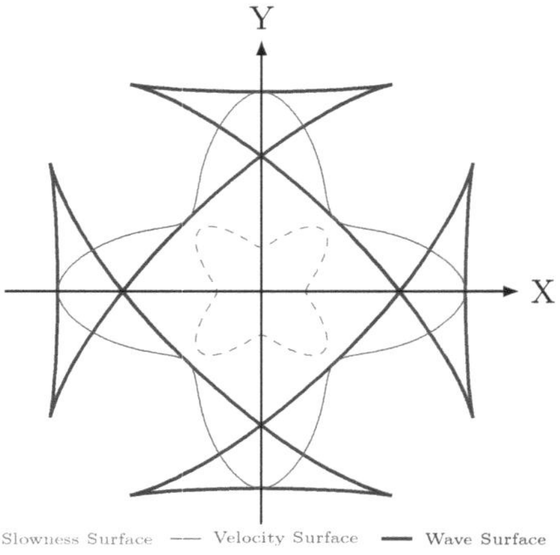

Figure 5.12 Relationships between slowness, velocity and wave surfaces for the T(||) wave in the XY section of TiO_2.

of the wave velocity as a function of wave-propagation direction, while wave surfaces are those constructed by connecting the end point of the energy velocity as a function of wave propagation direction.

Figure 5.12 shows the slowness, velocity and wave surfaces of the T(||) wave traveling in the XY section of TiO_2. The inner curve is the slowness surface (not to scale), the thin (red) curve possessing the same shape as the slowness surface but with a 45° rotation is the velocity surface, and the thick (blue) curve with sharp points is the wave surface. Since velocity and slowness have an inverse relationship, a velocity surface can be obtained directly by plotting the inverse of its corresponding slowness as a function wave propagation angle. Like slowness surfaces, there are three velocity surfaces in each section, one for the L wave and two for the T waves.

We now take a close look of the interdependence among these surfaces and the relationship between v and v_e. Referring to Figure 5.13(a), we first look at the wave propagating in the 30° direction. The vector $v_{(30°)}$ defines the magnitude of the phase velocity and wave propagation direction. The normal of the slowness surface (as marked by a dashed line) at the intersect of the $v_{(30°)}$ vector and the slowness surface defines the direction of the corresponding energy velocity, v_e. The magnitude of the energy velocity, $v_{e(30°)}$, can be determined, according to Eq. (5.6), from the

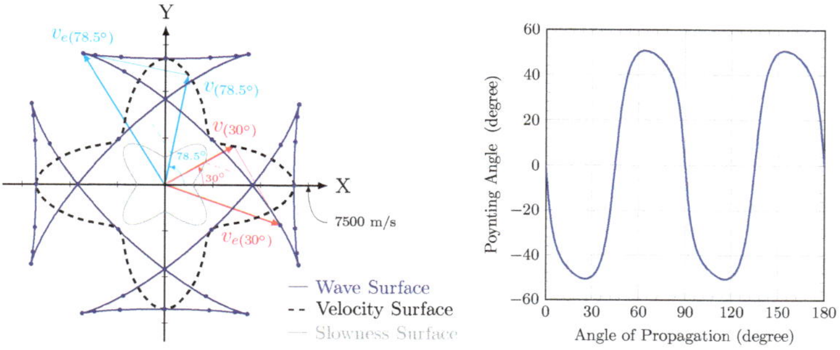

(a) Three surfaces

(b) Poynting angle

Figure 5.13 Energy velocity and Poynting angle of the T(||) wave in the XY section of TiO_2.

intersect of the perpendicular line at the end of the $v_{(30°)}$ vector and the radial line that is parallel to the dashed normal line. In the same way, for the wave propagating in the 78.5° direction, the $v_{e(78.5°)}$ vector can be found from the $v_{(78.5°)}$ vector. By considering all propagation directions and connecting the end points of the resulting energy velocity vectors, we can obtain the wave surface. The angle between the v and v_e vectors is the Poynting angle. Figure 5.13(b) shows the variation of Poynting angle as a function of propagation direction. Clearly, the Poynting angle is zero at 0°, 45°, 90°, and so on. In other word, only in these directions the T(⊥) wave is pure transverse wave and in all other directions it is quasi-transverse wave.

Figure 5.14(a) shows the slowness surface (inner curve, not to scale), velocity surface (dash line) and wave surface (solid line) of the T(⊥) wave traveling in the YZ section of $LiNbO_3$. Figure 5.14(b) shows the variation of Poynting angle as a function of propagation direction. Apparently, in the range shown, there are two directions in which the Poynting angle is zero, indicating that the T(⊥) wave is pure transverse wave at approximately 25° and 115°.

Figure 5.15(a) shows the slowness surface (inner curve, not to scale), velocity surface (dash line) and wave surface (solid line) of the T(||) wave traveling in the XY section of TeO_2. Figure 5.15(b) shows the corresponding Poynting angle as a function of propagation direction. Apparently, in the range shown, there are five directions in which the Poynting angle is zero, indicating that the T(||) wave is pure

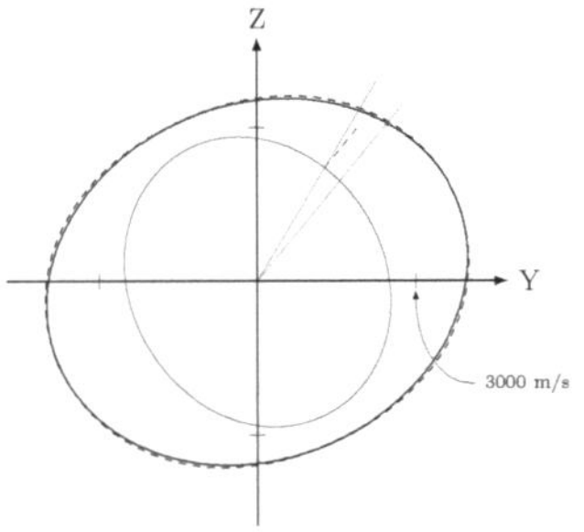

(a) Energy velocity

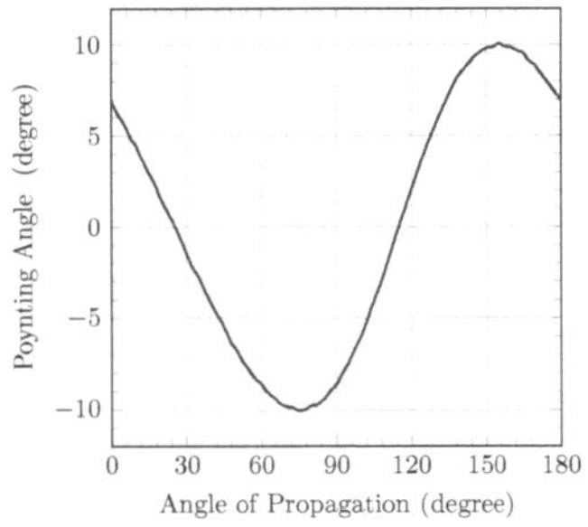

(b) Poynting angle

Figure 5.14 Energy velocity and Poynting angle of the T(⊥) wave in the YZ section of $LiNbO_3$.

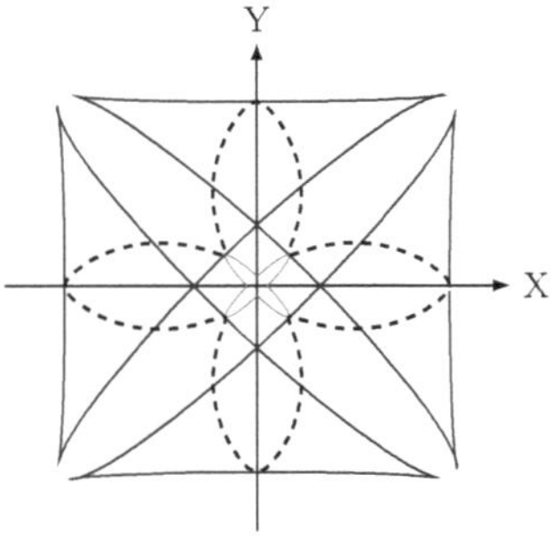

(a) Energy velocity

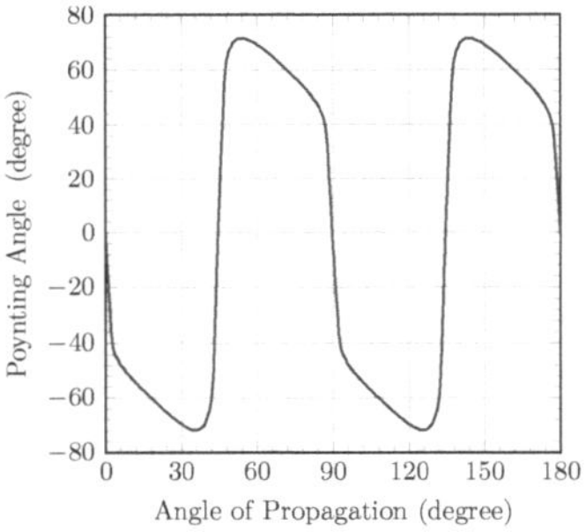

(b) Poynting angle

Figure 5.15 Energy velocity and Poynting angle of the T(||) wave in the XY section of TeO_2.

transverse wave along these directions, including 0°, 45°, 90°, 135° and 180°.

5.4 Bulk Acoustic Waves in Piezoelectric Solids

In Section 5.3, all slowness surfaces examined are either for non-piezoelectric crystal materials or for piezoelectric crystals without the consideration of the piezoelectric coupling effect. In this section, we will examine the characteristics of BAWs in piezoelectric crystals by considering the piezoelectric mechano-electric coupling effect.

5.4.1 Governing Equation for BAWs with Piezoelectric Coupling

As discussed earlier, Eq. (2.25) gives the set of governing equations for acoustic waves in piezoelectric solids, which we repeat here for convenience sake

$$[\nabla_s]\left([c][\nabla_s]^T\{u\} + [e]^T[\nabla]^T\{\phi\}\right) - \rho\frac{\partial^2}{\partial t^2}\{u\} = 0$$
$$[\nabla]\left([e][\nabla_s]^T\{u\} - [\epsilon][\nabla]^T\{\phi\}\right) = 0$$

In a matrix form, this set of equations can be expressed as

$$\begin{bmatrix} \frac{\partial}{\partial x} & 0 & 0 & 0 & \frac{\partial}{\partial z} & \frac{\partial}{\partial y} \\ 0 & \frac{\partial}{\partial y} & 0 & \frac{\partial}{\partial z} & 0 & \frac{\partial}{\partial x} \\ 0 & 0 & \frac{\partial}{\partial z} & \frac{\partial}{\partial y} & \frac{\partial}{\partial x} & 0 \end{bmatrix}$$
$$\times\left\{\begin{bmatrix} c_{11} & c_{12} & c_{13} & c_{14} & c_{15} & c_{16} \\ c_{12} & c_{22} & c_{23} & c_{24} & c_{25} & c_{26} \\ c_{13} & c_{23} & c_{33} & c_{34} & c_{35} & c_{36} \\ c_{14} & c_{24} & c_{34} & c_{44} & c_{45} & c_{46} \\ c_{15} & c_{25} & c_{35} & c_{45} & c_{55} & c_{56} \\ c_{16} & c_{26} & c_{36} & c_{46} & c_{56} & c_{66} \end{bmatrix}\begin{bmatrix} \frac{\partial}{\partial x} & 0 & 0 \\ 0 & \frac{\partial}{\partial y} & 0 \\ 0 & 0 & \frac{\partial}{\partial z} \\ 0 & \frac{\partial}{\partial z} & \frac{\partial}{\partial y} \\ \frac{\partial}{\partial z} & 0 & \frac{\partial}{\partial x} \\ \frac{\partial}{\partial y} & \frac{\partial}{\partial x} & 0 \end{bmatrix}\begin{Bmatrix} u_x \\ u_y \\ u_z \end{Bmatrix}\right.$$
$$\left. + \begin{bmatrix} e_{x1} & e_{y1} & e_{z1} \\ e_{x2} & e_{y2} & e_{z2} \\ e_{x3} & e_{y3} & e_{z3} \\ e_{x4} & e_{y4} & e_{z4} \\ e_{x5} & e_{y5} & e_{z5} \\ e_{x6} & e_{y6} & e_{z6} \end{bmatrix}\begin{Bmatrix} \frac{\partial}{\partial x} \\ \frac{\partial}{\partial y} \\ \frac{\partial}{\partial z} \end{Bmatrix}\{\phi\}\right\} = \rho\frac{\partial^2}{\partial t^2}\begin{Bmatrix} u_x \\ u_y \\ u_z \end{Bmatrix} \tag{5.7}$$

and

$$\begin{bmatrix} \dfrac{\partial}{\partial x} & \dfrac{\partial}{\partial y} & \dfrac{\partial}{\partial z} \end{bmatrix}$$

$$\times \left\{ \begin{bmatrix} e_{x1} & e_{x2} & e_{x3} & e_{x4} & e_{x5} & e_{x6} \\ e_{y1} & e_{y2} & e_{y3} & e_{y4} & e_{y5} & e_{y6} \\ e_{z1} & e_{z2} & e_{z3} & e_{z4} & e_{z5} & e_{z6} \end{bmatrix} \begin{bmatrix} \dfrac{\partial}{\partial x} & 0 & 0 \\ 0 & \dfrac{\partial}{\partial y} & 0 \\ 0 & 0 & \dfrac{\partial}{\partial z} \\ 0 & \dfrac{\partial}{\partial z} & \dfrac{\partial}{\partial y} \\ \dfrac{\partial}{\partial z} & 0 & \dfrac{\partial}{\partial x} \\ \dfrac{\partial}{\partial y} & \dfrac{\partial}{\partial x} & 0 \end{bmatrix} \begin{Bmatrix} u_x \\ u_y \\ u_z \end{Bmatrix} \right.$$

$$\left. - \begin{bmatrix} \epsilon_{xx} & \epsilon_{xy} & \epsilon_{xz} \\ \epsilon_{xy} & \epsilon_{yy} & \epsilon_{yz} \\ \epsilon_{xz} & \epsilon_{yz} & \epsilon_{zz} \end{bmatrix} \begin{Bmatrix} \dfrac{\partial}{\partial x} \\ \dfrac{\partial}{\partial y} \\ \dfrac{\partial}{\partial z} \end{Bmatrix} \{\phi\} \right\} = 0 \tag{5.8}$$

Referring to the discussions on the fields of BAWs in Section 5.2, we consider that the generated electrical potential (ϕ), like the particle displacements (u_x, u_y, y_z), is also proportional to

$$\phi \propto e^{i(\omega t - kr)} = e^{ik(vt-r)} \tag{5.9}$$

With this expression, we note that

$$\begin{Bmatrix} \dfrac{\partial}{\partial x} \\ \dfrac{\partial}{\partial y} \\ \dfrac{\partial}{\partial z} \end{Bmatrix} \{\phi\} = -ik \begin{Bmatrix} l_x \\ l_y \\ l_z \end{Bmatrix} \{\phi\}$$

where l_x, l_y and l_z are the same as those defined in Eq. (5.5). By taking advantage of the derivations in Section 5.2, we rewrite Eq. (5.7) as

$$\begin{bmatrix} l_x & 0 & 0 & 0 & l_z & l_y \\ 0 & l_y & 0 & l_z & 0 & l_x \\ 0 & 0 & l_z & l_y & l_x & 0 \end{bmatrix} \begin{bmatrix} c_{11} & c_{12} & c_{13} & c_{14} & c_{15} & c_{16} \\ c_{12} & c_{22} & c_{23} & c_{24} & c_{25} & c_{26} \\ c_{13} & c_{23} & c_{33} & c_{34} & c_{35} & c_{36} \\ c_{14} & c_{24} & c_{34} & c_{44} & c_{45} & c_{46} \\ c_{15} & c_{25} & c_{35} & c_{45} & c_{55} & c_{56} \\ c_{16} & c_{26} & c_{36} & c_{46} & c_{56} & c_{66} \end{bmatrix} \begin{bmatrix} l_x & 0 & 0 \\ 0 & l_y & 0 \\ 0 & 0 & l_z \\ 0 & l_z & l_y \\ l_z & 0 & l_x \\ l_y & l_x & 0 \end{bmatrix} \begin{Bmatrix} u_x \\ u_y \\ u_z \end{Bmatrix}$$

$$+ \begin{bmatrix} l_x & 0 & 0 & 0 & l_z & l_y \\ 0 & l_y & 0 & l_z & 0 & l_x \\ 0 & 0 & l_z & l_y & l_x & 0 \end{bmatrix} \begin{bmatrix} e_{x1} & e_{y1} & e_{z1} \\ e_{x2} & e_{y2} & e_{z2} \\ e_{x3} & e_{y3} & e_{z3} \\ e_{x4} & e_{y4} & e_{z4} \\ e_{x5} & e_{y5} & e_{z5} \\ e_{x6} & e_{y6} & e_{z6} \end{bmatrix} \begin{Bmatrix} l_x \\ l_y \\ l_z \end{Bmatrix} \{\phi\} - \rho v^2 \begin{bmatrix} 1 & 0 & 0 \\ 0 & 1 & 0 \\ 0 & 0 & 1 \end{bmatrix} \begin{Bmatrix} u_x \\ u_y \\ u_z \end{Bmatrix} = 0 \tag{5.10}$$

and Eq. (5.8) as

$$\begin{bmatrix} l_x & l_y & l_z \end{bmatrix} \begin{bmatrix} e_{x1} & e_{x2} & e_{x3} & e_{x4} & e_{x5} & e_{x6} \\ e_{y1} & e_{y2} & e_{y3} & e_{y4} & e_{y5} & e_{y6} \\ e_{z1} & e_{z2} & e_{z3} & e_{z4} & e_{z5} & e_{z6} \end{bmatrix} \begin{bmatrix} l_x & 0 & 0 \\ 0 & l_y & 0 \\ 0 & 0 & l_z \\ 0 & l_z & l_y \\ l_z & 0 & l_x \\ l_y & l_x & 0 \end{bmatrix} \begin{Bmatrix} u_x \\ u_y \\ u_z \end{Bmatrix}$$

$$- \begin{bmatrix} l_x & l_y & l_z \end{bmatrix} \begin{bmatrix} \epsilon_{xx} & \epsilon_{xy} & \epsilon_{xz} \\ \epsilon_{xy} & \epsilon_{yy} & \epsilon_{yz} \\ \epsilon_{xz} & \epsilon_{yz} & \epsilon_{zz} \end{bmatrix} \begin{Bmatrix} l_x \\ l_y \\ l_z \end{Bmatrix} \{\phi\} = 0 \tag{5.11}$$

Since the product of the three matrices in the second term in Eq. (5.11) is a 1×1 matrix, which is in fact a scalar, we denote it as

$|\alpha|$, and write

$$|\alpha| = \begin{bmatrix} l_x & l_y & l_z \end{bmatrix} \begin{bmatrix} \epsilon_{xx} & \epsilon_{xy} & \epsilon_{xz} \\ \epsilon_{xy} & \epsilon_{yy} & \epsilon_{yz} \\ \epsilon_{xz} & \epsilon_{yz} & \epsilon_{zz} \end{bmatrix} \begin{Bmatrix} l_x \\ l_y \\ l_z \end{Bmatrix}$$

then with Eq. (5.11) we can express the electrical potential, $\{\phi\}$, as a function of particle displacements:

$$\{\phi\} = \frac{1}{|\alpha|} \begin{bmatrix} l_x & l_y & l_z \end{bmatrix} \begin{bmatrix} e_{x1} & e_{x2} & e_{x3} & e_{x4} & e_{x5} & e_{x6} \\ e_{y1} & e_{y2} & e_{y3} & e_{y4} & e_{y5} & e_{y6} \\ e_{z1} & e_{z2} & e_{z3} & e_{z4} & e_{z5} & e_{z6} \end{bmatrix} \begin{bmatrix} l_x & 0 & 0 \\ 0 & l_y & 0 \\ 0 & 0 & l_z \\ 0 & l_z & l_y \\ l_z & 0 & l_x \\ l_y & l_x & 0 \end{bmatrix} \begin{Bmatrix} u_x \\ u_y \\ u_z \end{Bmatrix} \tag{5.12}$$

With the second term in Eq. (5.10) and Eq. (5.12) in mind, we introduce a 6×6 matrix $[c']$ and let

$$[c'] = \frac{1}{|\alpha|} \begin{bmatrix} e_{x1} & e_{y1} & e_{z1} \\ e_{x2} & e_{y2} & e_{z2} \\ e_{x3} & e_{y3} & e_{z3} \\ e_{x4} & e_{y4} & e_{z4} \\ e_{x5} & e_{y5} & e_{z5} \\ e_{x6} & e_{y6} & e_{z6} \end{bmatrix} \begin{Bmatrix} l_x \\ l_y \\ l_z \end{Bmatrix} \begin{bmatrix} l_x & l_y & l_z \end{bmatrix} \begin{bmatrix} e_{x1} & e_{x2} & e_{x3} & e_{x4} & e_{x5} & e_{x6} \\ e_{y1} & e_{y2} & e_{y3} & e_{y4} & e_{y5} & e_{y6} \\ e_{z1} & e_{z2} & e_{z3} & e_{z4} & e_{z5} & e_{z6} \end{bmatrix} \tag{5.13}$$

Recalling the following familiar equation:

$$\begin{bmatrix} \Gamma_{11} & \Gamma_{12} & \Gamma_{13} \\ \Gamma_{12} & \Gamma_{22} & \Gamma_{23} \\ \Gamma_{13} & \Gamma_{23} & \Gamma_{33} \end{bmatrix} = \begin{bmatrix} l_x & 0 & 0 & 0 & l_z & l_y \\ 0 & l_y & 0 & l_z & 0 & l_x \\ 0 & 0 & l_z & l_y & l_x & 0 \end{bmatrix} [c] \begin{bmatrix} l_x & 0 & 0 \\ 0 & l_y & 0 \\ 0 & 0 & l_z \\ 0 & l_z & l_y \\ l_z & 0 & l_x \\ l_y & l_x & 0 \end{bmatrix}$$

we write

$$\begin{bmatrix} \Gamma'_{11} & \Gamma'_{12} & \Gamma'_{13} \\ \Gamma'_{12} & \Gamma'_{22} & \Gamma'_{23} \\ \Gamma'_{13} & \Gamma'_{23} & \Gamma'_{33} \end{bmatrix} = \begin{bmatrix} l_x & 0 & 0 & 0 & l_z & l_y \\ 0 & l_y & 0 & l_z & 0 & l_x \\ 0 & 0 & l_z & l_y & l_x & 0 \end{bmatrix} [c'] \begin{bmatrix} l_x & 0 & 0 \\ 0 & l_y & 0 \\ 0 & 0 & l_z \\ 0 & l_z & l_y \\ l_z & 0 & l_x \\ l_y & l_x & 0 \end{bmatrix}$$

Putting all these relations into Eq. (5.10), we derive the following:

$$\left\{ \begin{bmatrix} \Gamma_{11} & \Gamma_{12} & \Gamma_{13} \\ \Gamma_{12} & \Gamma_{22} & \Gamma_{23} \\ \Gamma_{13} & \Gamma_{23} & \Gamma_{33} \end{bmatrix} + \begin{bmatrix} \Gamma'_{11} & \Gamma'_{12} & \Gamma'_{13} \\ \Gamma'_{12} & \Gamma'_{22} & \Gamma'_{23} \\ \Gamma'_{13} & \Gamma'_{23} & \Gamma'_{33} \end{bmatrix} - \rho v^2 \begin{bmatrix} 1 & 0 & 0 \\ 0 & 1 & 0 \\ 0 & 0 & 1 \end{bmatrix} \right\} \begin{Bmatrix} u_x \\ u_y \\ u_z \end{Bmatrix} = 0$$

which leads us to the following governing equation for BAWs in piezoelectric solids:

$$\begin{bmatrix} \Gamma_{11} & \Gamma_{12} & \Gamma_{13} \\ \Gamma_{12} & \Gamma_{22} & \Gamma_{23} \\ \Gamma_{13} & \Gamma_{23} & \Gamma_{33} \end{bmatrix} + \begin{bmatrix} \Gamma'_{11} & \Gamma'_{12} & \Gamma'_{13} \\ \Gamma'_{12} & \Gamma'_{22} & \Gamma'_{23} \\ \Gamma'_{13} & \Gamma'_{23} & \Gamma'_{33} \end{bmatrix} - \rho v^2 \begin{bmatrix} 1 & 0 & 0 \\ 0 & 1 & 0 \\ 0 & 0 & 1 \end{bmatrix} = 0 \quad (5.14)$$

This equation is commonly known as the *Stiffened Christoffel equation* in a matrix form, with the stiffening coming from piezoelectric coupling. Like the case of BAWs in elastic solids, this equation can be solved to find the velocity and slowness of three BAWs in piezoelectric solids.

The above matrix form governing equation will be useful for the reader who has an access to software that can handle linear algebra and matrix manipulation, to determine the characteristics of BAWs in piezoelectric solids.

For conceptual benefit and completeness of discussion, it may be advantageous to express this governing equation in a condensed form. From the second equation in (2.25) we have

$$[\nabla][e][\nabla_s]^T\{u\} - [\nabla][\epsilon][\nabla]^T\{\phi\} = 0$$

As discussed earlier, for bulk waves the term $[\nabla][\epsilon][\nabla]^T$ is a scalar. This means that electrical potential, $\{\phi\}$, is related to the particle displacement fields by the following expression:

$$\{\phi\} = \frac{[\nabla][e][\nabla_s]^T}{[\nabla][\epsilon][\nabla]^T}\{u\} \tag{5.15}$$

Plugging this $\{\phi\}$ expression into the first Equation in (2.25) with some rearrangements, we obtain

$$[\nabla_s]\left([c] + \frac{[e]^T[\nabla]^T[\nabla][e]}{[\nabla][\epsilon][\nabla]^T}\right)[\nabla_s]^T\{u\} - \rho\frac{\partial^2}{\partial t^2}\{u\} = 0 \tag{5.16}$$

in which the term in the parentheses is known as stiffened stiffness matrix, $[c]_{\text{Stiffened}}$:

$$[c]_{\text{Stiffened}} = [c] + \frac{[e]^T[\nabla]^T[\nabla][e]}{[\nabla][\epsilon][\nabla]^T} \tag{5.17}$$

It is now clear that mechano-electric coupling in a piezoelectric material is to stiffen its mechanical stiffness ($[c]$) by its piezoelectric properties ($[e], [\epsilon]$). As a result, it will lead to enhanced wave propagation performance in the material. Note that the second term in Eq. (5.17) is the same as the $[c']$ matrix in Eq. (5.13). This is to say that the $[c']$ matrix is a function of the wave propagation direction, as defined by (l_x, l_y, l_z); hence, the piezoelectric stiffening effect is, therefore, direction dependent.

Example 3: Use the matrix form *Stiffened Christoffel equation* given in (5.14) along with the piezoelectric stress constants and permittivity given below to resolve the problem given in Example 12 in Chapter 2.

$$[e] = \begin{bmatrix} 0 & 0 & 0 & 0 & 2.6 & -1.6 \\ -1.6 & 1.6 & 0 & 2.6 & 0 & 0 \\ 0 & 0 & 1.9 & 0 & 0 & 0 \end{bmatrix} \text{C/m}^2;$$

$$[\epsilon] = 10^{-11} \times \begin{bmatrix} 36.3 & 0 & 0 \\ 0 & 36.3 & 0 \\ 0 & 0 & 38.1 \end{bmatrix} \text{F/m}$$

Answer: Referring to Eq. (5.14), with the results from Example 1 early in this chapter we already know the first and third terms in the equation as follows

$$\begin{bmatrix} \Gamma_{11} & \Gamma_{12} & \Gamma_{13} \\ \Gamma_{12} & \Gamma_{22} & \Gamma_{23} \\ \Gamma_{13} & \Gamma_{23} & \Gamma_{33} \end{bmatrix} = 10^{11} \times \begin{bmatrix} 2.33 & 0 & 0 \\ 0 & 0.93 & -0.11 \\ 0 & -0.11 & 0.94 \end{bmatrix}$$

and

$$\rho v^2 \begin{bmatrix} 1 & 0 & 0 \\ 0 & 1 & 0 \\ 0 & 0 & 1 \end{bmatrix} = 7450v^2 \begin{bmatrix} 1 & 0 & 0 \\ 0 & 1 & 0 \\ 0 & 0 & 1 \end{bmatrix}$$

so we only need to determine the second term

$$\begin{bmatrix} \Gamma'_{11} & \Gamma'_{12} & \Gamma'_{13} \\ \Gamma'_{12} & \Gamma'_{22} & \Gamma'_{23} \\ \Gamma'_{13} & \Gamma'_{23} & \Gamma'_{33} \end{bmatrix}$$

Knowing

$$l_x = 1, l_y = 0, l_z = 0$$

we first calculate

$$|\alpha| = \begin{bmatrix} l_x & l_y & l_z \end{bmatrix} \begin{bmatrix} \epsilon_{xx} & \epsilon_{xy} & \epsilon_{xz} \\ \epsilon_{xy} & \epsilon_{yy} & \epsilon_{yz} \\ \epsilon_{xz} & \epsilon_{yz} & \epsilon_{zz} \end{bmatrix} \begin{Bmatrix} l_x \\ l_y \\ l_z \end{Bmatrix}$$

$$= 10^{-11} \times \begin{bmatrix} 1 & 0 & 0 \end{bmatrix} \begin{bmatrix} 36.3 & 0 & 0 \\ 0 & 36.3 & 0 \\ 0 & 0 & 38.1 \end{bmatrix} \begin{Bmatrix} 1 \\ 0 \\ 0 \end{Bmatrix} = 36.3 \times 10^{-11}$$

then

$$[c'] = \frac{1}{|\alpha|}\begin{bmatrix} e_{x1} & e_{y1} & e_{z1} \\ e_{x2} & e_{y2} & e_{z2} \\ e_{x3} & e_{y3} & e_{z3} \\ e_{x4} & e_{y4} & e_{z4} \\ e_{x5} & e_{y5} & e_{z5} \\ e_{x6} & e_{y6} & e_{z6} \end{bmatrix} \begin{Bmatrix} l_x \\ l_y \\ l_z \end{Bmatrix} \begin{bmatrix} l_x & l_y & l_z \end{bmatrix} \begin{bmatrix} e_{x1} & e_{x2} & e_{x3} & e_{x4} & e_{x5} & e_{x6} \\ e_{y1} & e_{y2} & e_{y3} & e_{y4} & e_{y5} & e_{y6} \\ e_{z1} & e_{z2} & e_{z3} & e_{z4} & e_{z5} & e_{z6} \end{bmatrix}$$

$$= \frac{1}{36.3 \times 10^{-11}} \begin{bmatrix} 0 & -1.6 & 0 \\ 0 & -1.6 & 0 \\ 0 & 0 & 1.9 \\ 0 & 2.6 & 0 \\ 2.6 & 0 & 0 \\ -1.6 & 0 & 0 \end{bmatrix}$$

$$\times \begin{Bmatrix} 1 \\ 0 \\ 0 \end{Bmatrix} \begin{bmatrix} 1 & 0 & 0 \end{bmatrix} \begin{bmatrix} 0 & 0 & 0 & 0 & 2.6 & -1.6 \\ -1.6 & 1.6 & 0 & 2.6 & 0 & 0 \\ 0 & 0 & 1.9 & 0 & 0 & 0 \end{bmatrix}$$

$$= 10^{11} \begin{bmatrix} 0 & 0 & 0 & 0 & 0 & 0 \\ 0 & 0 & 0 & 0 & 0 & 0 \\ 0 & 0 & 0 & 0 & 0 & 0 \\ 0 & 0 & 0 & 0 & 0 & 0 \\ 0 & 0 & 0 & 0 & 0.18 & -0.11 \\ 0 & 0 & 0 & 0 & -011 & 0.071 \end{bmatrix}$$

With that we further calculate

$$\begin{bmatrix} \Gamma'_{11} & \Gamma'_{12} & \Gamma'_{13} \\ \Gamma'_{12} & \Gamma'_{22} & \Gamma'_{23} \\ \Gamma'_{13} & \Gamma'_{23} & \Gamma'_{33} \end{bmatrix} = \begin{bmatrix} l_x & 0 & 0 & 0 & l_z & l_y \\ 0 & l_y & 0 & l_z & 0 & l_x \\ 0 & 0 & l_z & l_y & l_x & 0 \end{bmatrix} [c'] \begin{bmatrix} l_x & 0 & 0 \\ 0 & l_y & 0 \\ 0 & 0 & l_z \\ 0 & l_z & l_y \\ l_z & 0 & l_x \\ l_y & l_x & 0 \end{bmatrix}$$

$$= 10^{11} \times \begin{bmatrix} 1 & 0 & 0 & 0 & 0 & 0 \\ 0 & 0 & 0 & 0 & 0 & 1 \\ 0 & 0 & 0 & 0 & 1 & 0 \end{bmatrix} \begin{bmatrix} 0 & 0 & 0 & 0 & 0 & 0 \\ 0 & 0 & 0 & 0 & 0 & 0 \\ 0 & 0 & 0 & 0 & 0 & 0 \\ 0 & 0 & 0 & 0 & 0 & 0 \\ 0 & 0 & 0 & 0 & 0.18 & -0.11 \\ 0 & 0 & 0 & 0 & -0.11 & 0.071 \end{bmatrix} \begin{bmatrix} 1 & 0 & 0 \\ 0 & 0 & 0 \\ 0 & 0 & 0 \\ 0 & 0 & 0 \\ 0 & 0 & 1 \\ 0 & 1 & 0 \end{bmatrix}$$

$$= 10^{11} \times \begin{bmatrix} 0 & 0 & 0 \\ 0 & 0.071 & -0.11 \\ 0 & -0.11 & 0.18 \end{bmatrix}$$

Putting the three terms back in *Stiffened Christoffel equation* given in (5.14) we have

$$10^{11} \times \begin{bmatrix} 2.33 & 0 & 0 \\ 0 & 0.93 & -0.11 \\ 0 & -0.11 & 0.94 \end{bmatrix} + 10^{11} \times \begin{bmatrix} 0 & 0 & 0 \\ 0 & 0.071 & -0.11 \\ 0 & -0.11 & 0.18 \end{bmatrix} - 7450v^2 \begin{bmatrix} 1 & 0 & 0 \\ 0 & 1 & 0 \\ 0 & 0 & 1 \end{bmatrix} = 0$$

After some rearrangements, we obtain

$$\begin{bmatrix} 2.33 \times 10^{11} - 7450v^2 & 0 & 0 \\ 0 & 1.01 \times 10^{11} - 7450v^2 & -0.22 \times 10^{11} \\ 0 & -0.22 \times 10^{11} & 1.12 \times 10^{11} - 7450v^2 \end{bmatrix} = 0$$

This governing matrix equation is exactly the same as that obtained in Example 12 in Chapter 2 as expected.

The three roots of this matrix equation are

$$v_{L(x)} = 5592.4 \text{ m/s}, \quad v_{T(y)} = 3338.1 \text{ m/s}, \quad v_{T(z)} = 4171.8 \text{ m/s}$$

representing the wave velocity of the L wave polarizing in the x direction, and T waves polarizing in the y and z directions, respectively.

5.4.2 Slowness Curves Affected by Piezoelectric Coupling

In this section, we use a case study to examine how piezoelectric coupling affects the slowness curves.

Example 4: Solve the *Stiffened Christoffel equation* using lithium niobate ($LiNbO_3$) as the material.

Answer: Since $LiNbO_3$ is of trigonal Class 3m crystals, referring to Table 4.5, we have the following material constants:

$$[c] = 10^{10} \times \begin{bmatrix} 20.3 & 5.3 & 7.5 & 0.9 & 0 & 0 \\ 5.3 & 20.3 & 7.5 & -0.9 & 0 & 0 \\ 7.5 & 7.5 & 24.5 & 0 & 0 & 0 \\ 0.9 & -0.9 & 0 & 6.0 & 0 & 0 \\ 0 & 0 & 0 & 0 & 6.0 & 0.9 \\ 0 & 0 & 0 & 0 & 0.9 & 7.5 \end{bmatrix} \text{N/m}^2; \;\; \rho = 4700 \text{ Kg/m}^3$$

and

$$[e] = \begin{bmatrix} 0 & 0 & 0 & 0 & 3.7 & -2.5 \\ -2.5 & 2.5 & 0 & 3.7 & 0 & 0 \\ 0.2 & 0.2 & 1.3 & 0 & 0 & 0 \end{bmatrix} \text{C/m}^2;$$

$$[\epsilon] = 10^{-11} \times \begin{bmatrix} 38.9 & 0 & 0 \\ 0 & 38.9 & 0 \\ 0 & 0 & 25.7 \end{bmatrix} \text{F/m}$$

For the direction of wave propagation, we select the simplest case with Euler angles of $(0°, 0°, 0°)$, which is the case illustrated in Figure 5.3(a). By plugging these angles into Eq. (3.14), we have

$$[a] = \begin{bmatrix} 1 & 0 & 0 \\ 0 & 1 & 0 \\ 0 & 0 & 1 \end{bmatrix} \times \begin{bmatrix} 1 & 0 & 0 \\ 0 & 1 & 0 \\ 0 & 0 & 1 \end{bmatrix} \times \begin{bmatrix} 1 & 0 & 0 \\ 0 & 1 & 0 \\ 0 & 0 & 1 \end{bmatrix} = \begin{bmatrix} 1 & 0 & 0 \\ 0 & 1 & 0 \\ 0 & 0 & 1 \end{bmatrix}$$

From Eq. (5.5), we know $l_x = 1, ly = 0$ and $l_z = 0$; hence, we write

$$[\nabla_s] = \begin{bmatrix} 1 & 0 & 0 & 0 & 0 & 0 \\ 0 & 0 & 0 & 0 & 0 & 1 \\ 0 & 0 & 0 & 0 & 1 & 0 \end{bmatrix} \text{ or } [\nabla_s]^T = \begin{bmatrix} 1 & 0 & 0 \\ 0 & 0 & 0 \\ 0 & 0 & 0 \\ 0 & 0 & 0 \\ 0 & 0 & 1 \\ 0 & 1 & 0 \end{bmatrix};$$

$$[\nabla] = \begin{bmatrix} 1 & 0 & 0 \end{bmatrix} \text{ or } [\nabla]^T = \begin{bmatrix} 1 \\ 0 \\ 0 \end{bmatrix}$$

with these expressions, we calculate the following (note the units conversion relationships of C=FV and C=Nm/V)

$$|\alpha| = 10^{-11} \times \begin{bmatrix} 1 & 0 & 0 \end{bmatrix} \begin{bmatrix} 38.9 & 0 & 0 \\ 0 & 38.9 & 0 \\ 0 & 0 & 25.7 \end{bmatrix} \begin{bmatrix} 1 \\ 0 \\ 0 \end{bmatrix} = 3.89 \times 10^{-10} \text{ (F/m)}$$

$$[c'] = \frac{10^{10}}{3.89} \begin{bmatrix} 0 & -2.5 & 0.2 \\ 0 & 2.5 & 0.2 \\ 0 & 0 & 1.3 \\ 0 & 3.7 & 0 \\ 3.7 & 0 & 0 \\ -2.5 & 0 & 0 \end{bmatrix} \begin{bmatrix} 1 \\ 0 \\ 0 \end{bmatrix} \begin{bmatrix} 1 & 0 & 0 \end{bmatrix} \begin{bmatrix} 0 & 0 & 0 & 0 & 3.7 & -2.5 \\ -2.5 & 2.5 & 0 & 3.7 & 0 & 0 \\ 0.2 & 0.2 & 1.3 & 0 & 0 & 0 \end{bmatrix}$$

$$= 10^{10} \times \begin{bmatrix} 0 & 0 & 0 & 0 & 0 & 0 \\ 0 & 0 & 0 & 0 & 0 & 0 \\ 0 & 0 & 0 & 0 & 0 & 0 \\ 0 & 0 & 0 & 0 & 3.52 & -2.38 \\ 0 & 0 & 0 & 0 & -2.38 & 1.61 \end{bmatrix} \text{ (N/m}^2\text{)}$$

$$\begin{bmatrix} \Gamma_{11} & \Gamma_{12} & \Gamma_{13} \\ \Gamma_{12} & \Gamma_{22} & \Gamma_{23} \\ \Gamma_{13} & \Gamma_{23} & \Gamma_{33} \end{bmatrix} = 10^{10} \times \begin{bmatrix} 1 & 0 & 0 & 0 & 0 & 0 \\ 0 & 0 & 0 & 0 & 0 & 1 \\ 0 & 0 & 0 & 0 & 1 & 0 \end{bmatrix} \begin{bmatrix} 20.3 & 5.3 & 7.5 & 0.9 & 0 & 0 \\ 5.3 & 20.3 & 7.5 & -0.9 & 0 & 0 \\ 7.5 & 7.5 & 24.5 & 0 & 0 & 0 \\ 0.9 & -0.9 & 0 & 6.0 & 0 & 0 \\ 0 & 0 & 0 & 0 & 6.0 & 0.9 \\ 0 & 0 & 0 & 0 & 0.9 & 7.5 \end{bmatrix} \begin{bmatrix} 1 & 0 & 0 \\ 0 & 0 & 0 \\ 0 & 0 & 0 \\ 0 & 0 & 0 \\ 0 & 0 & 1 \\ 0 & 1 & 0 \end{bmatrix}$$

$$= 10^{10} \times \begin{bmatrix} 20.3 & 0 & 0 \\ 0 & 7.5 & 0.9 \\ 0 & 0.9 & 6.0 \end{bmatrix} \text{ (N/m}^2\text{)}$$

$$\begin{bmatrix} \Gamma'_{11} & \Gamma'_{12} & \Gamma'_{13} \\ \Gamma'_{12} & \Gamma'_{22} & \Gamma'_{23} \\ \Gamma'_{13} & \Gamma'_{23} & \Gamma'_{33} \end{bmatrix} = 10^{10} \times \begin{bmatrix} 1 & 0 & 0 & 0 & 0 & 0 \\ 0 & 0 & 0 & 0 & 0 & 1 \\ 0 & 0 & 0 & 0 & 1 & 0 \end{bmatrix} \begin{bmatrix} 0 & 0 & 0 & 0 & 0 & 0 \\ 0 & 0 & 0 & 0 & 0 & 0 \\ 0 & 0 & 0 & 0 & 0 & 0 \\ 0 & 0 & 0 & 0 & 3.52 & -2.38 \\ 0 & 0 & 0 & 0 & -2.38 & 1.61 \end{bmatrix} \begin{bmatrix} 1 & 0 & 0 \\ 0 & 0 & 0 \\ 0 & 0 & 0 \\ 0 & 0 & 0 \\ 0 & 0 & 1 \\ 0 & 1 & 0 \end{bmatrix}$$

$$= 10^{10} \times \begin{bmatrix} 0 & 0 & 0 \\ 0 & 1.61 & -2.38 \\ 0 & -2.38 & 3.51 \end{bmatrix} \ (\mathrm{N/m^2})$$

and

$$\begin{bmatrix} \Gamma_{11} & \Gamma_{12} & \Gamma_{13} \\ \Gamma_{12} & \Gamma_{22} & \Gamma_{23} \\ \Gamma_{13} & \Gamma_{23} & \Gamma_{33} \end{bmatrix} + \begin{bmatrix} \Gamma'_{11} & \Gamma'_{12} & \Gamma'_{13} \\ \Gamma'_{12} & \Gamma'_{22} & \Gamma'_{23} \\ \Gamma'_{13} & \Gamma'_{23} & \Gamma'_{33} \end{bmatrix} = 10^{10} \times \begin{bmatrix} 20.3 & 0 & 0 \\ 0 & 9.10 & -1.48 \\ 0 & -1.48 & 9.52 \end{bmatrix} \ (\mathrm{N/m^2})$$

With the above expressions, by Eq. (5.4), we obtain the non-stiffened Christoffel equation as

$$10^{10} \times \begin{bmatrix} 20.3 & 0 & 0 \\ 0 & 7.5 & 0.9 \\ 0 & 0.9 & 6.0 \end{bmatrix} - 4700\, v^2 \begin{bmatrix} 1 & 0 & 0 \\ 0 & 1 & 0 \\ 0 & 0 & 1 \end{bmatrix} = 0$$

and by Eq. (5.14), the stiffened Christoffel equation

$$10^{10} \times \begin{bmatrix} 20.3 & 0 & 0 \\ 0 & 9.10 & -1.48 \\ 0 & -1.48 & 9.52 \end{bmatrix} - 4700\, v^2 \begin{bmatrix} 1 & 0 & 0 \\ 0 & 1 & 0 \\ 0 & 0 & 1 \end{bmatrix} = 0$$

Solving these equations, we find the three roots for the velocity (and slowness) of the three BAWs in each case, and the results are listed in Table 5.2 given below.

Table 5.2 Solutions for the velocity and slowness values

Piezoelectric	Velocity (m/s $\times 10^3$)			Slowness (s/m $\times 10^{-4}$)		
coupling	L	T(⊥)	T(‖)	L	T(⊥)	T(‖)
No	6.5720	3.4451	4.1054	1.5216	2.9026	2.4358
Yes	6.5720	4.0792	4.7947	1.5216	2.4534	2.0856

It is seen here that in this particular propagation direction, piezoelectric coupling has no effect on the L wave, but it has caused an increase in velocity for the T(⊥) wave from approximately 3445 to 4079 m/s (or a decrease in slowness from 2.9026 to 2.4534 s/m$\times 10^{-4}$), and an increase in the velocity for the T(||) wave from 4105 to 4795 m/s (or a decrease in slowness from 2.4358 to 2.0856 s/m$\times 10^{-4}$).

Expanding on this case study by considering more angles for the direction of propagation in the XY section plane, we can find the corresponding slowness curves. Figure 5.16(a) shows the obtained results with and without considering the piezoelectric coupling effect. Note in these plots, thick line curves represent the cases with the piezoelectric effect, and thin lines curves the cases without the piezoelectric effect. It is obvious that the size of the T(⊥) wave in solid blue lines is decreased drastically, indicating a strong piezoelectric coupling effect. For the T(||) wave in dashed brown lines, the coupling effect is also strong, but due to the almost overlapping of the unstiffened case (thin dashed brown line) with the stiffened T(⊥) wave (thick blue line), the change is not visually discernible in these slowness curves. However, the slowness values listed in Table 5.2 offer a much clear view of the coupling effect, and the overlapping of the slowness of the unstiffened T(||) wave with that of the stiffened T(⊥) wave is evident by the fact that the two slowness values, 2.4358 and 2.4534, are so close to each other. This observation suggests that a different way to evaluate the piezoelectric coupling effect is necessary.

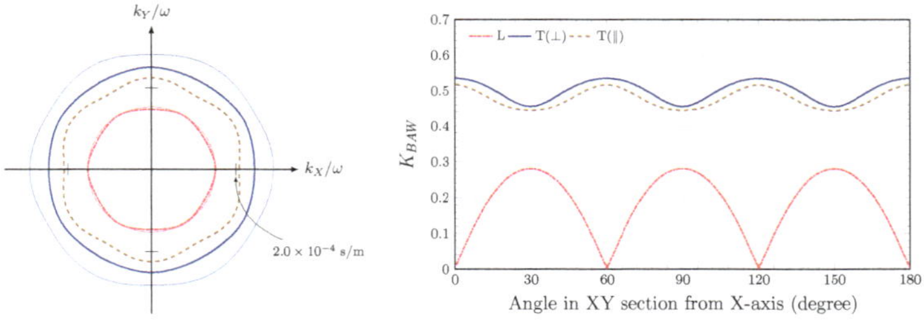

(a) Slowness curve (b) Electromechanical coupling coefficient

Figure 5.16 Slowness curves and electromechanical coupling coefficient of $LiNbO_3$ in XY section.

5.4.3 Electromechanical Coupling Coefficient K for BAWs

We now introduce a better way to quantify the piezoelectric coupling effect using a variable called *electromechanical coupling coefficient K*. For BAWs, it is defined as follows:

$$K_{BAW} = \sqrt{1 - (\frac{v_{\text{unstiffened}}}{v_{\text{stiffened}}})^2} \quad (5.18)$$

where $v_{\text{unstiffened}}$ and $v_{\text{stiffened}}$ are the wave velocity when the piezoelectric coupling effect is ignored and considered, respectively. Note that we use a subscript of BAW here for K for BAWs to distinguish it from that for surface acoustic waves (SAWs) because that is determined differently as discussed in Chapter 6.

By applying Eq. (5.18) to the slowness curves of $LiNbO_3$ given in Figure 5.16(a), and plotting the K_{BAW} values as a function of propagation angle from 0° to 180° for each of the three waves, we obtain a set of K_{BAW} curves as shown in Figure 5.16(b).

Apparently, for the case with Euler angles of $(0^\circ, 0^\circ, 0^\circ)$, which corresponds to the values at 0° angle in the Figure, we can see that the K_{BAW} value for the L wave is zero, but it is approximately 0.54 and 0.52 for the T($\perp$) and T($\|$) waves, respectively. As the angle increases, the K_{BAW} value rises and falls, suggesting the piezoelectric coupling effect is indeed direction dependent. Of the angular range shown here, the highest K_{BAW} value for the L wave is reached at 30°, 90° and 150°, while a peak is reached at 0°, 60°, 120° and 180° for the two T waves.

Figure 5.17 shows the slowness curves with and without the piezoelectric coupling effect and the corresponding variation of electromechanical coupling coefficient for $LiNbO_3$ in the XZ section. In this section, all three waves show some piezoelectric coupling effect, with stronger effect exhibiting in the T($\perp$) and T($\|$) waves, where K_{BAW} values can reach as high as 0.54 and 0.52, respectively. The L wave also exhibits a good coupling effect with K_{BAW} values as high as $\approx$0.37. Figure 5.18 shows the slowness curves with and without the piezoelectric coupling effect and the corresponding variation of electromechanical coupling coefficient for $LiNbO_3$ in the YZ section. In this section, however, the T($\perp$) wave seems not affected at all by the piezoelectric coupling effect, but the L and T($\|$) waves both exhibit strong piezoelectric effect.

Of the commonly used piezoelectric materials, $LiNbO_3$ likely has the highest electromechanical coupling effect. This fact becomes more evident if we express the *electromechanical coupling coefficient K* in the following squared form:

$$K_{BAW}^2 = [1 - (\frac{v_{\text{unstiffened}}}{v_{\text{stiffened}}})^2] \times 100\% \quad (5.19)$$

Though this way of quantifying the piezoelectric coupling effect does not change the meaning of the *electromechanical coupling coefficient K*, it does, however, highlight the strong piezoelectric coupling effect of some materials. As we can see in Figure 5.19, the K_{BAW}^2 values of $LiNbO_3$ can reach as high as 30% in the XY and XZ sections, and about 40% in the YZ section.

Comparing with $LiNbO_3$, some other piezoelectric materials can be regarded as having poor piezoelectric coupling effect. Among

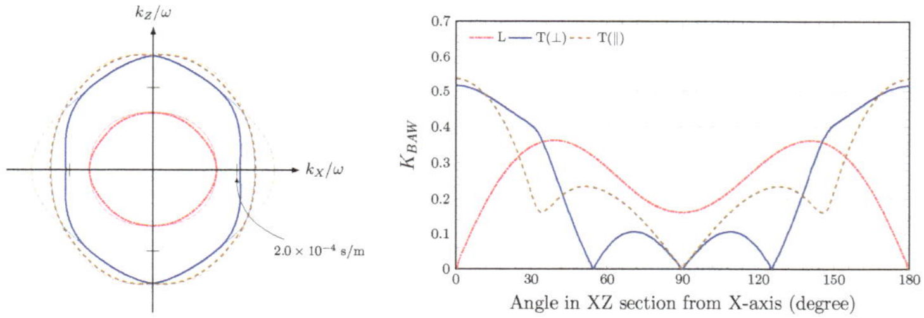

(a) Slowness curve (b) Electromechanical coupling coefficient

Figure 5.17 Slowness curves and electromechanical coupling coefficient of $LiNbO_3$ in XZ section.

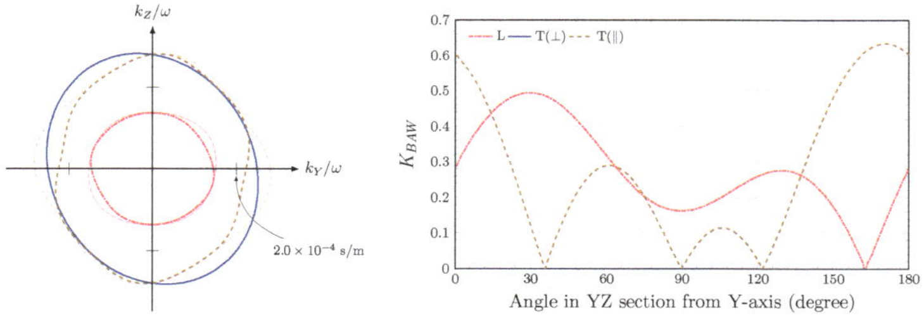

(a) Slowness curve (b) Electromechanical coupling coefficient

Figure 5.18 Slowness curves and electromechanical coupling coefficient of $LiNbO_3$ in YZ section.

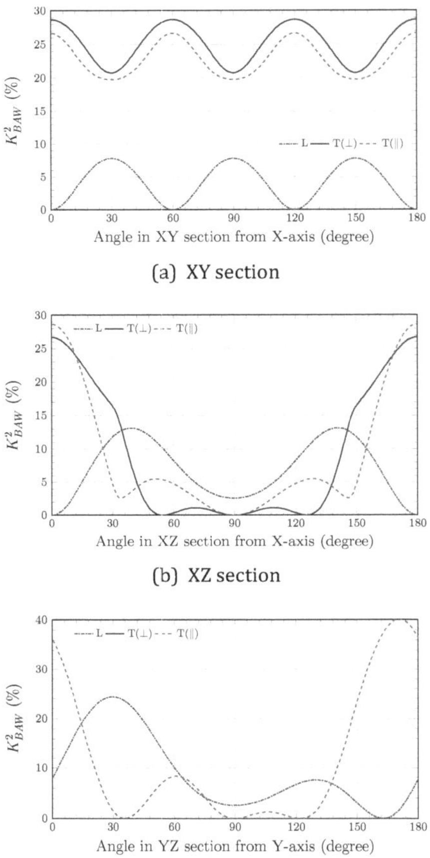

(a) XY section

(b) XZ section

(c) YZ section

Figure 5.19 K^2_{BAW} curves for $LiNbO_3$ in three orthogonal sections.

them are gallium arsenide, langasites, and quartz, to name just a few. In the case of quartz (SiO_2), its K^2_{BAW} values are typically less than 2%. Figure 5.20 shows its slowness curves and the corresponding K^2_{BAW} curves in the XY, XZ and YZ sections, respectively. From these plots we can see that the K^2_{BAW} values for the two T waves in the XZ and YZ sections are below 0.4%.

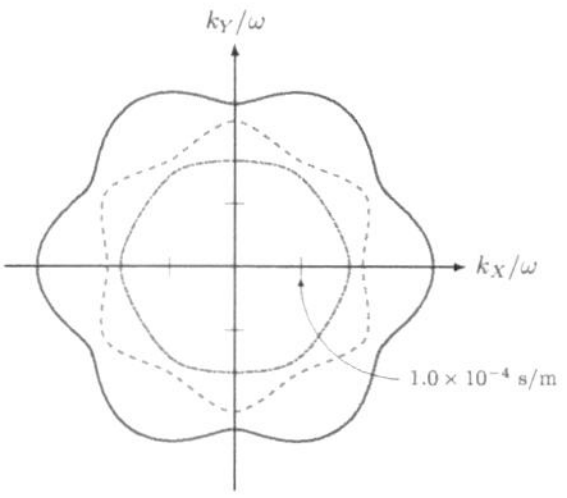

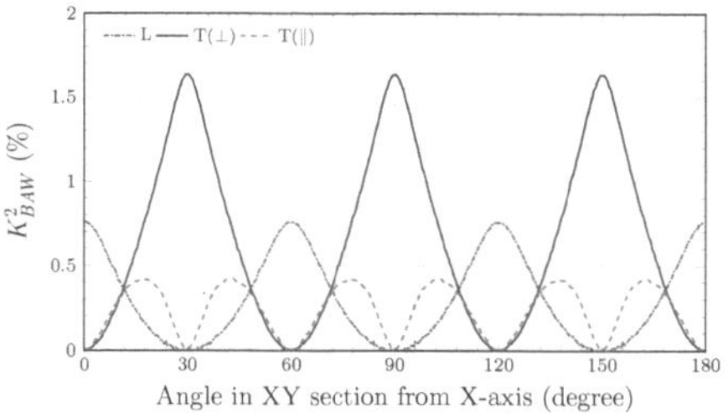

(a) Slowness curve in XY

(b) Coupling coefficient in XY

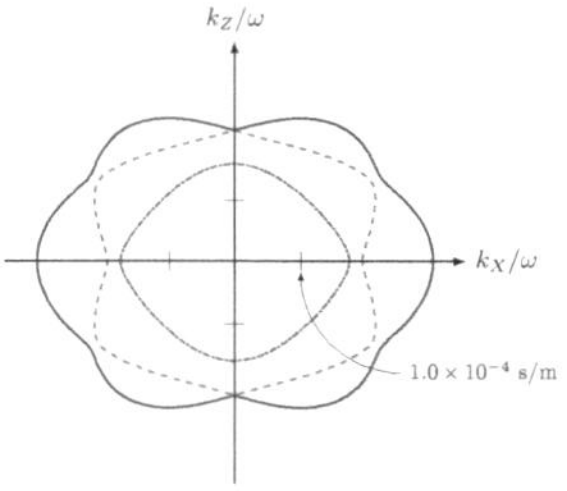

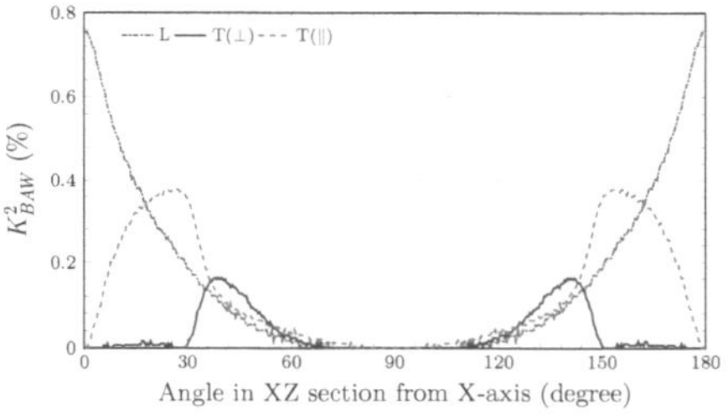

(c) Slowness curve in XZ

(d) Coupling coefficient in XZ

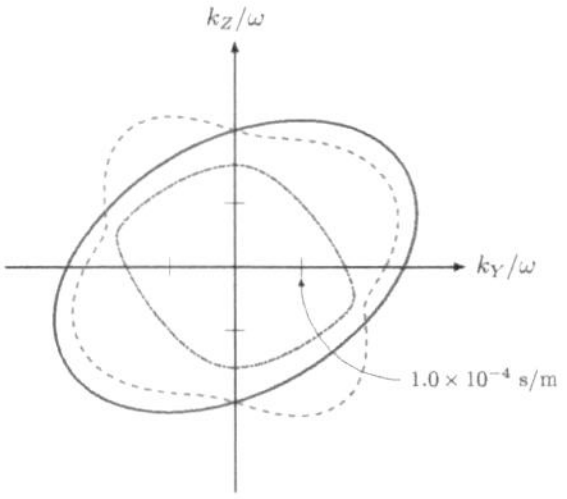

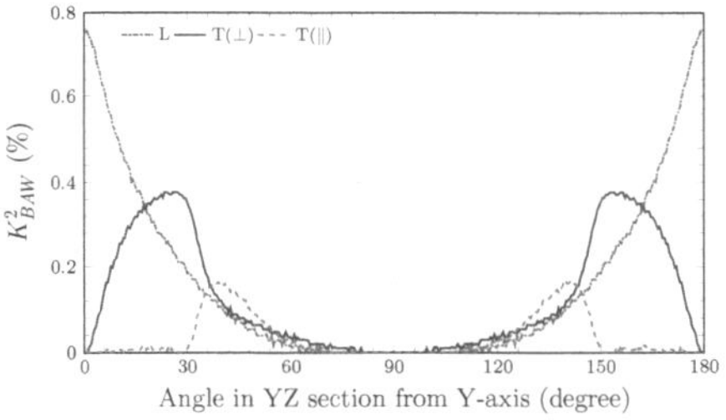

(e) Slowness curve in YZ

(f) Coupling coefficient in YZ

Figure 5.20 Slowness curves and electromechanical coupling coefficient of SiO_2 in three orthogonal sections.

5.4.4 A Few Other Crystals with Strong Piezoelectric Coupling Effect

In this section, a few more examples of piezoelectric crystal materials possessing relatively strong electromechanical coupling effect are listed.

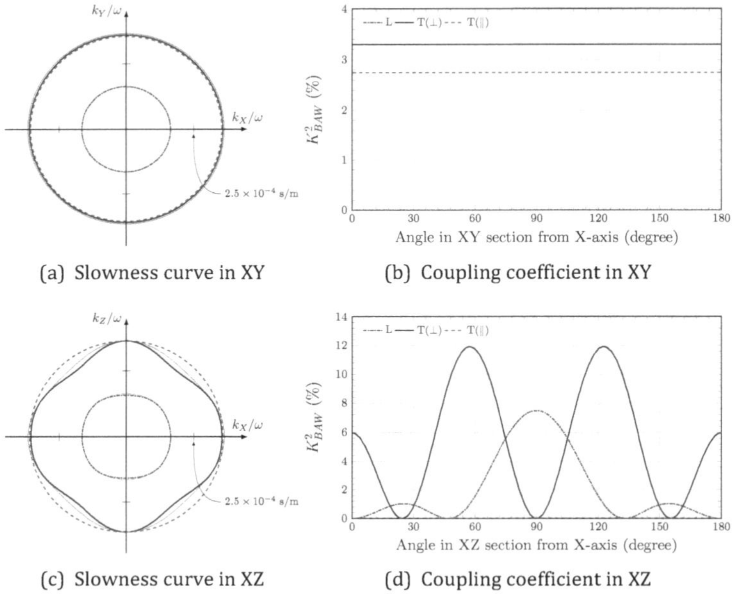

(a) Slowness curve in XY (b) Coupling coefficient in XY

(c) Slowness curve in XZ (d) Coupling coefficient in XZ

Figure 5.21 Slowness curves and electromechanical coupling coefficient of ZnO in three orthogonal sections.

Figure 5.21 shows the slowness curves and the corresponding K^2_{BAW} curves for hexagonal crystal (Class 6mm) zinc oxide (ZnO). Interestingly, in the XY section, K^2_{BAW} values for the two T wave curves remain at their respective constant levels, and that for the L wave is zero. In the XZ and YZ sections, K^2_{BAW} values for the L and T($\perp$) waves are relatively high, with the coupling for the T($\perp$) wave reaching close to 12% and that for the L wave close to 8%, but the coupling coefficient for the T($\|$) wave is zero.

Figure 5.22 shows the slowness curves with (thick line) and without (thin lines) the piezoelectric coupling effect and the corresponding variation of electromechanical coupling coefficient (K^2_{BAW}) for trigonal crystal (Class 3m) lithium tantalate ($LiTaO_3$). Clearly, the K^2_{BAW} values of $LiTaO_3$ can be as high as almost 20%. Due to the same trigonal (Class 3m) crystal structure of $LiTaO_3$ as $LiNbO_3$, we can see some similarity between the two materials, especially in the YZ section, though the piezoelectric coupling effect is generally higher in $LiNbO_3$ than in $LiTaO_3$ in all sections.

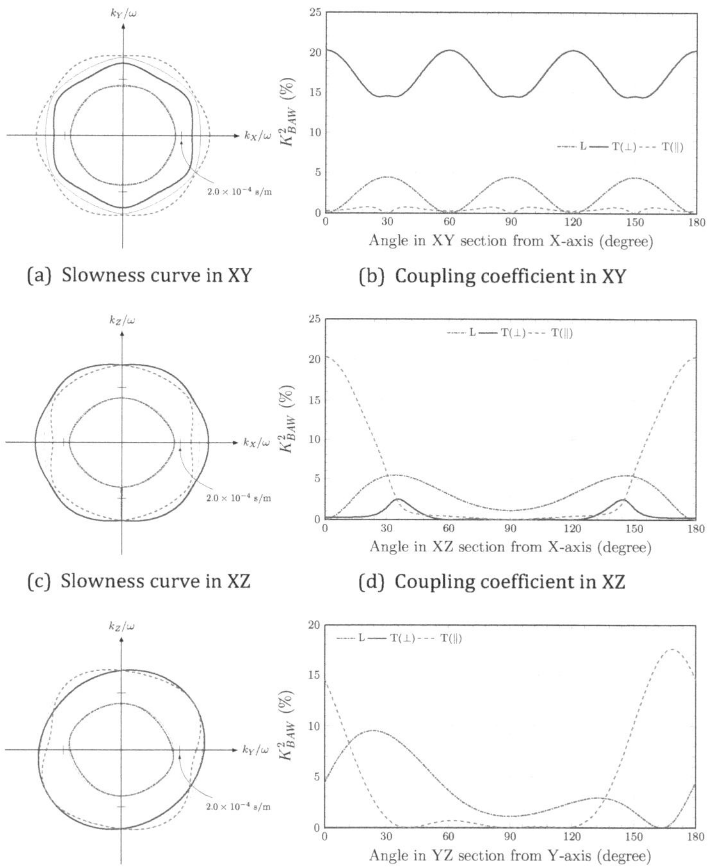

(a) Slowness curve in XY (b) Coupling coefficient in XY

(c) Slowness curve in XZ (d) Coupling coefficient in XZ

(e) Slowness curve in YZ (f) Coupling coefficient in YZ

Figure 5.22 Slowness curves and electromechanical coupling coefficient of $LiTaO_3$ in three orthogonal sections.

Figure 5.23 shows the slowness curves and the corresponding K^2_{BAW} curves for hexagonal crystal (Class 6mm) cadmium sulfate (CdS). Due to its same hexagonal (Class 6mm) crystal structure as ZnO, CdS exhibits similarity characteristics in both the slowness curves and the K^2_{BAW} curves as ZnO, especially in the XZ and YZ sections. Piezoelectric coupling effect is lower in CdS than in ZnO. In the XZ and YZ sections (note that results in the YZ section are not shown because they look just like the results in the XZ section), K^2_{BAW}

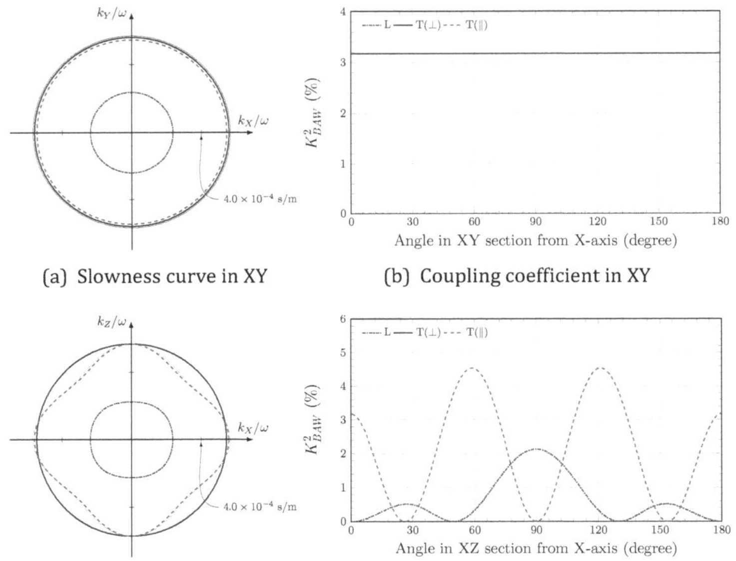

(a) Slowness curve in XY (b) Coupling coefficient in XY

(c) Slowness curve in XZ (d) Coupling coefficient in XZ

Figure 5.23 Slowness curves and electromechanical coupling coefficient of CdS in three orthogonal sections.

values for the L and T(⊥) waves reach slightly over 2% and 4.5%, respectively. Also, the two T waves appear to be switching positions in the two materials, that is, the coupling coefficient for the T(||) wave in ZnO remains zero but that for the T(⊥) wave is zero.

Figure 5.24 shows the slowness curves and the corresponding K^2_{BAW} curves for tetragonal crystal (Class 4mm) barium titanate oxide ($BaTiO_3$). In the XY section, K^2_{BAW} values for the T(⊥) wave are very high (about 32%) and those for the L and T(||) waves are zero. In XZ and YZ sections (note that results in the YZ section are not shown because they look the same as in the XZ section), K^2_{BAW} values for the T(⊥) wave are zero, but K^2_{BAW} values the T(||) wave reach as high as approximately 32%, and that for L reach as high as over 10%.

Figure 5.25 shows the slowness curves and the corresponding K^2_{BAW} curves for cubic crystal (Class 23) bismuth germanium oxide ($Bi_{12}GeO_{20}$). Note that because the results in the XY and XZ sections appear the same, only the results in the XZ section are shown, Moreover, The letter D here represents the direction defined by the

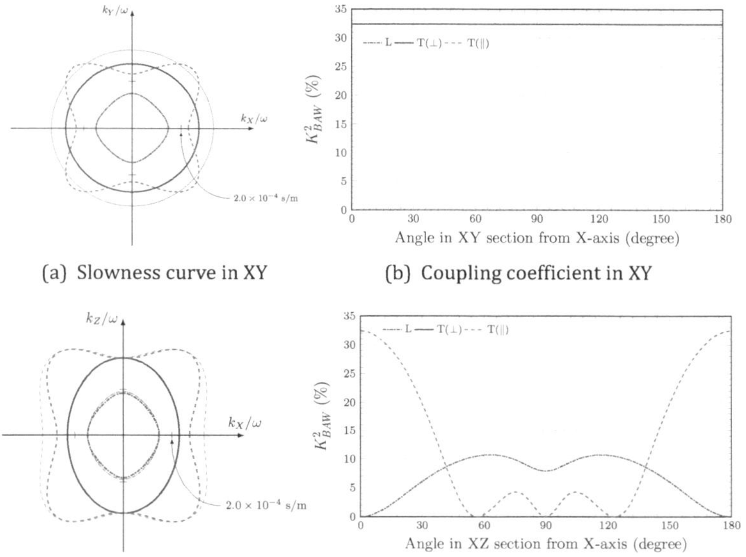

(a) Slowness curve in XY

(b) Coupling coefficient in XY

(c) Slowness curve in XZ

(d) Coupling coefficient in XZ

Figure 5.24 Slowness curves and electromechanical coupling coefficient of $BaTiO_3$ in three orthogonal sections.

diagonal line in the XY section. Clearly, in all three sections, K^2_{BAW} values can be as high as 10% such as the T(||) wave in the XY and XZ sections, and the T(⊥) wave in the DZ section.

5.4.5 Energy Velocity, Wave Velocity and Poynting Angle

When the piezoelectric effect is considered, the relationship between the energy velocity and phase velocity remain the same as when the effect is ignored, which we have discussed in good detail in Section 5.3.4. This means that the energy velocity as well as wave surface can be determined once the slowness surface is known. Figure 5.26 shows an example, in which the slowness surface (inner curve, not to scale), velocity surface (dash line) and wave surface (solid line) of the T(||) wave traveling in the YZ section of $LiNbO_3$ are plotted (Figure 5.26(a)), and the corresponding Poynting angle as a function of propagation direction for the waves is given in Figure 5.26(b). The Poynting angle is found to vary within the

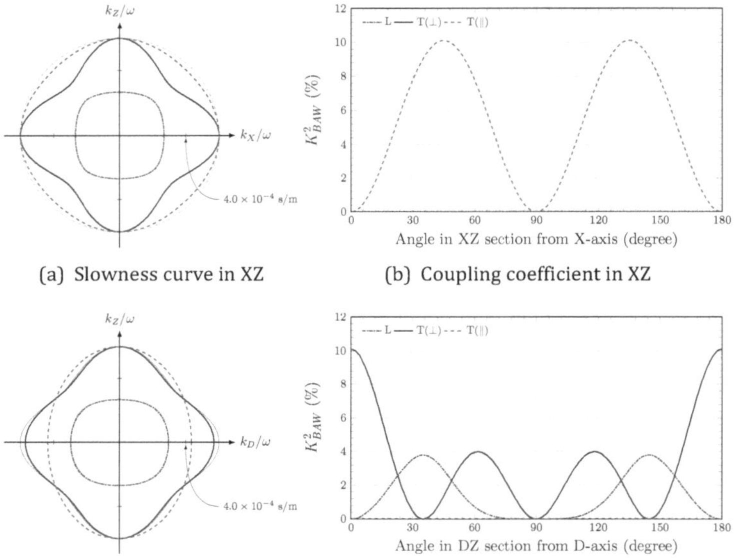

(a) Slowness curve in XZ

(b) Coupling coefficient in XZ

(c) Slowness curve in DZ

(d) Coupling coefficient in DZ

Figure 5.25 Slowness curves and electromechanical coupling coefficient of $Bi_{12}GeO_{20}$ in three selected sections.

bounds of approximately $\pm 16°$, and in the propagation direction ranging from $0°$ to $180°$ shown here, there exist two directions, at approximately $84°$ and $168°$, where the Poynting angle is zero, suggesting that the T(∥) wave is a pure transverse wave along these two directions in the YZ section of $LiNbO_3$. Figure 5.26(b) also shows the variation of Poynting angle when the piezoelectric coupling effect is not considered. Apparently, the difference is quite obvious. For the T(⊥) wave, since the piezoelectric coupling effect is zero in the YZ section (see Figures 5.18 and 5.19) the Poynting angle remains the same as in Figure 5.14(b).

Figure 5.27 shows another example, in which the slowness surface (inner curve, not to scale), velocity surface (dash line) and wave surface (solid line) of the T(⊥) wave traveling in the XY section of $LiTaO_3$ are plotted (Figure 5.27(a)), and the corresponding Poynting angle as a function of propagation direction is given in Figure 5.27(b). In this case, the Poynting angle varies within the bounds of approximately $\pm 15°$, and within the range from $0°$ to $180°$,

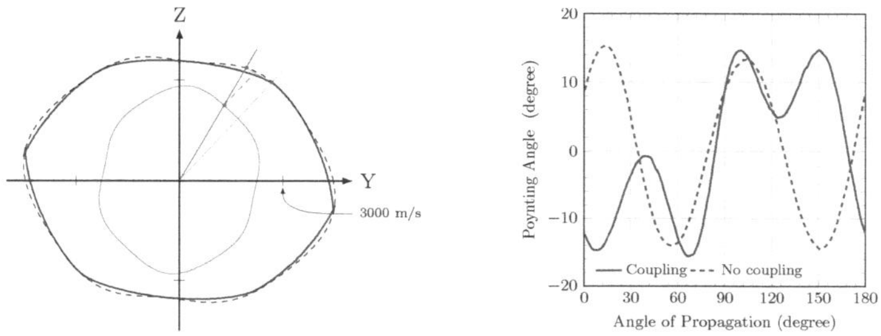

(a) Energy velocity (b) Poynting angle

Figure 5.26 Energy velocity and Poynting angle of the T(||) wave in the YZ section of $LiNbO_3$.

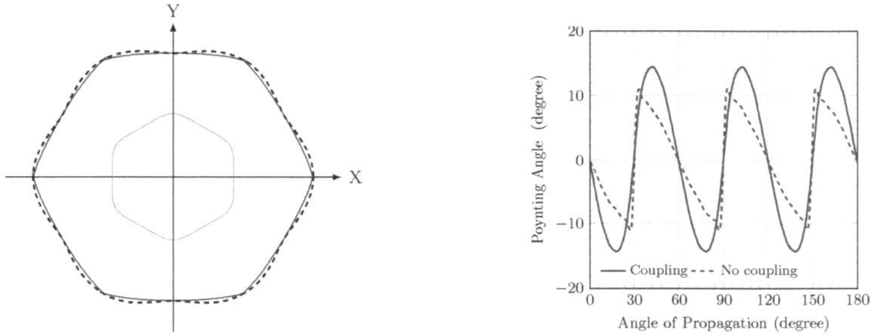

(a) Energy velocity (b) Poynting angle

Figure 5.27 Energy velocity and Poynting angle of the T($\perp$) wave in the XY section of $LiTaO_3$.

there exist seven directions, starting from 0° with a 30° increment all the way to 180°, where the Poynting angle is zero, suggesting that the T($\perp$) wave in the XY section of $LiTaO_3$ is a pure transverse wave along these directions. Figure 5.27(b) also shows the variation of Poynting angle when the piezoelectric coupling effect is ignored.

5.5 Study Problems

1. With the material properties given in Section 4.9 for barium titanate crystal, find the wave velocity, slowness and the K^2_{BAW} values for BAWs traveling at 60° angle in the XZ section plane.

2. For CdS, find the wave velocity, slowness and the K^2_{BAW} values for BAWs traveling at 90° angle in the XZ section plane.
3. For ZnO, find the wave velocity, slowness and the K^2_{BAW} values for BAWs traveling at 60° angle in the XZ section plane.
4. For $LiTaO_3$, find the wave velocity, slowness and the K^2_{BAW} values for BAWs traveling at 24° angle in the YZ section plane.
5. For $LiTaO_3$, find the wave velocity, slowness and the K^2_{BAW} values for BAWs traveling at 30° angle in the XY section plane.
6. For $Bi_{12}GeO_{20}$, find the wave velocity, slowness and the K^2_{BAW} values for BAWs traveling at 45° angle in the XY section plane.
7. For Si, find the wave velocity, slowness and the K^2_{BAW} values for BAWs traveling at 27° angle in the YZ section plane.
8. For Si, find the wave velocity, slowness and the K^2_{BAW} values for BAWs traveling at 30° angle in the XY section plane.
9. For $LiNbO_3$, find the wave velocity, slowness and the K^2_{BAW} values for BAWs traveling at 40° angle in the XZ section plane.

Chapter 6

Surface Acoustic Waves

As we have seen in Chapter 5, because BAWs travel in the interior of a solid material through particulate vibration occurring in wave-front planes that cut through the entire material, mathematically speaking, this allows us to write their wave fields using simple plane functions. SAWs, on the other hand, travel near the surface of the material with particulate vibration confined to a narrow region of the material within a distance of only a few wavelengths away from the surface. Their wave fields cannot be represented by plane functions. In this chapter, we will discuss the polarization and propagation of SAWs near the surfaces of various elastic and piezoelectric solids.

6.1 Fixed and Rotated Coordinates Systems

In dealing with BAWs, even we select a direction for wave propagation, we always refer to the fixed coordinates system, namely the XYZ-system, as the system for reference. In dealing with SAWs, however, since the existence of a boundary surface (e.g., the boundary of a semi-infinite solid material) will alter the wave propagation behavior, it is necessary to use the rotated coordinates system, in addition to the fixed one, for referencing purposes.

Bulk and Surface Acoustic Waves: Fundamentals, Devices, and Applications
Guigen Zhang

ISBN 978-981-4877-73-2 (Hardcover), 978-1-003-25662-5 (eBook)
www.jennystanford.com

This becomes more relevant when we design SAW devices using crystals with different orientations (or cuts) as defined by a set of Euler angles (ϕ, θ, ψ). In this chapter, we will use the *xyz*-system and the XYZ-system simultaneously as the rotated and fixed systems, respectively, for referencing purposes. As illustrated in Figure 6.1, one advantage of using the rotated coordinates system is that no matter what crystal cuts we choose, we will always have the *z* axis represent the direction of surface normal. This is very important because in dealing with SAWs we need to know where this direction points at such that proper mathematical functions can be developed to capture the essence of SAWs, that is, the displacements of polarized particles decay with the distance deep into the material from the surface.

With the rotated *xyz* coordinates system, we will always use the *x* direction as the wave propagation direction. We can then set the following orientation convention: The wave polarizing in the *x* direction is the longitudinal wave (or the L wave), the wave polarizing in the *z* direction normal to the surface is the transverse vertical wave (or the TV wave), and the wave polarizing in the *y* direction parallel with the surface is the transverse horizontal wave (or the TH wave). In Figure 6.1, we use the same line types and colors as in Chapter 5 to mark these sub waves. Specifically, by using the surface as the reference plane, the wave polarizing parallel with the surface (i.e., horizontally in the *xy* plane) is the TH wave (represented by the brown dashed line), and the wave polarizing perpendicular to the

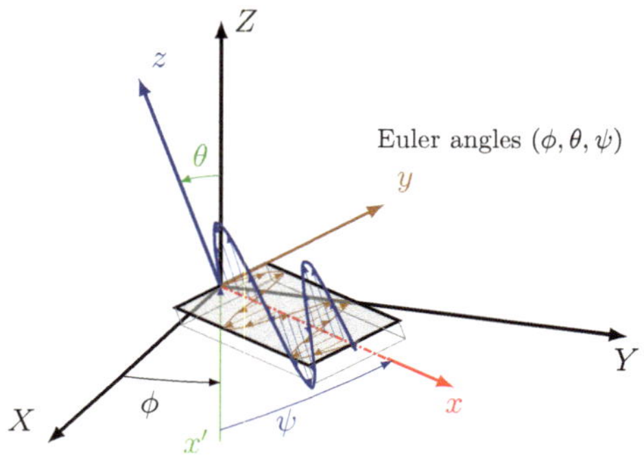

Figure 6.1 Illustration of polarization and propagation in surface acoustic waves.

surface the TV wave (represented by blue solid line), aside from the L wave polarizing in the wave propagation direction (marked by the red dash-dotted line).

Referring to the discussions in Chapter 3, once the orientation of a crystal cut is selected by a set of Euler angles (ϕ, θ, ψ), we can determine the transformation matrix by using Eq. (3.14):

$$[a] = \begin{bmatrix} a_{11} & a_{12} & a_{13} \\ a_{21} & a_{22} & a_{23} \\ a_{31} & a_{32} & a_{33} \end{bmatrix}$$

$$= \begin{bmatrix} \cos\psi & \sin\psi & 0 \\ -\sin\psi & \cos\psi & 0 \\ 0 & 0 & 1 \end{bmatrix} \times \begin{bmatrix} 1 & 0 & 0 \\ 0 & \cos\theta & \sin\theta \\ 0 & -\sin\theta & \cos\theta \end{bmatrix} \times \begin{bmatrix} \cos\phi & \sin\phi & 0 \\ -\sin\phi & \cos\phi & 0 \\ 0 & 0 & 1 \end{bmatrix}$$

$$= \begin{bmatrix} \cos\phi\cos\psi - \cos\theta\sin\phi\sin\psi & \cos\psi\sin\phi + \cos\phi\cos\theta\sin\psi & \sin\psi\sin\theta \\ -\cos\phi\sin\psi - \cos\psi\cos\theta\sin\phi & \cos\phi\cos\psi\cos\theta - \sin\phi\sin\psi & \cos\psi\sin\theta \\ \sin\phi\sin\theta & -\cos\phi\sin\theta & \cos\theta \end{bmatrix}$$

With this, we can calculate the $[M]$ matrix (see Eq. (3.4))

$$[M] = \begin{bmatrix} a_{11}^2 & a_{12}^2 & a_{13}^2 & 2a_{12}a_{13} & 2a_{13}a_{11} & 2a_{11}a_{12} \\ a_{21}^2 & a_{22}^2 & a_{23}^2 & 2a_{22}a_{23} & 2a_{23}a_{21} & 2a_{21}a_{22} \\ a_{31}^2 & a_{32}^2 & a_{33}^2 & 2a_{32}a_{33} & 2a_{33}a_{31} & 2a_{31}a_{32} \\ a_{21}a_{31} & a_{22}a_{32} & a_{23}a_{33} & a_{22}a_{33} + a_{23}a_{32} & a_{21}a_{33} + a_{23}a_{31} & a_{22}a_{31} + a_{21}a_{32} \\ a_{31}a_{11} & a_{32}a_{12} & a_{33}a_{13} & a_{12}a_{33} + a_{13}a_{32} & a_{13}a_{31} + a_{11}a_{33} & a_{11}a_{32} + a_{12}a_{31} \\ a_{11}a_{21} & a_{12}a_{22} & a_{13}a_{23} & a_{12}a_{23} + a_{13}a_{22} & a_{13}a_{21} + a_{11}a_{23} & a_{11}a_{22} + a_{12}a_{21} \end{bmatrix}$$

Then, with the known material properties of a piezoelectric crystal material given in the original XYZ system, denoted as $[c_0]$, $[e_0]$ and $[\epsilon_0]$ here, we apply Eqs. (3.8) (3.10) and (3.12) to obtain the material properties as viewed in the rotated *xyz* system, as follows:

$$[c_{xyz}] = [M][c_0][M]^T, \quad [e_{xyz}] = [a][e_0][M]^T, \quad [\epsilon_{xyz}] = [a][\epsilon_0][a]^T \quad (6.1)$$

6.2 Propagation of SAWs in Isotropic Solids

To better understand the difference between BAWs and SAWs, it is beneficial to start with Rayleigh waves in an isotropic material. In this section, we will discuss how this Rayleigh type of surface acoustic waves is generated using an isotropic material and demonstrate how the waves vanish with the depth into the material from its surface.

6.2.1 Governing Eqs. for SAWs in Isotropic Solids

Surface acoustic waves are governed by the same wave equations as those for bulk acoustic waves, as discussed in Section 5.2. Since we are dealing with SAWs in isotropic elastic materials, the direction of wave propagation is no longer relevant as the material will have the same property in all directions. Thus, referring to the coordinates systems shown in Figure 6.1, we can simply regard the two coordinates systems as overlapping with each other without any rotation.

We now consider a surface wave propagating along the x direction and express the particle displacement fields as proportional to

$$\{u\} \propto e^{-ikbz} e^{i(\omega t - kx)} = e^{-ikbz} e^{ik(vt - x)} \tag{6.2}$$

in which v, ω and k are the same variables as defined before, namely, v is the wave phase velocity ($v = \omega/k$), k is the wavenumber ($k = 2\pi/\lambda$), and b is a constant in a complex form. Note that having a negative sign in the exponent allows us to mentally regard that a positive z value is on the piezoelectric substrate side.

Apparently, this wave field function is not planar. It is expressed explicitly as a function of z because it is necessary to have this feature in the function to capture the very essence of SAWs, that is, the displacements of polarized particles decay as z increases in depth. For this to happen, the constant b needs to be a negative imaginary number (i.e., $b = -ai$, with a being a positive real number, where $i^2 = -1$) such that when z increases, we have

$$e^{-ikbz} = e^{-ik(-ai)z} = e^{-akz} \to 0$$

With the expression given in Eq. (6.2), we express the three components of particle displacements of surface waves as follows:

$$\begin{Bmatrix} u_x \\ u_y \\ u_z \end{Bmatrix} = \begin{Bmatrix} {}^\circ u_x \\ {}^\circ u_y \\ {}^\circ u_z \end{Bmatrix} e^{-ikbz} e^{ik(vt-x)} \tag{6.3}$$

where ${}^\circ u_x$,${}^\circ u_y$ and ${}^\circ u_z$ are the magnitudes of the three components of particles displacements, respectively.

With Eq. (6.3) we note that taking a partial derivative of these expressions with respect to either x, y, z or t is equivalent to multiplying a constant factor as listed below

$$\frac{\partial}{\partial x} = -ik, \quad \frac{\partial}{\partial y} = 0, \quad \frac{\partial}{\partial z} = -ikb, \quad \frac{\partial^2}{\partial t^2} = -k^2 v^2$$

By applying these notions to the matrix form wave equations, given in Eq. (5.1), we write the following matrix equation for SAWs:

$$\left\{ \begin{bmatrix} 1 & 0 & 0 & 0 & b & 0 \\ 0 & 0 & 0 & b & 0 & 1 \\ 0 & 0 & b & 0 & 1 & 0 \end{bmatrix} [c] \begin{bmatrix} 1 & 0 & 0 \\ 0 & 0 & 0 \\ 0 & 0 & b \\ 0 & b & 0 \\ b & 0 & 1 \\ 0 & 1 & 0 \end{bmatrix} \begin{Bmatrix} {}^\circ u_x \\ {}^\circ u_y \\ {}^\circ u_z \end{Bmatrix} - \rho v^2 \begin{Bmatrix} {}^\circ u_x \\ {}^\circ u_y \\ {}^\circ u_z \end{Bmatrix} \right\} \times e^{-ikbz} e^{ik(vt-x)} = 0$$

Consider the following stiffness matrix for an isotropic elastic material:

$$[c] = \begin{bmatrix} c_{11} & c_{12} & c_{12} & 0 & 0 & 0 \\ c_{12} & c_{11} & c_{12} & 0 & 0 & 0 \\ c_{12} & c_{12} & c_{11} & 0 & 0 & 0 \\ 0 & 0 & 0 & c_{44} & 0 & 0 \\ 0 & 0 & 0 & 0 & c_{44} & 0 \\ 0 & 0 & 0 & 0 & 0 & c_{44} \end{bmatrix}$$

where $c_{44} = (c_{11} - c_{12})/2$, we can simplify the above matrix equation into

$$\begin{bmatrix} c_{11} + c_{44}b^2 - \rho v^2 & 0 & (c_{11} - c_{44})b \\ 0 & c_{44}b^2 + c_{44} - \rho v^2 & 0 \\ (c_{11} - c_{44})b & 0 & c_{11}b^2 + c_{44} - \rho v^2 \end{bmatrix} \begin{Bmatrix} {}^{\circ}u_x \\ {}^{\circ}u_y \\ {}^{\circ}u_z \end{Bmatrix} e^{-ikbz} e^{ik(vt-x)} = 0$$

or simply

$$\begin{bmatrix} c_{11} + c_{44}b^2 - \rho v^2 & 0 & (c_{11} - c_{44})b \\ 0 & c_{44}b^2 + c_{44} - \rho v^2 & 0 \\ (c_{11} - c_{44})b & 0 & c_{11}b^2 + c_{44} - \rho v^2 \end{bmatrix} \begin{Bmatrix} {}^{\circ}u_x \\ {}^{\circ}u_y \\ {}^{\circ}u_z \end{Bmatrix} = 0 \tag{6.4}$$

This is the set of governing equations for SAWs polarizing and propagating in isotropic elastic materials.

Judging from the layout of the square matrix in Eq. (6.4), it is apparent that this equation can be decoupled into two independent subgroups of waves. The first subgroup is for two coupled wave components polarizing along the x and z directions, respectively, and the second one in the y direction. We will discuss the two subgroups of waves separately.

6.2.2 Rayleigh Waves

The waves with components polarizing in the x and z directions are known as Rayleigh waves. By removing the second column and second row from the matrix given in Eq. (6.4), we have the following reduced governing equation for Rayleigh waves:

$$\begin{bmatrix} c_{11} + c_{44}b^2 - \rho v^2 & (c_{11} - c_{44})b \\ (c_{11} - c_{44})b & c_{11}b^2 + c_{44} - \rho v^2 \end{bmatrix} \begin{Bmatrix} {}^{\circ}u_x \\ {}^{\circ}u_z \end{Bmatrix} = 0 \tag{6.5}$$

whose solutions can be found by setting the determinant of its characteristic matrix to zero, namely,

$$\det\begin{bmatrix} c_{11}+c_{44}b^2-\rho v^2 & (c_{11}-c_{44})b \\ (c_{11}-c_{44})b & c_{11}b^2+c_{44}-\rho v^2 \end{bmatrix}=0 \tag{6.6}$$

For a given isotropic material, Eq. (6.6) provides two roots as solutions for the parameter b in terms of phase velocity (v). Here we denote these two roots as b_j ($j=1,3$). Subsequently, for each b_j ($j=1,3$) we can find a corresponding null polarization vector that satisfies Eq. (6.5):

$$\begin{Bmatrix} {}^{\circ}u_x^1 \\ {}^{\circ}u_z^1 \end{Bmatrix}, \quad \begin{Bmatrix} {}^{\circ}u_x^3 \\ {}^{\circ}u_z^3 \end{Bmatrix}$$

These null polarization vectors contain the normalized magnitudes of the particle displacements of the two partial waves. Since these b values and null polarization vectors are expressed as functions of v, solving for v requires us to consider the stress boundary conditions at the surface. This is another way to understand why the existence of a surface and its stress boundary conditions will affect the polarization and propagation of acoustic waves near the surface.

With the obtained two b values and their corresponding null polarization vectors, we can then express particle displacements as the summation of two partial waves using the following equations:

$$\begin{Bmatrix} u_x \\ u_y \\ u_z \end{Bmatrix}=\begin{Bmatrix} {}^{\circ}u_x^1\alpha_1 e^{-ikb_1z}+{}^{\circ}u_x^3\alpha_3 e^{-ikb_3z} \\ 0 \\ {}^{\circ}u_z^1\alpha_1 e^{-ikb_1z}+{}^{\circ}u_z^3\alpha_3 e^{-ikb_3z} \end{Bmatrix} e^{ik(vt-x)} \tag{6.7}$$

where α_1 and α_3 are constants that adjust the contributions of the two partial waves and they can be found once the phase velocity of the waves, v, is determined.

With the field expressions given in Eq. (6.7), along with Eqs. (2.6) and (2.15), we write the stress vector as

$$\{T\} = \begin{Bmatrix} \sigma_x \\ \sigma_y \\ \sigma_z \\ \tau_{yz} \\ \tau_{xz} \\ \tau_{xy} \end{Bmatrix} = -ik[c]\left\{ \begin{bmatrix} 1 & 0 & 0 \\ 0 & 0 & 0 \\ 0 & 0 & b_1 \\ 0 & b_1 & 0 \\ b_1 & 0 & 1 \\ 0 & 1 & 0 \end{bmatrix} \begin{Bmatrix} {}^{\circ}u_x^1 \alpha_1 e^{-ikb_1 z} \\ 0 \\ {}^{\circ}u_z^1 \alpha_1 e^{-ikb_1 z} \end{Bmatrix} \right.$$

$$\left. + \begin{bmatrix} 1 & 0 & 0 \\ 0 & 0 & 0 \\ 0 & 0 & b_3 \\ 0 & b_3 & 0 \\ b_3 & 0 & 1 \\ 0 & 1 & 0 \end{bmatrix} \begin{Bmatrix} {}^{\circ}u_x^3 \alpha_3 e^{-ikb_3 z} \\ 0 \\ {}^{\circ}u_z^3 \alpha_3 e^{-ikb_3 z} \end{Bmatrix} \right\} e^{ik(vt-x)}$$

which produces the following expression (note that $c_{12} = c_{11} - 2c_{44}$):

$$\begin{Bmatrix} \sigma_x \\ \sigma_y \\ \sigma_z \\ \tau_{yz} \\ \tau_{xz} \\ \tau_{xy} \end{Bmatrix}$$
$$= -ik \begin{Bmatrix} c_{11}({}^{\circ}u_x^1\alpha_1 e^{-ikb_1 z} + {}^{\circ}u_x^3\alpha_3 e^{-ikb_3 z}) + c_{12}({}^{\circ}u_z^1\alpha_1 b_1 e^{-ikb_1 z} + {}^{\circ}u_z^3\alpha_3 b_3 e^{-ikb_3 z}) \\ c_{12}({}^{\circ}u_x^1\alpha_1 e^{-ikb_1 z} + {}^{\circ}u_x^3\alpha_3 e^{-ikb_3 z}) + c_{12}({}^{\circ}u_z^1\alpha_1 b_2 e^{-ikb_1 z} + {}^{\circ}u_z^3\alpha_3 b_3 e^{-ikb_3 z}) \\ c_{12}({}^{\circ}u_x^1\alpha_1 e^{-ikb_1 z} + {}^{\circ}u_x^3\alpha_3 e^{-ikb_3 z}) + c_{11}({}^{\circ}u_z^1\alpha_1 b_1 e^{-ikb_1 z} + {}^{\circ}u_z^3\alpha_3 b_3 e^{-ikb_3 z}) \\ 0 \\ c_{44}({}^{\circ}u_x^1\alpha_1 b_1 e^{-ikb_1 z} + b_3\,{}^{\circ}u_x^3\alpha_3 b_3 e^{-ikb_3 z}) + c_{44}({}^{\circ}u_z^1\alpha_1 e^{-ikb_1 z} + {}^{\circ}u_z^3\alpha_3 e^{-ikb_3 z}) \\ 0 \end{Bmatrix} e^{ik(vt-x)} \quad (6.8)$$

If the surface of interest is of zero stresses, then the three stress components associated with z can be set to zero, thus we have the following zero-stress conditions:

$$\sigma_{z|z=0} = \tau_{yz|z=0} = \tau_{xz|z=0} = 0$$

By applying this zero-stress boundary condition to Eqs. (6.8) at the surface (i.e., letting $z = 0$), we can determine the phase velocity of the waves.

Example 1: Use an isotropic elastic material with $c_{11} = 10.7\times10^{10}$ N/m^2, $c_{44} = 2.8\times10^{10}$ N/m^2, and $\rho = 2700$ kg/m^3 to examine how Rayleigh waves are formed along a surface and determine the phase velocity of the waves.

Answer: By substituting the given material constants into Eq. (6.5), we have

$$10^{10} \times \begin{bmatrix} 2.8b^2 + 10.7 - 2.7 \times 10^{-7}v^2 & 7.9b \\ 7.9b & 10.7b^2 + 2.8 - 2.7 \times 10^{-7}v^2 \end{bmatrix} \begin{Bmatrix} {}^{\circ}u_x \\ {}^{\circ}u_z \end{Bmatrix} = 0$$

By setting the following determinant to zero

$$\det \begin{bmatrix} 2.8b^2 + 10.7 - 2.7 \times 10^{-7}v^2 & 7.9b \\ 7.9b & 10.7b^2 + 2.8 - 2.7 \times 10^{-7}v^2 \end{bmatrix} = 0$$

we find two roots with negative imaginary part that satisfy the wave decaying requirement as depth (z) increases

$$b_1 = -\sqrt{\rho v^2/c_{11} - 1} = -\sqrt{2.523 \times 10^{-8}v^2 - 1}$$

$$b_3 = -\sqrt{\rho v^2/c_{44} - 1} = -\sqrt{9.643 \times 10^{-8}v^2 - 1}$$

With these two b values, we can find their corresponding null polarization vector

$$\begin{Bmatrix} {}^{\circ}u_x^1 \\ {}^{\circ}u_z^1 \end{Bmatrix} = \begin{Bmatrix} 1 \\ \sqrt{\rho v^2/c_{11} - 1} \end{Bmatrix} = \begin{Bmatrix} 1 \\ \sqrt{2.523 \times 10^{-8} v^2 - 1} \end{Bmatrix}$$

$$\begin{Bmatrix} {}^{\circ}u_x^3 \\ {}^{\circ}u_z^3 \end{Bmatrix} = \begin{Bmatrix} 1 \\ \dfrac{-1}{\sqrt{\rho v^2/c_{44} - 1}} \end{Bmatrix} = \begin{Bmatrix} 1 \\ \dfrac{-1}{\sqrt{9.643 \times 10^{-8} v^2 - 1}} \end{Bmatrix}$$

Substituting these b values and null polarization vectors into Eq. (6.8), along with letting $z = 0$ for surface stresses, we find σ_x, σ_z and τ_{xz} in terms of α_1, α_3 and v as follows:

$$\sigma_{x|_{z=0}} = \left[-5.6 \times 10^{10} - 1287v^2\right] \alpha_1 - \left[5.6 \times 10^{10}\right] \alpha_3$$

$$\sigma_{z|_{z=0}} = \left[5.6 \times 10^{10} - 2700v^2\right] \alpha_1 + \left[5.6 \times 10^{10}\right] \alpha_3$$

$$\tau_{xz|_{z=0}} = 2.8 \times 10^{10} \left[2\alpha_1 \sqrt{2.523 \times 10^{-8} v^2 - 1} + \alpha_3 \left(\sqrt{9.643 \times 10^{-8} v^2 - 1} - \frac{1}{\sqrt{9.643 \times 10^{-8} v^2 - 1}}\right)\right]$$

By applying the zero-stress boundary conditions of

$$\sigma_{z|_{z=0}} = \tau_{xz|_{z=0}} = 0$$

we obtain the following two equations:

$$\left[5.6 \times 10^{10} - 2700v^2\right] \alpha_1 + \left[5.6 \times 10^{10}\right] \alpha_3 = 0$$

$$2\left(\sqrt{2.523 \times 10^{-8} v^2 - 1}\right)\alpha_1 + \left[\sqrt{9.643 \times 10^{-8} v^2 - 1} - \frac{1}{\sqrt{9.643 \times 10^{-8} v^2 - 1}}\right] \alpha_3 = 0$$

After separating the two constants, we write the above equations in a matrix form as

$$\begin{bmatrix} \frac{5.6 \times 10^{10} - 2700v^2}{2\sqrt{2.523 \times 10^{-8}v^2 - 1}} & \frac{5.6 \times 10^{10}}{\sqrt{9.643 \times 10^{-8}v^2 - 1} - \frac{1}{\sqrt{9.643 \times 10^{-8}v^2 - 1}}} \end{bmatrix} \begin{Bmatrix} \alpha_1 \\ \alpha_3 \end{Bmatrix} = 0$$

For the two constants (α_1, α_3) having real value solutions, the determinant of the characteristic matrix of the boundary-satisfying equations must be zero; hence,

$$\det \begin{bmatrix} \frac{5.6 \times 10^{10} - 2700v^2}{2\sqrt{2.523 \times 10^{-8}v^2 - 1}} & \frac{5.6 \times 10^{10}}{\sqrt{9.643 \times 10^{-8}v^2 - 1} - \frac{1}{\sqrt{9.643 \times 10^{-8}v^2 - 1}}} \end{bmatrix} = 0$$

This leads to a fourth-order polynomial equation with wave phase velocity, v, as its sole variable. Of the four roots, the only one acceptable as the solution for the Rayleigh waves is the real and positive root, that is,

$$v = 2997.8 \text{ m/s}$$

To put this in perspectives, for the given isotropic material, the phase velocity for the three bulk waves can be found by setting $b = 0$ in the governing Eq. (6.4). Doing so we find

$$v_L = \sqrt{\frac{c_{11}}{\rho}} = 6295.2 \text{ m/s}, \quad v_{T(\perp)} = v_{T(\parallel)} = \sqrt{\frac{c_{44}}{\rho}} = 3220.3 \text{ m/s}$$

Apparently, the phase velocity of surface Rayleigh waves is typically less than that of bulk waves.

Knowing the v value, we can now find the following:

$$b_1 = -0.879i, \; b_3 = -0.365i; \; {}^{\circ}u_x^1 = 1, \; {}^{\circ}u_z^1 = -0.878i,$$
$${}^{\circ}u_x^3 = 1, \; {}^{\circ}u_z^3 = -2.738i$$

Also, with the obtained v along with the zero-stress boundary condition, we can express α_3 in terms of α_1 as

$$\alpha_3 = -0.567\alpha_1$$

By plugging all these obtained values into Eq. (6.7), we find the displacement fields. After normalizing the field magnitude with u_z at the surface (i.e., $u_{z|z=0}$), and with $k = 2\pi/\lambda$, we express the displacement fields as follows:

$$\{u\} = \begin{Bmatrix} u_x \\ u_y \\ u_z \end{Bmatrix} = \begin{Bmatrix} -0.8428e^{-2.295z/\lambda} + 1.487e^{-5.525z/\lambda} \\ 0 \\ \left(2.308e^{-2.295z/\lambda} - 1.308e^{-5.525z/\lambda}\right) i \end{Bmatrix} e^{ik(vt-x)}$$

6.2.3 Displacements and Stresses Caused by Rayleigh Waves

It is now clear that in isotropic solids, Rayleigh waves are the coupled outcome of two wave components near the surface, with one component polarizing in the longitudinal x direction (the L wave component) and the other in the z direction (the TV wave component). Moreover, the fact that the u_x field is of a real number while the u_z field is of an imaginary number (i.e., it is multiplied by the imaginary index i) suggests that the two fields have a $\pi/2$, or $90°$, phase difference, indicating that the trajectories of the particle motions of Rayleigh waves are ellipses.

Figure 6.2 shows the magnitude of the two displacement fields as a function of depth into the material away from the surface. It is apparent that particle polarization of both wave components decays to almost zero at a depth of about two times the wavelength ($z = 2\lambda$).

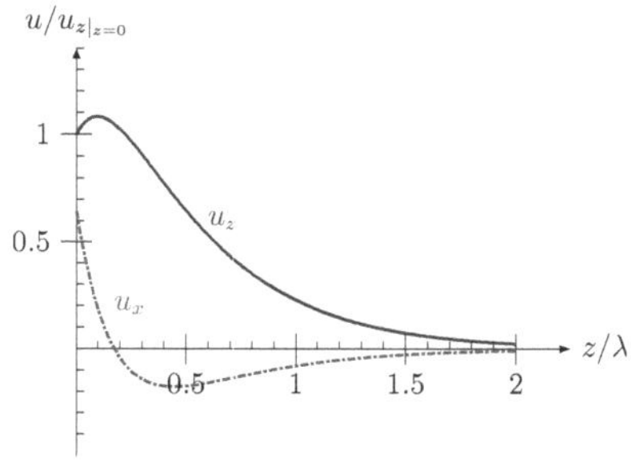

Figure 6.2 Relative magnitudes of particle displacements caused by Rayleigh waves.

The decaying profiles for u_x and u_z, however, are different: u_z increases a bit initially and then decreases gradually to zero beyond a depth of 2λ, but u_x decreases rather drastically and it changes its sign (or direction of displacement) at a depth less than $z = 0.2\lambda$ before it bends backward and decays to zero slowly from the negative side.

By plugging all the obtained values into the stress expressions, we obtain the induced stresses normalized by σ_x at the surface. Similarly, τ_{xz} has a phase shift of $\pi/2$, or 90°, with respect to σ_x and σ_z.

$$\begin{Bmatrix} \sigma_x \\ \sigma_z \\ \tau_{xz} \end{Bmatrix} = \begin{Bmatrix} -0.886e^{-2.295z/\lambda} + 1.886e^{-5.525z/\lambda} \\ 0.886e^{-2.295z/\lambda} - 0.886e^{-5.525z/\lambda} \\ \left(1.374e^{-2.295z/\lambda} - 1.374e^{-5.525z/\lambda}\right) i \end{Bmatrix} e^{ik(vt-x)}$$

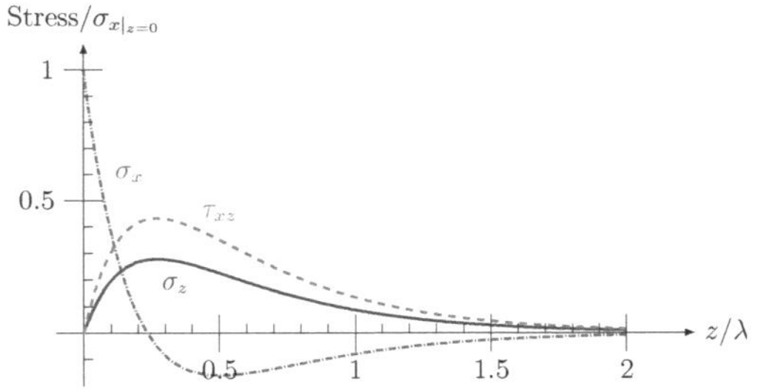

Figure 6.3 Relative magnitudes of stresses induced by Rayleigh waves.

Figure 6.3 sketches the relative magnitude of the three stress components as a function of depth. As a confirmation, the stresses almost vanish at a depth twice the wavelength ($z = 2\lambda$). Moreover, the decaying profiles for σ_x, σ_z and τ_{xz} are different also: both σ_z and τ_{xz} are zero at the surface, but they reach their respective peak values at around $z = 0.25\lambda$ and then decrease gradually to zero; σ_x, on the other hand, has its highest value at the surface and it decrease rapidly, changes its sign (from tension to compression) at a depth of about $z = 0.23\lambda$, and then decays to zero slowly.

Figure 6.4 shows a 3D illustration of the particle displacements induced by the polarization of Rayleigh waves. Here the deformed meshes are sketched in a way that both the x and z displacements are in the right proportion to the displacement fields as a function of depth as shown in Figure 6.2. Apparently, displacements in both x

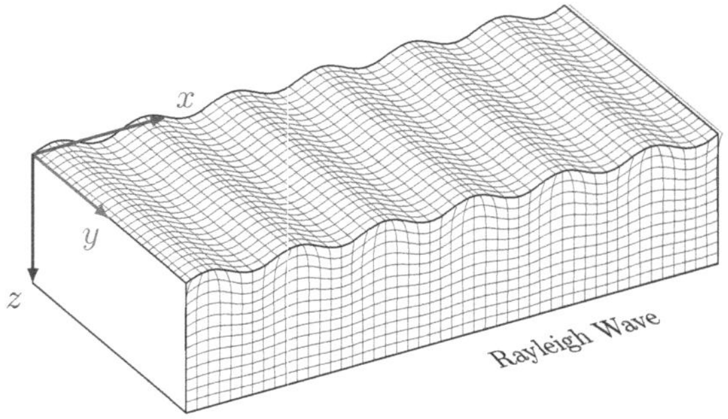

Figure 6.4 Illustration of particle displacements generated by surface Rayleigh waves.

and z directions vanish when the depth reaches about 2λ. Moreover, the u_x component changes direction from moving rightward to leftward, or vice versa, within a depth of 0.2λ from the surface. This illustration is a quantitative depiction of why we call such waves surface waves: their particle displacements caused by wave polarization are confined within an extremely narrow region close to the surface with a depth of about two times the wavelength.

6.2.4 Transverse Wave

Now let us take a quick look of the other uncoupled wave in Eq. (6.4), namely the wave polarizing in the y direction. We write its reduced governing equation as follows:

$$(c_{44}b^2 + c_{44} - \rho v^2)\{^\circ u_y\} = 0 \tag{6.9}$$

Using the same concepts discussed earlier, we know that this wave has one b value and with only one component $(^\circ u_y^2)$ in its null polarization vector; hence, we express its wave field as

$$\begin{Bmatrix} u_x \\ u_y \\ u_z \end{Bmatrix} = \begin{Bmatrix} 0 \\ ^\circ u_y^2 \alpha_2 e^{-ikb_2 z} \\ 0 \end{Bmatrix} e^{ik(vt-x)}$$

in which α_2 is a constant. Applying this displacement field to the stress vector, we find the two non-zero terms at the surface, as

$$\tau_{yz} = c_{44}{}^{\circ}u_y^2 b\alpha_2 e^{-ikb_2z}, \ \tau_{xy} = c_{44}{}^{\circ}u_y^2 \alpha_2 e^{-ikb_2z}$$

Since the surface (with the z axis as its normal) is considered as having zero stresses, we have $\tau_{yz} = 0$. By setting

$$\tau_{yz} = c_{44}{}^{\circ}u_y^2 b\alpha_2 e^{-ikb_2z} = 0$$

it becomes clear that

$$b = 0$$

So the resulting wave displacement is a function of x only, independent of z, as given below:

$$\begin{Bmatrix} u_x \\ u_y \\ u_z \end{Bmatrix} = \begin{Bmatrix} 0 \\ {}^{\circ}u_y^2\alpha_2 \\ 0 \end{Bmatrix} e^{ik(vt-x)} \tag{6.10}$$

Apparently, this wave is not a surface wave, but a bulk plane wave: it is a transverse plane wave traveling in a semi-infinite isotropic solid material with particle displacements occurring in a plane parallel with the surface.

Example 2: Use the same material given in Example 1 to find the phase velocity of the transverse waves predicted by Eq. (6.10).

Answer: With $b = 0$, from Eq. (6.9) we find its phase velocity as

$$v = \sqrt{\frac{c_{44}}{\rho}} = 3220.3 \text{ m/s}$$

which is the same as that for the two transverse bulk waves.

6.3 Propagation of SAWs in Piezoelectric Solids

6.3.1 Governing Equations for SAWs in Piezoelectric Solids

For SAWs traveling in piezoelectric materials, we refer to Eq. (2.25), or its expanded matrix forms given in equations (5.7) and (5.8), for their governing equations. As a conceptual reminder, we list Eq. (2.25) here again,

$$\begin{aligned}[\nabla_s]\left([c][\nabla_s]^T\{u\} + [e]^T[\nabla]^T\{\phi\}\right) - \rho\frac{\partial^2}{\partial t^2}\{u\} &= 0\\ [\nabla]\left([e][\nabla_s]^T\{u\} - [\epsilon][\nabla]^T\{\phi\}\right) &= 0\end{aligned}$$

Using the *xyz* coordinates system shown in Figure 6.1, we consider surface waves as propagating along the *x* direction and express both the particle displacement and electrical potential fields as proportional to

$$\{u\}, \{\phi\} \propto e^{-ikbz}e^{i(\omega t-kx)} = e^{-ikbz}e^{ik(vt-x)} \tag{6.11}$$

in which v, ω, k and b are the same as defined in Section 6.2.1, and only b values with negative imaginary part should be selected to ensure that as z increases, $e^{-ikbz} \Longrightarrow 0$.

With this expression we write the three components of particle displacements and electrical potential of the surface waves as follows:

$$\begin{Bmatrix} u_x \\ u_y \\ u_z \\ \phi \end{Bmatrix} = \begin{Bmatrix} {}^{\circ}u_x \\ {}^{\circ}u_y \\ {}^{\circ}u_z \\ {}^{\circ}\phi \end{Bmatrix} e^{-ikbz}e^{ik(vt-x)} \tag{6.12}$$

where ${}^{\circ}u_x, {}^{\circ}u_y, {}^{\circ}u_z$ and ${}^{\circ}\phi$ are the magnitudes of the three components of particles displacements and electrical potential, respectively. Similarly, with these expressions, we note that

$$\frac{\partial}{\partial x} = -ik,\ \frac{\partial}{\partial y} = 0,\ \frac{\partial}{\partial z} = -ikb,\ \frac{\partial^2}{\partial t^2} = -k^2v^2$$

By applying these notions to the matrix form wave equations, given in equations (5.7) and (5.8), and eliminating the common factor, $-k^2$, we derive the following two matrix equations for SAWs:

$$\left\{ \begin{bmatrix} 1 & 0 & 0 & 0 & b & 0 \\ 0 & 0 & 0 & b & 0 & 1 \\ 0 & 0 & b & 0 & 1 & 0 \end{bmatrix} [c_{xyz}] \begin{bmatrix} 1 & 0 & 0 \\ 0 & 0 & 0 \\ 0 & 0 & b \\ 0 & b & 0 \\ b & 0 & 1 \\ 0 & 1 & 0 \end{bmatrix} \begin{Bmatrix} {}^{\circ}u_x \\ {}^{\circ}u_y \\ {}^{\circ}u_z \end{Bmatrix} \right.$$

$$\left. + \begin{bmatrix} 1 & 0 & 0 & 0 & b & 0 \\ 0 & 0 & 0 & b & 0 & 1 \\ 0 & 0 & b & 0 & 1 & 0 \end{bmatrix} [e_{xyz}]^T \begin{Bmatrix} 1 \\ 0 \\ b \end{Bmatrix} \{{}^{\circ}\phi\} - \rho v^2 \begin{Bmatrix} {}^{\circ}u_x \\ {}^{\circ}u_y \\ {}^{\circ}u_z \end{Bmatrix} \right\} e^{-ikbz} e^{ik(vt-x)} = 0$$

and

$$\left\{ \begin{bmatrix} 1 & 0 & b \end{bmatrix} [e_{xyz}] \begin{bmatrix} 1 & 0 & 0 \\ 0 & 0 & 0 \\ 0 & 0 & b \\ 0 & b & 0 \\ b & 0 & 1 \\ 0 & 1 & 0 \end{bmatrix} \begin{Bmatrix} {}^{\circ}u_x \\ {}^{\circ}u_y \\ {}^{\circ}u_z \end{Bmatrix} \right.$$

$$\left. - \begin{bmatrix} 1 & 0 & b \end{bmatrix} [\epsilon_{xyz}] \begin{Bmatrix} 1 \\ 0 \\ b \end{Bmatrix} \{{}^{\circ}\phi\} \right\} e^{-ikbz} e^{ik(vt-x)} = 0$$

By introducing the following three expressions

$$\begin{bmatrix} \Gamma_{11} & \Gamma_{12} & \Gamma_{13} \\ \Gamma_{12} & \Gamma_{22} & \Gamma_{23} \\ \Gamma_{13} & \Gamma_{23} & \Gamma_{33} \end{bmatrix} = \begin{bmatrix} 1 & 0 & 0 & 0 & b & 0 \\ 0 & 0 & 0 & b & 0 & 1 \\ 0 & 0 & b & 0 & 1 & 0 \end{bmatrix} [c_{xyz}] \begin{bmatrix} 1 & 0 & 0 \\ 0 & 0 & 0 \\ 0 & 0 & b \\ 0 & b & 0 \\ b & 0 & 1 \\ 0 & 1 & 0 \end{bmatrix}$$

$$\begin{Bmatrix} \gamma_1 \\ \gamma_2 \\ \gamma_3 \end{Bmatrix} = \begin{bmatrix} 1 & 0 & 0 & 0 & b & 0 \\ 0 & 0 & 0 & b & 0 & 1 \\ 0 & 0 & b & 0 & 1 & 0 \end{bmatrix} [e_{xyz}]^T \begin{Bmatrix} 1 \\ 0 \\ b \end{Bmatrix} \quad \text{and} \quad \{\epsilon_4\} = \begin{bmatrix} 1 & 0 & b \end{bmatrix} [\epsilon_{xyz}] \begin{Bmatrix} 1 \\ 0 \\ b \end{Bmatrix}$$

we can simplify the above two matrix equations into

$$\left\{ \begin{bmatrix} \Gamma_{11} - \rho v^2 & \Gamma_{12} & \Gamma_{13} \\ \Gamma_{12} & \Gamma_{22} - \rho v^2 & \Gamma_{23} \\ \Gamma_{13} & \Gamma_{23} & \Gamma_{33} - \rho v^2 \end{bmatrix} \begin{Bmatrix} {}^\circ u_x \\ {}^\circ u_y \\ {}^\circ u_z \end{Bmatrix} + \begin{Bmatrix} \gamma_1 \\ \gamma_2 \\ \gamma_3 \end{Bmatrix} \{{}^\circ \phi\} \right\} e^{-ikbz} e^{ik(vt-x)} = 0$$

and

$$\left\{ \begin{Bmatrix} \gamma_1 \\ \gamma_2 \\ \gamma_3 \end{Bmatrix}^T \begin{Bmatrix} {}^\circ u_x \\ {}^\circ u_y \\ {}^\circ u_z \end{Bmatrix} - \{\epsilon_4\}\{{}^\circ \phi\} \right\} e^{-ikbz} e^{ik(vt-x)} = 0$$

These two equations can be further condensed into

$$\begin{bmatrix} \Gamma_{11} - \rho v^2 & \Gamma_{12} & \Gamma_{13} & \gamma_1 \\ \Gamma_{12} & \Gamma_{22} - \rho v^2 & \Gamma_{23} & \gamma_2 \\ \Gamma_{13} & \Gamma_{23} & \Gamma_{33} - \rho v^2 & \gamma_3 \\ \gamma_1 & \gamma_2 & \gamma_3 & -\epsilon_4 \end{bmatrix} \begin{Bmatrix} {}^\circ u_x \\ {}^\circ u_y \\ {}^\circ u_z \\ {}^\circ \phi \end{Bmatrix} e^{-ikbz} e^{ik(vt-x)} = 0$$

or

$$\begin{bmatrix} \Gamma_{11}-\rho v^2 & \Gamma_{12} & \Gamma_{13} & \gamma_1 \\ \Gamma_{12} & \Gamma_{22}-\rho v^2 & \Gamma_{23} & \gamma_2 \\ \Gamma_{13} & \Gamma_{23} & \Gamma_{33}-\rho v^2 & \gamma_3 \\ \gamma_1 & \gamma_2 & \gamma_3 & -\epsilon_4 \end{bmatrix} \begin{Bmatrix} {}^{\circ}u_x \\ {}^{\circ}u_y \\ {}^{\circ}u_z \\ {}^{\circ}\phi \end{Bmatrix} = 0 \qquad (6.13)$$

This is the set of governing equations for SAWs, with which we can find the solutions by setting the determinant of its characteristic matrix to zero, namely,

$$\det \begin{bmatrix} \Gamma_{11}-\rho v^2 & \Gamma_{12} & \Gamma_{13} & \gamma_1 \\ \Gamma_{12} & \Gamma_{22}-\rho v^2 & \Gamma_{23} & \gamma_2 \\ \Gamma_{13} & \Gamma_{23} & \Gamma_{33}-\rho v^2 & \gamma_3 \\ \gamma_1 & \gamma_2 & \gamma_3 & -\epsilon_4 \end{bmatrix} = 0 \qquad (6.14)$$

6.3.2 Displacement and Potential Fields

For a given piezoelectric material and a preselected wave propagation direction, Eq. (6.14) provides four roots as solutions for the parameter b in terms of wave phase velocity (v). Here we denote these four roots as b_j $(j = 1, 2, 3, 4)$. Subsequently, for each b_j $(j = 1, 2, 3, 4)$ we can find a corresponding null polarization vector that satisfies Eq. (6.13)

$$\begin{Bmatrix} {}^{\circ}u_x^j \\ {}^{\circ}u_y^j \\ {}^{\circ}u_z^j \\ {}^{\circ}\phi^j \end{Bmatrix}; \; j = 1, 2, 3, 4$$

This null polarization vector contains the normalized magnitudes of the particle displacements and potential of the surface waves.

Again, because these b values and null polarization vectors are expressed as functions of the wave phase velocity (v), this leaves us with v as the unknown variable. To solve for v, we need to consider the mechanical (in terms of stresses or strains) and electrical (in terms of potential) boundary conditions at the surface, because the existence of a surface and its mechanical and electrical boundary conditions will alter the polarization and propagation of acoustic waves near the surface.

With the obtained four b values and their corresponding null polarization vectors, we further express the three particle displacement fields and the potential field in linear superposition of four partial waves using the following equations:

$$u_x = \sum_{j=1}^{4} {}^{\circ}u_x^j \alpha_j e^{-ikb_j z} e^{ik(vt-x)}, \quad u_y = \sum_{j=1}^{4} {}^{\circ}u_y^j \alpha_j e^{-ikb_j z} e^{ik(vt-x)}$$

$$u_z = \sum_{j=1}^{4} {}^{\circ}u_z^j \alpha_j e^{-ikb_j z} e^{ik(vt-x)}, \quad \phi = \sum_{j=1}^{4} {}^{\circ}\phi^j \alpha_j e^{-ikb_j z} e^{ik(vt-x)}$$

where α_j ($j = 1, 2, 3, 4$) are constants, which will be determined with the consideration of boundary conditions once v is known. In a full expansion, the above expressions are

$$u_x = \left({}^{\circ}u_x^1 \alpha_1 e^{-ikb_1 z} + {}^{\circ}u_x^2 \alpha_2 e^{-ikb_2 z} + {}^{\circ}u_x^3 \alpha_3 e^{-ikb_3 z} + {}^{\circ}u_x^4 \alpha_4 e^{-ikb_4 z} \right) e^{ik(vt-x)}$$

$$u_y = \left({}^{\circ}u_y^1 \alpha_1 e^{-ikb_1 z} + {}^{\circ}u_y^2 \alpha_2 e^{-ikb_2 z} + {}^{\circ}u_y^3 \alpha_3 e^{-ikb_3 z} + {}^{\circ}u_y^4 \alpha_4 e^{-ikb_4 z} \right) e^{ik(vt-x)}$$

$$u_z = \left({}^{\circ}u_z^1 \alpha_1 e^{-ikb_1 z} + {}^{\circ}u_z^2 \alpha_2 e^{-ikb_2 z} + {}^{\circ}u_z^3 \alpha_3 e^{-ikb_3 z} + {}^{\circ}u_z^4 \alpha_4 e^{-ikb_4 z} \right) e^{ik(vt-x)}$$

$$\phi = \left({}^{\circ}\phi^1 \alpha_1 e^{-ikb_1 z} + {}^{\circ}\phi^2 \alpha_2 e^{-ikb_2 z} + {}^{\circ}\phi^3 \alpha_3 e^{-ikb_3 z} + {}^{\circ}\phi^4 \alpha_4 e^{-ikb_4 z} \right) e^{ik(vt-x)} \tag{6.15}$$

6.3.3 Mechanical and Electrical Conditions at the Surface

By substituting these field expressions into Eq. (2.24), we find the stress and electrical displacement vectors as

$$\{T\} = \begin{Bmatrix} \sigma_x \\ \sigma_y \\ \sigma_z \\ \tau_{yz} \\ \tau_{xz} \\ \tau_{xy} \end{Bmatrix} = -ik[c_{xyz}] \sum_{j=1}^{4} \left[\begin{bmatrix} 1 & 0 & 0 \\ 0 & 0 & 0 \\ 0 & 0 & b_j \\ 0 & b_j & 0 \\ b_j & 0 & 1 \\ 0 & 1 & 0 \end{bmatrix} \begin{Bmatrix} {}^{\circ}u_x^j \alpha_j e^{-ikb_jz} \\ {}^{\circ}u_y^j \alpha_j e^{-ikb_jz} \\ {}^{\circ}u_z^j \alpha_j e^{-ikb_jz} \end{Bmatrix} \right] e^{ik(vt-x)}$$

$$- ik[e_{xyz}]^T \sum_{j=1}^{4} \left[\begin{Bmatrix} 1 \\ 0 \\ b_j \end{Bmatrix} {}^{\circ}\phi^j \alpha_j e^{-ikb_jz} \right] e^{ik(vt-x)} \tag{6.16}$$

and

$$\{D\} = \begin{Bmatrix} D_x \\ D_y \\ D_z \end{Bmatrix} = -ik[e_{xyz}] \sum_{j=1}^{4} \left[\begin{bmatrix} 1 & 0 & 0 \\ 0 & 0 & 0 \\ 0 & 0 & b_j \\ 0 & b_j & 0 \\ b_j & 0 & 1 \\ 0 & 1 & 0 \end{bmatrix} \begin{Bmatrix} {}^{\circ}u_x^j \alpha_j e^{-ikb_jz} \\ {}^{\circ}u_y^j \alpha_j e^{-ikb_jz} \\ {}^{\circ}u_z^j \alpha_j e^{-ikb_jz} \end{Bmatrix} \right] e^{ik(vt-x)}$$

$$+ ik[\epsilon_{xyz}]^T \sum_{j=1}^{4} \left[\begin{Bmatrix} 1 \\ 0 \\ b_j \end{Bmatrix} {}^{\circ}\phi^j \alpha_j e^{-ikb_jz} \right] e^{ik(vt-x)} \tag{6.17}$$

If the surface of interest is not subjected to any mechanical stresses, the three stress components associated with z (the direction of the surface normal) are zero, thus we have the following zero-stress conditions:

$$\sigma_z|_{z=0} = \tau_{yz}|_{z=0} = \tau_{xz}|_{z=0} = 0$$

Additionally, a proper electrical boundary condition for the electrical displacement must be met. If the surface is at a state of electrically free floating, the following continuity condition for the electrical displacement (referred to as free-floating surface, or simply free surface, hereafter) should be satisfied:

$$D_{z|_{z=0}} = k\epsilon_{\text{vacuum}}\phi_{|_{z=0}}$$

where ϵ_{vacuum} is the permittivity of the vacuum with $\epsilon_{\text{vacuum}} = 8.854 \times 10^{-12}$ F/m^2.

If the surface is coated with a thin metallic layer (note that the thickness is much smaller than the wavelength) and grounded electrically, then the following zero-potential condition (referred to as metallized surface hereafter) must be met:

$$\phi_{|_{z=0}} = 0$$

6.3.3.1 Zero-Stress and Free-Floating Surfaces

Put it another way, when the surface of interest is with zero-stresses mechanically and free-floating electrically, the following four boundary conditions must be satisfied:

$$\sigma_{z|_{z=0}} = \tau_{yz|_{z=0}} = \tau_{xz|_{z=0}} = 0, \quad D_{z|_{z=0}} = \epsilon_{\text{vacuum}}k\phi_{|_{z=0}} \tag{6.18}$$

6.3.3.2 Zero-Stress and Metallized Surfaces

When the surface is with zero-stresses and metallized, the following four boundary conditions must be met:

$$\sigma_{z|_{z=0}} = \tau_{yz|_{z=0}} = \tau_{xz|_{z=0}} = 0, \quad \phi_{|_{z=0}} = 0 \tag{6.19}$$

By applying the boundary conditions given in equations (6.18) and (6.19), respectively, to equations (6.16) and (6.17) we first find a 4×4 boundary-satisfying matrix equation in terms of wave phase velocity (v) after substitution of the expressions of the four b values which are determined from solving the governing Eq. (6.14). By setting the determinant of this boundary-satisfying matrix to zero, we can solve for the phase velocity for the surface waves of interest.

6.4 Common Crystal Cuts for SAW Applications

Before we discuss how the governing equations and boundary conditions presented in the previous section are used to determine the type and phase velocity of SAWs in piezoelectric materials, let us first define some commonly utilized crystal cuts for the design and fabrication of SAW devices. Here the term *crystal cuts* carries the meaning of not only selecting a proper orientation but also isolating a small piece from a large structure of a crystal material as illustrated in Figure 6.5. For SAW device applications, these small pieces typically possess at least one large surface intended for SAW generation. The direction of the surface normal in the original coordinates system (i.e., the XYZ system) is commonly used to denote the cuts. This is necessary because the *xyz* system will rotate with the cut orientation. When a small cut piece has the normal of its intended surface in the Z direction, this cut piece is called a Z-cut piece. By simple extension, there are X-cut and Y-cut crystals. There are other cuts too, including rotated X and rotated Y cuts, among others. In the following subsections, we will discuss some of the most common ones using pictorial illustrations, as well as associating them with Euler angles and the three wave components.

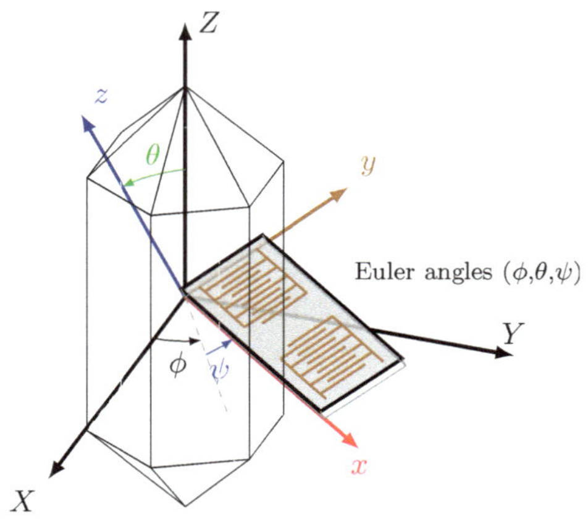

Figure 6.5 Illustration of a small piece of crustal slab cut from a crystal bar.

6.4.1 Z-Cut Crystals

As shown in Figure 6.6, a Z-cut crystal has its surface normal, i.e., the *z* axis, point to the Z direction. Recall the orientation convention discussed in Section 6.1 in which the *x* direction is designated as the wave propagation direction. As the *x* axis sweeps around about the Z axis in the XY plane, we can see that all resulting crystal cuts will have their surface normal in the Z direction; hence, they are all Z-cut crystals. So, Z-cut crystals can be represented by Euler angles $(\phi_{XY}, 0°, 0°)$, with ϕ_{XY} varying from 0° to 360° in the XY plane staring from the X axis.

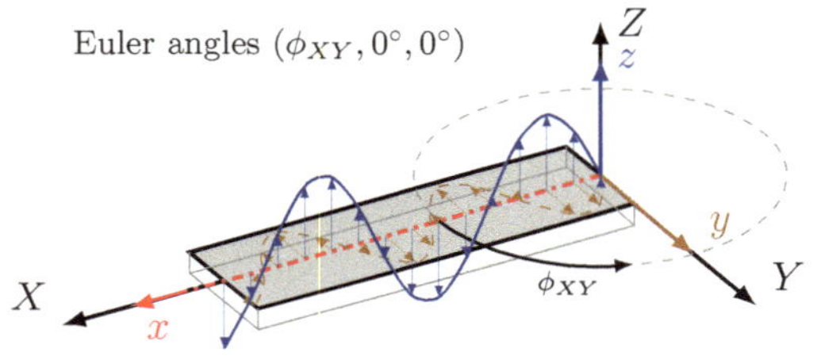

Figure 6.6 Z-cut crystals with Euler angles $(\phi_{XY}, 0°, 0°)$.

In a conventional way, we follow the right-hand rule for determining the sign for the rotation angle: with the thumb pointing to the *z* direction, and the rotation direction pointed by the four fingers is given a positive sign, and the opposite rotation will be given a negative sign. With this convention in mind, of the Z cuts, two common crystal selections are (1) Z-cut X-propagation with Euler angles $(0°, 0°, 0°)$, and (2) Z-cut Y-propagation with Euler angles $(90°, 0°, 0°)$. The three wave components, namely, the L wave (red dash-dotted line), the TH wave (brown dashed line) and the TV wave (blue solid line) are illustrated accordingly. For inconvenience sake, we will shorten the names for these cuts using two letters, with the first one representing the cut orientation and the second the intended propagation direction. So these two common cuts can be called ZX cut and ZY cut, respectively.

6.4.2 Y-Cut Crystals

Figure 6.7 shows a Y-cut crystal with its surface normal of z axis pointing in the Y direction. This cut is obtained with Euler angles of $(0°, -90°, 0°)$. By sweeping the x axis around about the Y axis in the XZ plane, we have all other Y-cut crystals, thus Y-cut crystals can be represented using Euler angles $(0°, -90°, \psi_{XZ})$, with ψ_{XZ} varying from 0° to 360° in the XZ plane starting from the X axis. In Y cuts, two common crystal selections are (1) YX cut with Euler angles $(0°, -90°, 0°)$, and (2) YZ cut with Euler angles $(0°, -90°, -90°)$.

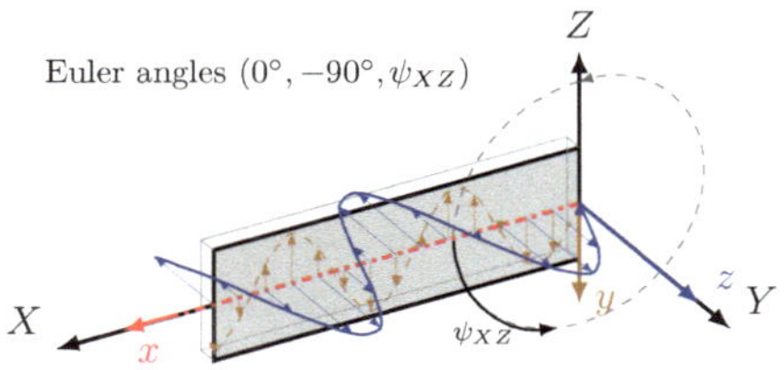

Figure 6.7 Y-cut crystals with Euler angles $(0°, -90°, \psi_{XY})$.

It is worth noting that once a Y-cut crystal is made, it makes no difference if the upper or lower surface is used for SAW device applications. This suggests that all Y-cut crystals can be equally described by Euler angles $(0°, 90°, \psi_{XZ})$.

6.4.3 X-Cut Crystals

Figure 6.8 shows a X-cut crystal with its surface normal of z axis pointing in the X direction. This cut is obtained with Euler angles $(90°, 90°, 0°)$.

By sweeping the x axis around about the X axis in the YZ plane, we have all other X-cut crystals, thus X-cut crystals can be represented using Euler angles $(90°, 90°, \psi_{YZ})$, with ψ_{YZ} varying from 0° to 360° in the XZ plane starting from the Y axis. Similarly, in X cuts, two common crystal selections are (1) XY cut with Euler angles $(90°, 90°, 0°)$, and (2) XZ cut with Euler angles $(90°, 90°, 90°)$.

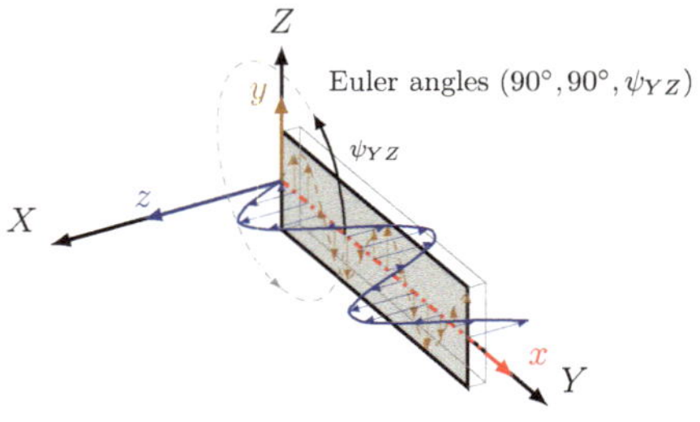

Figure 6.8 X-cut crystals with Euler angles $(90°, 90°, \psi_{XY})$.

6.4.4 Rotated Y-Cut Crystals with X(||) Wave Propagation

Aside from these cuts named after the axes of the fixed coordinates system, a different way of selecting crystal cuts is to fix the propagation direction while adjusting the cut angles. Figure 6.9 shows one such example, where a cut crystal with $(0°, -90°, 0°)$ has the wave propagation direction fixed parallel with the X axis (i.e., having the *x* axis pointing to the X direction). In this case, adjusting the cut angles is achieved by rotating the crystal about the X (or *x*) axis. As rotation takes place, the angle (θ_{YZ}) will change from 0° to 360° in the YZ plane starting from the Y axis. Since the resulting cuts have Euler angles of $(0°, -90° + \theta_{YZ}, 0°)$, this kind of cuts is termed rotated YX(||) cut, or Rt.-YX(||) as an abbreviation expression.

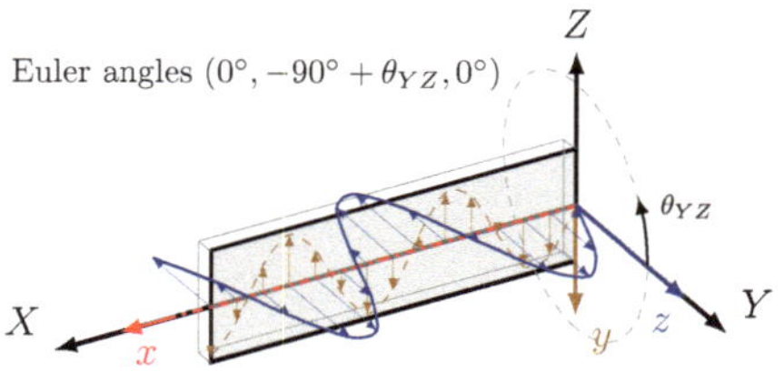

Figure 6.9 Rotated Y-cut crystals having wave propagation set parallel with the X axis with Euler angles $(0°, -90° + \theta_{XY}, 0°)$.

Comparing with the Y and Z cuts, the 0° case here is the same as the YX cut, and the 90° case is the same as the ZX cut. Very often we write the angle of rotation in front of Y to mark such cuts. For example, the 128° rotated cut can be expresses as 128° Rt.-YX(||) cut. Thus, the set of Euler angles of $(0°, 38°, 0°)$ will yield this crystal cut (see Figure 6.13(a)).

6.4.5 Rotated Y-Cut Crystals with X($\perp$) Wave Propagation

Figure 6.10 shows another type of Rt.-Y cuts, in which a crystal cut with Euler angles $(0°, -90°, -90°)$ has the wave propagation direction set at the Z direction, or more precisely, in a direction perpendicular to the X axis. In this case, adjusting the cut angles is achieved by rotating the crystal about the X axis, and as rotation takes place, the angle (θ_{YZ}) will change from 0° to 360° in the YZ plane staring from the Y axis. Since all the cuts resulting from such a rotation will have its *x* direction perpendicular to the X axis, this kind of cuts is known as Rt.-YX($\perp$) cut and all these cuts can be represented using Euler angles $(0°, -90° + \theta_{YZ}, -90°)$. Comparing with Y cuts, the 0° case here is the same as the YZ cut.

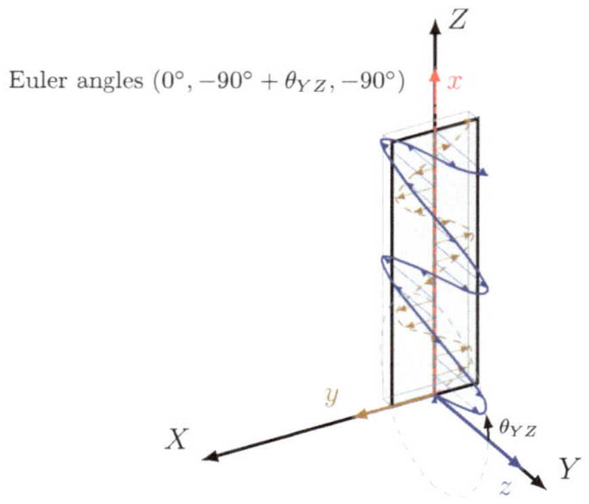

Figure 6.10 Rotated Y-cut crystals having wave propagation set perpendicular to in the X axis with Euler angles $(0°, -90° + \theta_{XY}, -90°)$.

6.4.6 45° Rotated Y-Cut Crystals

Figure 6.11 shows a series of 45° rotated Y cuts, in which a crystal cut with Euler angle $(0°, -45°, 0°)$ has the wave propagation direction pointing in the X direction. For these cuts, adjusting the cut angles is achieved by rotating the crystal about the *z* axis, and as rotation takes place, the angle (ψ_{xy}) will change from 0° to 360° in the *xy* plane (the plane perpendicular to the *z* axis) staring from the X axis. Since all the cuts resulting from such a rotation will have its *z* direction form a

45° angle with the Y axis, this kind of cuts is known as 45° Rt.-Y cuts. Comparing with the Rt.-YX(||) cuts, the 0° case here is the same as the 45° Rt.-Y X(||) cut, and comparing with the Rt.-YX(⊥) cuts, the –90° case here is the same as the 45° Rt.-YX(⊥) cut.

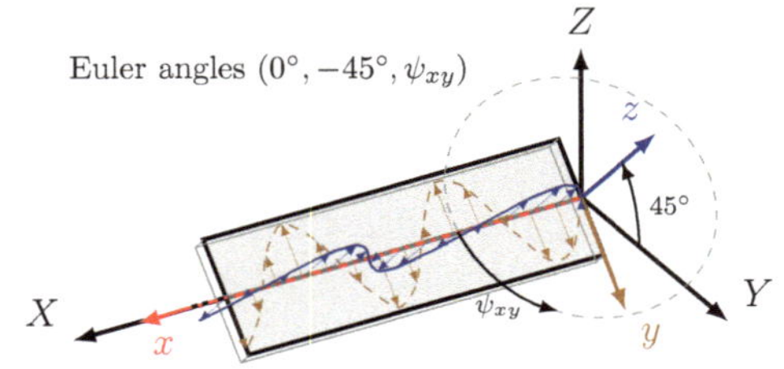

Figure 6.11 45° Rotated Y-cut crystals with Euler angles $(0°, -45°, \psi_{xy})$.

6.4.7 Rotated X-Cut Crystals with Y(⊥) Wave Propagation

Figure 6.12 shows a series of rotated X cuts, in which a crystal cut with Euler angles $(90°, 90°, 90°)$ has the wave propagation direction set in the Z direction, or more precisely, in a direction perpendicular to the Y axis. In this case, adjusting the cut angles is achieved by rotating the crystal about the Y axis, and as rotation takes place, the angle (θ_{XZ}) will change from 0° to 360° in the XZ plane staring from

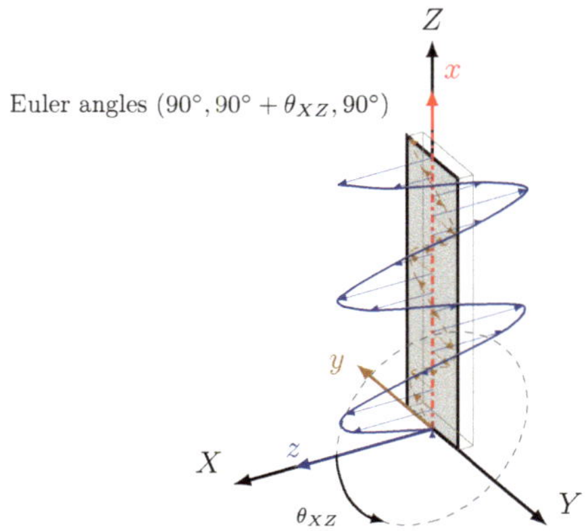

Figure 6.12 Rotated X-cut crystals with Euler angles defined as $(90°, 90° + \theta_{XY}, 90°)$.

the X axis. Since all the cuts resulting from such a rotation will have its x direction perpendicular to the Y axis, this kind of cuts is known as Rt.XY($\perp$) cut and all these cuts can be represented using Euler angles $(90°, 90° + \theta_{XZ}, 90°)$. Comparing with X cuts, the $0°$ case here is the same as the X-cut $90°$ case.

Example 3: Referring to the discussions and illustrations in Section 6.4, sketch the following crystal cuts defined by the given Euler angles:

1. $(0°, 38°, 0°)$
2. $(90°, 135°, 90°)$
3. $(0°, -54.75°, 0°)$
4. $(90°, 90°, 112°)$

Answer: To be able to tell which cuts these Euler angles represent, it is helpful to examine the given angles in reference to the common cuts discussed in Section 6.4. Here we will do this one by one. For case (1), since it has a non-zero value for the second angle, it is of the Rt.-Y cut group. Because the $0°$ situation of the Rt.-YX(||) cuts has $-90°$ for the second angle (see Figure 6.7), it is clear this cut is the $128°$ Rt.-YX(||) cut. For case (2), it belongs to the Rt.-XY($\perp$) cut group (see Figure 6.12), and it is the $45°$ Rt.XY($\perp$) cut. For case (3), like case (1) it is of Rt.-YX(||) cut group and it is $35.25°$ Rt.-YX(||) cut. In quartz, this cut is known as the AT cut. For case (4), it is of the X-cut group at $112°$ angle. Figure 6.13 shows the sketches for these four crystal cuts. This exercise highlights the fact that a rotation angle may or may not be the same as the corresponding angle in the expression of Euler angles.

6.5 Decoupling SAWs with Specific Cut Orientations

As we have seen in Section 6.2.1, the governing equation for SAWs traveling in isotropic materials (see Eq. 6.4) consists of two independent systems of waves, one contains surface Rayleigh waves

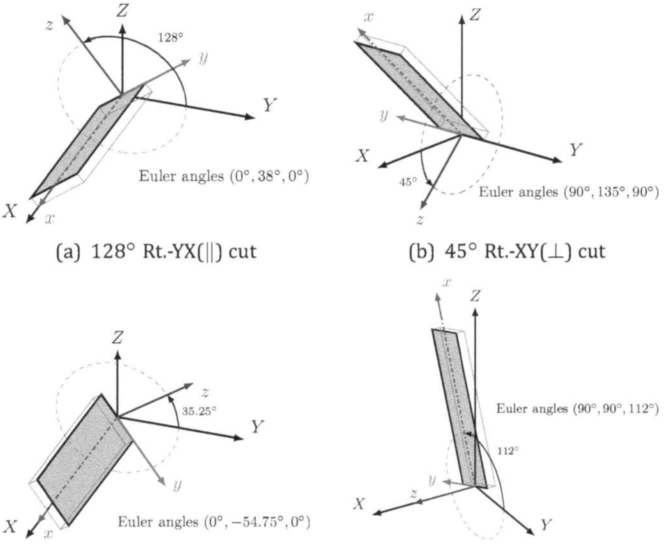

(a) 128° Rt.-YX(||) cut

(b) 45° Rt.-XY(⊥) cut

(c) 35.25° Rt.-YX(||) cut

(d) X 112° cut

Figure 6.13 Four selected cuts with rotation angles shown and Euler angles given.

and the other a transverse bulk wave. In piezoelectric materials, depending on the type of crystals and the selected cut orientation, the governing equation can also be decoupled into independent sub-systems of waves. In some situations, SAWs are decoupled into Rayleigh waves and a piezoelectric stiffened transverse wave, and in others, piezoelectric stiffened Rayleigh waves and a non-piezoelectric transverse wave. In this Section, we will discuss some of these cases.

6.5.1 Cuts with *xz*-Plane Perpendicular to Axes of Rotation of Even-Fold

For piezoelectric crystals possessing two-fold, four-fold and six-fold rotational symmetry about an axis, when the *xz* plane (sometimes, it is called sagittal plane) of a crystal cut is perpendicular to this axis of rotation, the governing equation for SAWs, as given in (6.13), will take the following form:

$$\begin{bmatrix} \Gamma_{11}-\rho v^2 & 0 & \Gamma_{13} & 0 \\ 0 & \Gamma_{22}-\rho v^2 & 0 & \gamma_2 \\ \Gamma_{13} & 0 & \Gamma_{33}-\rho v^2 & 0 \\ 0 & \gamma_2 & 0 & -\epsilon_4 \end{bmatrix} \begin{Bmatrix} {}^{\circ}u_x \\ {}^{\circ}u_y \\ {}^{\circ}u_z \\ {}^{\circ}\phi \end{Bmatrix} = 0 \qquad (6.20)$$

This means that SAWs traveling in this type of cut crystals will decouple into two independent systems of waves, consisting of

$$\begin{bmatrix} \Gamma_{11}-\rho v^2 & \Gamma_{13} \\ \Gamma_{13} & \Gamma_{33}-\rho v^2 \end{bmatrix} \begin{Bmatrix} {}^{\circ}u_x \\ {}^{\circ}u_z \end{Bmatrix} = 0 \qquad (6.21)$$

and

$$\begin{bmatrix} \Gamma_{22}-\rho v^2 & \gamma_2 \\ \gamma_2 & -\epsilon_4 \end{bmatrix} \begin{Bmatrix} {}^{\circ}u_y \\ {}^{\circ}\phi \end{Bmatrix} = 0 \qquad (6.22)$$

Referring to Eq. (6.5), it is clear that the waves governed by Eq. (6.21) are non-piezoelectric Rayleigh waves, and the wave governed by Eq. (6.22) is a piezoelectric stiffened TH wave, which is commonly known as Bleustein–Gulyaev *transverse horizontal* wave, to honor its discovery by them in 1968.

To demonstrate this decoupling phenomenon in piezoelectric crystals with two-fold, four-fold and six-fold rotational symmetry, we provide below the 4×4 $[\Gamma]$ matrix of selected cases determined using the procedures discussed in Section 6.3.1. Cases listed include orthorhombic (two-fold), tetragonal (four-fold), hexagonal (six-fold), trigonal (two-fold) and cubic (two- and four-fold) crystals at the respective cuts specified.

6.5.1.1 Two-fold symmetry

Orthorhombic 2mm crystals of XY cut (symmetry axis: Z):

$$[\Gamma] = \begin{bmatrix} c_{22} + c_{66}b^2 - \rho v^2 & 0 & c_{12}b + c_{66}b & 0 \\ 0 & c_{44} + c_{55}b^2 - \rho v^2 & 0 & e_{x5}b^2 + e_{y4} \\ c_{12}b + c_{66}b & 0 & c_{11}b^2 + c_{66} - \rho v^2 & 0 \\ 0 & e_{x5}b^2 + e_{y4} & 0 & -\epsilon_{xx}b^2 - \epsilon_{yy} \end{bmatrix}$$

Orthorhombic 2mm crystals of YX cut (symmetry axis: Z):

$$\begin{bmatrix} c_{11} + c_{66}b^2 - \rho v^2 & 0 & c_{12}b + c_{66}b & 0 \\ 0 & c_{44}b^2 + c_{55} - \rho v^2 & 0 & e_{y4}b^2 + e_{x5} \\ c_{12}b + c_{66}b & 0 & c_{22}b^2 + c_{66} - \rho v^2 & 0 \\ 0 & e_{y4}b^2 + e_{x5} & 0 & -\epsilon_{yy}b^2 - \epsilon_{xx} \end{bmatrix}$$

Trigonal 32 crystals of YZ cut (symmetry axis: X):

$$\begin{bmatrix} c_{33} + c_{44}b^2 - \rho v^2 & 0 & c_{14}b^2 + (c_{13} + c_{44})b & 0 \\ 0 & c_{44} - 2c_{14}b + \dfrac{(c_{11} - c_{12})b^2}{2} - \rho v^2 & 0 & e_{x1}b^2 - (e_{x4}b \\ c_{14}b^2 + (c_{13} + c_{44})b & 0 & c_{11}b^2 + c_{44} + 2c_{14}b - \rho v^2 & 0 \\ 0 & e_{x1}b^2 - (e_{x4}b & 0 & -\epsilon_{xx}b^2 - \epsilon_{zz} \end{bmatrix}$$

Trigonal 32 crystals of ZY cut (symmetry axis: X):

$$\begin{bmatrix} c_{11} - 2c_{14}b + c_{44}b^2 - \rho v^2 & 0 & (c_{13} + c_{44})b - c_{14} & 0 \\ 0 & c_{44}b^2 + 2c_{14}b + \dfrac{c_{11} - c_{12}}{2} - \rho v^2 & 0 & e_{x1} + e_{x4}b \\ (c_{13} + c_{44})b - c_{14} & 0 & c_{33}b^2 + c_{44} - \rho v^2 & 0 \\ 0 & e_{x1} + e_{x4}b & 0 & -\epsilon_{xx} - \epsilon_{zz}b^2 \end{bmatrix}$$

Tetragonal 422 crystals of ZY cut (symmetry axis: X) and ZX cut (symmetry axis: Y):

$$\begin{bmatrix} c_{11}+c_{44}b^2-\rho v^2 & 0 & (c_{13}+c_{44})b & 0 \\ 0 & c_{44}b^2+c_{66}-\rho v^2 & 0 & e_{x4}b \\ (c_{13}+c_{44})b & 0 & c_{33}b^2+c_{44}-\rho v^2 & 0 \\ 0 & e_{x4}b & 0 & -\epsilon_{zz}b^2-\epsilon_{xx} \end{bmatrix}$$

Tetragonal 422 crystals of YZ cut (symmetry axis: X) and XZ cut (symmetry axis: Y):

$$\begin{bmatrix} c_{33}+c_{44}b^2-\rho v^2 & 0 & (c_{13}+c_{44})b & 0 \\ 0 & c_{44}+c_{66}b^2-\rho v^2 & 0 & -e_{x4}b \\ (c_{13}+c_{44})b & 0 & c_{11}b^2+c_{44}-\rho v^2 & 0 \\ 0 & -e_{x4}b & 0 & -\epsilon_{xx}b^2-\epsilon_{zz} \end{bmatrix}$$

Cubic 23 and $\bar{4}3$m crystals of XY cut (symmetry axis: Z), ZX cut (symmetry axis: Y) and YZ cut (symmetry axis: X):

$$\begin{bmatrix} c_{11}+c_{44}b^2-\rho v^2 & 0 & (c_{12}+c_{44})b & 0 \\ 0 & c_{44}(1+b^2)-\rho v^2 & 0 & 2e_{x4}b \\ (c_{12}+c_{44})b & 0 & c_{11}b^2+c_{44}-\rho v^2 & 0 \\ 0 & 2e_{x4}b & 0 & -\epsilon_{xx}(1+b^2) \end{bmatrix}$$

Cubic 23 and $\bar{4}3$m crystals of YX cut (symmetry axis: Z), XZ cut (symmetry axis: Y) and ZY cut (symmetry axis: X):

$$\begin{bmatrix} c_{11}+c_{44}b^2-\rho v^2 & 0 & (c_{12}+c_{44})b & 0 \\ 0 & c_{44}(1+b^2)-\rho v^2 & 0 & -2e_{x4}b \\ (c_{12}+c_{44})b & 0 & c_{11}b^2+c_{44}-\rho v^2 & 0 \\ 0 & -2e_{x4}b & 0 & -\epsilon_{xx}(1+b^2) \end{bmatrix}$$

6.5.1.2 Four-fold symmetry

Tetragonal 4mm crystals of XY cut (symmetry axis: Z) and YX cut (symmetry axis: Z):

$$\begin{bmatrix} c_{11}+c_{66}b^2-\rho v^2 & 0 & c_{12}b+c_{66}b & 0 \\ 0 & c_{44}(1+b^2)-\rho v^2 & 0 & e_{x5}(1+b^2) \\ c_{12}b+c_{66}b & 0 & c_{11}b^2+c_{66}-\rho v^2 & 0 \\ 0 & e_{x5}(b^2+1) & 0 & -\epsilon_{xx}(1+b^2) \end{bmatrix}$$

Tetragonal 422 and 4/mmm crystals of XY cut (symmetry axis: Z) and YX cut (symmetry axis: Z):

$$\begin{bmatrix} c_{11}+c_{66}b^2-\rho v^2 & 0 & (c_{12}+c_{66})b & 0 \\ 0 & c_{44}(1+b^2)-\rho v^2 & 0 & 0 \\ (c_{12}+c_{66})b & 0 & c_{11}b^2+c_{66}-\rho v^2 & 0 \\ 0 & 0 & 0 & -\epsilon_{xx}(1+b^2) \end{bmatrix}$$

In these two cuts of Class 422 tetragonal crystals, piezoelectric effect disappears due to crystal symmetry, thus the resulting $[\Gamma]$ matrix is the same as that for Class 4/mmm tetragonal crystals in the same cuts. Also, this matrix can be obtained from the one for Class 4mm crystals after ignoring the piezoelectric constants (see the case above). For the same reason, due to the presence of four-fold symmetry in Class m3m non-piezoelectric cubic crystals, their $[\Gamma]$ matrices will be the same as those of the cubic cases given above after ignoring the piezoelectric constants.

6.5.1.3 Six-fold symmetry

Hexagonal 6mm crystals of XY and YX cuts (symmetry axis: Z):

$$\begin{bmatrix} c_{11}+\frac{(c_{11}-c_{12})b^2}{2}-\rho v^2 & 0 & \dfrac{(c_{11}+c_{12})b}{2} & 0 \\ 0 & c_{44}(1+b^2)-\rho v^2 & 0 & e_{x5}(1+b^2) \\ \dfrac{(c_{11}+c_{12})b}{2} & 0 & c_{11}b^2+\dfrac{c_{11}-c_{12}}{2}-\rho v^2 & 0 \\ 0 & e_{x5}(1+b^2) & 0 & -\epsilon_{xx}(1+b^2) \end{bmatrix}$$

6.5.2 Cuts with *xz*-Plane Parallel with Mirror Planes

For piezoelectric crystals possessing mirror-plane symmetry, when the *xz* plane of a crystal cut is parallel with the mirror plane, the governing equation for SAWs, as given in Eq. (6.13), will take the following form:

$$\begin{bmatrix} \Gamma_{11}-\rho v^2 & 0 & \Gamma_{13} & \gamma_1 \\ 0 & \Gamma_{22}-\rho v^2 & 0 & 0 \\ \Gamma_{13} & 0 & \Gamma_{33}-\rho v^2 & \gamma_3 \\ \gamma_1 & 0 & \gamma_3 & -\epsilon_4 \end{bmatrix} \begin{Bmatrix} {}^\circ u_x \\ {}^\circ u_y \\ {}^\circ u_z \\ {}^\circ \phi \end{Bmatrix} = 0 \qquad (6.23)$$

This points to the fact that SAWs traveling in this type of cut crystals will decouple into two different independent systems of waves, namely,

$$\begin{bmatrix} \Gamma_{11}-\rho v^2 & \Gamma_{13} & \gamma_1 \\ \Gamma_{13} & \Gamma_{33}-\rho v^2 & \gamma_3 \\ \gamma_1 & \gamma_3 & -\epsilon_4 \end{bmatrix} \begin{Bmatrix} {}^\circ u_x \\ {}^\circ u_z \\ {}^\circ \phi \end{Bmatrix} = 0 \qquad (6.24)$$

and

$$[\Gamma_{22} - \rho v^2]\{^\circ u_y\} = 0 \tag{6.25}$$

The waves governed by Eq. (6.24) are piezoelectric stiffened Rayleigh waves, and the wave governed by Eq. (6.25) is a non-piezoelectric TH wave.

We list below some example cases for these situations with the 4×4 $[\Gamma]$ matrix. Cases listed include orthorhombic, trigonal, tetragonal, and hexagonal crystals at the respective cuts specified.

6.5.2.1 Single Mirror Plane

Orthorhombic 2mm crystals of XZ cut (mirror plane: XZ):

$$\begin{bmatrix} c_{33} + c_{55}b^2 - \rho v^2 & 0 & (c_{13} + c_{55})b & e_{x5}b^2 + e_{z3} \\ 0 & c_{44} + c_{66}b^2 - \rho v^2 & 0 & 0 \\ (c_{13} + c_{55})b & 0 & c_{11}b^2 + c_{55} - \rho v^2 & (e_{x5} + e_{z1})b \\ e_{x5}b^2 + e_{z3} & 0 & (e_{x5} + e_{z1})b & -\epsilon_{xx}b^2 - \epsilon_{zz} \end{bmatrix}$$

Orthorhombic 2mm crystals of ZX cut (mirror plane: XZ):

$$\begin{bmatrix} c_{11} + c_{55}b^2 - \rho v^2 & 0 & (c_{13} + c_{55})b & (e_{x5} + e_{z1})b \\ 0 & c_{44}b^2 + c_{66} - \rho v^2 & 0 & 0 \\ (c_{13} + c_{55})b & 0 & c_{33}b^2 + c_{55} - \rho v^2 & e_{z3}b^2 + e_{x5} \\ (e_{x5} + e_{z1})b & 0 & e_{z3}b^2 + e_{x5} & -\epsilon_{zz}b^2 - \epsilon_{xx} \end{bmatrix}$$

Trigonal 3m crystals of YZ cut (mirror plane: YZ):

$$\begin{bmatrix} c_{33} + c_{44}b^2 - \rho v^2 & 0 & c_{44}b + (c_{13} + c_{14}b) & e_{x5}b^2 + e_{z3} \\ 0 & c_{44} - 2c_{14}b + \dfrac{(c_{11} - c_{12})b^2}{2} - \rho v^2 & 0 & 0 \\ c_{14}b^2 + (c_{13} + c_{44})b & 0 & c_{44} + 2c_{14}b + c_{11}b^2 - \rho v^2 & (e_{z1} + e_{x5})b - e_{y2}b^2 \\ e_{x5}b^2 + e_{z3} & 0 & (e_{z1} + e_{x5})b - e_{y2}b^2 & -\epsilon_{xx}b^2 - \epsilon_{zz} \end{bmatrix}$$

Trigonal 3m crystals of ZY cut (mirror plane: YZ):

$$\begin{bmatrix} c_{11} - 2c_{14}b + c_{44}b^2 - \rho v^2 & 0 & (c_{13} + c_{44})b - c_{14} & e_{y2} + (e_{x5} + e_{z1})b \\ 0 & \dfrac{c_{11} - c_{12}}{2} + 2c_{14}b + c_{44}b^2 - \rho v^2 & 0 & 0 \\ (c_{13} + c_{44})b - c_{14} & 0 & c_{33}b^2 + c_{44} - \rho v^2 & e_{z3}b^2 + e_{x5} \\ e_{y2} + (e_{x5} + e_{z1})b & 0 & e_{z3}b^2 + e_{x5} & -\epsilon_{zz}b^2 - \epsilon_{xx} \end{bmatrix}$$

Tetragonal 4mm crystals of XZ cut (mirror plane: XZ):

$$\begin{bmatrix} c_{33}+c_{44}b^2-\rho v^2 & 0 & (c_{13}+c_{44})b & e_{x5}b^2+e_{z3} \\ 0 & c_{44}+c_{66}b^2-\rho v^2 & 0 & 0 \\ (c_{13}+c_{44})b & 0 & c_{11}b^2+c_{44}-\rho v^2 & (e_{x5}+e_{z1})b \\ e_{x5}b^2+e_{z3} & 0 & (e_{x5}+e_{z1})b & -\epsilon_{xx}b^2-\epsilon_{zz} \end{bmatrix}$$

Tetragonal 4mm crystals of ZX cut (mirror plane: XZ):

$$\begin{bmatrix} c_{11}+c_{44}b^2-\rho v^2 & 0 & (c_{13}+c_{44})b & (e_{x5}+e_{z1})b \\ 0 & c_{44}b^2+c_{66}-\rho v^2 & 0 & 0 \\ (c_{13}+c_{44})b & 0 & c_{33}b^2+c_{44}-\rho v^2 & e_{z3}b^2+e_{x5} \\ (e_{x5}+e_{z1})b & 0 & e_{z3}b^2+e_{x5} & -\epsilon_{zz}b^2-\epsilon_{xx} \end{bmatrix}$$

Hexagonal 6mm crystals of XZ cut (mirror plane: XZ):

$$\begin{bmatrix} c_{33}+c_{44}b^2-\rho v^2 & 0 & (c_{13}+c_{44})b & e_{x5}b^2+e_{z3} \\ 0 & \dfrac{(c_{11}-c_{12})b^2}{2}+c_{44}-\rho v^2 & 0 & 0 \\ (c_{13}+c_{44})b & 0 & c_{11}b^2+c_{44}-\rho v^2 & (e_{x5}+e_{z1})b \\ e_{x5}b^2+e_{z3} & 0 & (e_{x5}+e_{z1})b & -\epsilon_{xx}b^2-\epsilon_{zz} \end{bmatrix}$$

Hexagonal 6mm crystals of ZX cut (mirror plane: XZ):

$$\begin{bmatrix} c_{11}+c_{44}b^2-\rho v^2 & 0 & (c_{13}+c_{44})b & (e_{x5}+e_{z1})b \\ 0 & \dfrac{c_{11}/2-c_{12}}{2}+c_{44}b^2-\rho v^2 & 0 & 0 \\ (c_{13}+c_{44})b & 0 & c_{33}b^2+c_{44}-\rho v^2 & e_{z3}b^2+e_{x5} \\ (e_{x5}+e_{z1})b & 0 & e_{z3}b^2+e_{x5} & -\epsilon_{zz}b^2-\epsilon_{xx} \end{bmatrix}$$

6.5.2.2 Multiple Mirror Planes

Tetragonal 4/mmm crystals of ZX cut (mirror plane: XZ) and ZY cut (mirror plane: YZ):

$$\begin{bmatrix} c_{11}+c_{44}b^2-\rho v^2 & 0 & (c_{13}+c_{44})b & 0 \\ 0 & c_{44}b^2+c_{66}-\rho v^2 & 0 & 0 \\ (c_{13}+c_{44})b & 0 & c_{33}b^2+c_{44}-\rho v^2 & 0 \\ 0 & 0 & 0 & -\epsilon_{zz}b^2-\epsilon_{xx} \end{bmatrix}$$

Tetragonal 4/mmm crystals of XZ cut (mirror plane: XZ) and YZ cut (mirror plane: YZ):

$$\begin{bmatrix} c_{33}+c_{44}b^2-\rho v^2 & 0 & (c_{13}+c_{44})b & 0 \\ 0 & c_{44}+c_{66}b^2-\rho v^2 & 0 & 0 \\ (c_{13}+c_{44})b & 0 & c_{11}b^2+c_{44}-\rho v^2 & 0 \\ 0 & 0 & 0 & -\epsilon_{xx}b^2-\epsilon_{zz} \end{bmatrix}$$

Note that the $[\Gamma]$ matrices for the non-piezoelectric Class 4/mmm crystals are the same as those for Class 4mm crystals in the same respective cuts after ignoring the piezoelectric constants.

From the provided examples listed above, it appears that for some piezoelectric crystals, different cuts and propagation directions may lead to different decoupling of surface waves for possible generation of different types of sub-waves using one material. Though in practice, it may not be as simple as making such a judgment based on the appearance of $[\Gamma]$ matrices. Nevertheless, knowing this information is important for picking the right materials for the right applications. In designing SAWs devices to meet certain application criteria, it is important to select not only the right piezoelectric material but also the right crystal cut, because in certain applications, Rayleigh waves may be desirable, but in others, transverse horizontal wave might be a better choice.

6.6 Characterization of SAWs in Piezoelectric Solids

With these most commonly used crystal cuts introduced, we will refer to them in case studies for solving the governing matrix equations given in Eq. (6.14) along with boundary conditions at the surface. To help the reader develop a better understanding, we will go through some quantitative analyses, step-by-step, to solve the phase velocity, displacement fields and potential, and to provide sketches of the obtained results to show how the particle displacements, electric field, and stress fields distribute along the depth direction from the surface. Some of the materials selected for these analyses piezoelectric and some are non-piezoelectric.

6.6.1 Non-Piezoelectric Rayleigh Waves

To show how the concepts discussed earlier are put to use to solve governing equations with gradual ascending complexity, we begin by using crystals in a specific cut such that the generated SAWs will decouple into non-piezoelectric Rayleigh waves and piezoelectric stiffened transverse wave.

Example 4: For hexagonal Class 6mm crystal zinc oxide (ZnO), using the piezoelectric material constants given in Section 4.9 along with the selection of YX cut, i.e., with Euler angles $(0^\circ, -90^\circ, 0^\circ)$, to find the wave velocity, particle displacement and potential field generated by the waves.

Answer: From the discussions in Section 6.5.1, we know that for ZnO in YX-cut application, its governing matrix is reduced into Rayleigh waves as governed by Eq. (6.21) and a piezoelectric stiffened transverse wave by Eq. (6.22). We first examine the Rayleigh waves. Referring to the six-fold symmetry case, with substitution of the material constants given in Section 4.9, we have the governing matrix for Rayleigh waves as follows:

$$10^{10} \times \begin{bmatrix} 4.39b^2 + 20.9 - 5.68 \times 10^{-7}v^2 & -16.51b \\ -16.51b & 20.9b^2 + 4.39 - 5.68 \times 10^{-7}v^2 \end{bmatrix} \begin{Bmatrix} {}^\circ u_x \\ {}^\circ u_z \end{Bmatrix} = 0$$

By setting the following determinant to zero

$$\det\begin{bmatrix} 4.39b^2 + 20.9 - 5.68 \times 10^{-7}v^2 & -16.51b \\ -16.51b & 20.9b^2 + 4.39 - 5.68 \times 10^{-7}v^2 \end{bmatrix} = 0$$

we find two roots with negative imaginary part that satisfy the wave decaying requirement as depth (z) increases (note here that c_{11} and c_{66} are the constants in the original coordinates system)

$$b_1 = -\sqrt{\rho v^2/c_{66} - 1} = -\sqrt{1.29 \times 10^{-7}v^2 - 1}$$

$$b_3 = -\sqrt{\rho v^2/c_{11} - 1} = -\sqrt{2.72 \times 10^{-8}v^2 - 1}$$

Knowing these two b expressions, we can find the two corresponding null polarization vectors

$$\begin{Bmatrix} {}^\circ u_x^1 \\ {}^\circ u_z^1 \end{Bmatrix} = \begin{Bmatrix} 1 \\ \dfrac{-1}{\sqrt{1.29 \times 10^{-7}v^2 - 1}} \end{Bmatrix} \text{ and } \begin{Bmatrix} {}^\circ u_x^3 \\ {}^\circ u_z^3 \end{Bmatrix} = \begin{Bmatrix} 1 \\ \sqrt{2.72 \times 10^{-8}v^2 - 1} \end{Bmatrix}$$

Referring to Eq. (6.15), this type of decoupling allows us to regard $b_2 = b_4 = 0$ and $\alpha_2 = \alpha_4 = 0$; hence, we express the displacement fields as

$$\begin{Bmatrix} u_x \\ u_y \\ u_z \end{Bmatrix} = \begin{Bmatrix} {}^\circ u_x^1 \alpha_1 e^{-ikb_1 z} + {}^\circ u_x^3 \alpha_3 e^{-ikb_3 z} \\ 0 \\ {}^\circ u_z^1 \alpha_1 e^{-ikb_1 z} + {}^\circ u_z^3 \alpha_3 e^{-ikb_3 z} \end{Bmatrix} e^{ik(vt-x)}$$

and with Eq. (6.16), we have

$$\{T\} = \begin{Bmatrix} \sigma_x \\ \sigma_y \\ \sigma_z \\ \tau_{yz} \\ \tau_{xz} \\ \tau_{xy} \end{Bmatrix} = -ik[c_{xyz}]\left\{ \begin{bmatrix} 1 & 0 & 0 \\ 0 & 0 & 0 \\ 0 & 0 & b_1 \\ 0 & b_1 & 0 \\ b_1 & 0 & 1 \\ 0 & 1 & 0 \end{bmatrix} \begin{Bmatrix} {}^{\circ}u_x^1 \alpha_1 e^{-ikb_1 z} \\ 0 \\ {}^{\circ}u_z^1 \alpha_1 e^{-ikb_1 z} \end{Bmatrix} \right.$$

$$\left. + \begin{bmatrix} 1 & 0 & 0 \\ 0 & 0 & 0 \\ 0 & 0 & b_3 \\ 0 & b_3 & 0 \\ b_3 & 0 & 1 \\ 0 & 1 & 0 \end{bmatrix} \begin{Bmatrix} {}^{\circ}u_x^3 \alpha_3 e^{-ikb_3 z} \\ 0 \\ {}^{\circ}u_z^3 \alpha_3 e^{-ikb_3 z} \end{Bmatrix} \right\} e^{ik(vt-x)}$$

Substituting the two b expressions and null polarization vectors into stress vector and applying zero-stress boundary conditions of $\sigma_z = 0$ and $\tau_{xz} = 0$, we can set up two equations in terms of constants α_1 and α_3 as a function of v, as

$$8.79c_1 + 8.79c_3 - 5676 \times 10^{-10} c_3 v^2 = 0$$

and

$$\frac{3.44 \times 10^8 v^2 - 5.32 \times 10^{15}}{\sqrt{473v^2 - 3.66 \times 10^9}} c_1 - \frac{1.645 \times 10^8 v^2 - 6.06 \times 10^{15}}{\sqrt{473v^2 - 3.66 \times 10^9}} c_3 = 0$$

After variable separation by the two constants, we obtain the following 2×2 boundary-satisfying matrix equation:

$$\begin{bmatrix} 8.79 & +8.79 - 5676 \times 10^{-10} v^2 = 0 \\ \dfrac{3.44 \times 10^8 v^2 - 5.32 \times 10^{15}}{\sqrt{473v^2 - 3.66 \times 10^9}} & \dfrac{1.645 \times 10^8 v^2 - 6.06 \times 10^{15}}{\sqrt{473v^2 - 3.66 \times 10^9}} \end{bmatrix} \begin{Bmatrix} c_1 \\ c_3 \end{Bmatrix} = 0$$

By setting the determinant of this boundary-satisfying matrix to zero, we solve for the sole variable, v, and find (note that the two sub-waves have the same phase velocity)

$$v = 2608.68 \text{ m/s}$$

Knowing this phase velocity, we can further find

$$b_1 = -0.348i, b_3 = -0.903i, \alpha_1 = 1, \alpha_3 = -1.784,$$
$$^{\circ}u_x^3 = -2.873i, \ ^{\circ}u_z^3 = -0.903i$$

By substituting these values into the displacement expressions along with normalization with respect to u_z at the surface, we obtain the displacement fields for Rayleigh waves propagating in the YX-cut ZnO as

$$\begin{Bmatrix} u_x \\ u_y \\ u_z \end{Bmatrix} = \begin{Bmatrix} 1.413e^{-5.673z/\lambda} - 0.792e^{-2.187z/\lambda} \\ 0 \\ (2.2756e^{-2.187z/\lambda} - 1.275e^{-5.673z/\lambda})i \end{Bmatrix} e^{ik(vt-x)}$$

Like ZnO, hexagonal Class 6mm crystal cadmium sulfate (CdS) possesses similar behavior and decoupled Rayleigh waves can be generated in YX cut. By taking the same steps, we can characterize Rayleigh waves in CdS. Figure 6.14 shows the magnitudes of the displacements caused by Rayleigh waves in ZnO and CdS. Apparently, aside from the difference in phase velocity values, the profiles of the displacement fields caused by Rayleigh waves in these two materials are very similar.

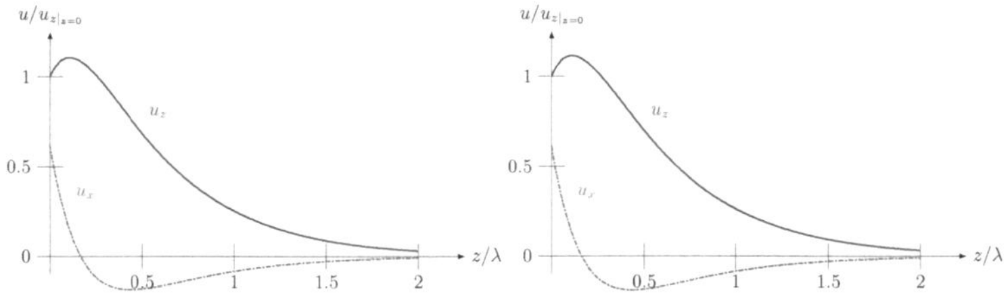

(a) ZnO: v=2608.68 m/s (b) CdS: v=1722.29 m/s

Figure 6.14 Magnitudes of particle displacements caused by Rayleigh waves in YX-cut hexagonal crystals in ZnO and CdS.

6.6.2 Influence of Anisotropy on Rayleigh Waves

Rayleigh waves are known to be influenced by the degree of anisotropy of a material. The *anisotropy factor* given below is typically used as a means to quantify such anisotropy.

$$A = \frac{2c_{66}}{c_{11} - c_{12}} \tag{6.26}$$

The particle displacement profiles of all decoupled Rayleigh waves we have seen so far, including the cases for isotropic material, ZnO and CdS, are of a similar shape. This is so because all these materials have $A = 1$. In the following example, we will see how this anisotropic factor affects the profiles of particle displacements.

Example 5: Use three anisotropic materials, namely, Class $\bar{4}3$m cubic crystal Gallium Arsenide (GaAs), Class m3m cubic crystal Silicon (Si), and Class 4/mmm tetragonal crystal titanium oxide (TiO_2), to show the effect of anisotropy factor on particle displacement profiles induced by Rayleigh waves.

Answer: By the discussion in Section 6.5.1, we know that with the selection of Y-cut X-propagation, the governing matrix equation for SAWs in these crystals will reduce to Eq. (6.21), so all these cases can be solved using the same procedures as for the ZnO case discussed earlier.

For GaAs, the governing matrix, phase velocity of the Rayleigh waves, and the b and α values are, respectively,

$$\det \begin{bmatrix} 5.94 \times 10^{10} b^2 + 11.88 \times 10^{10} - 5307v^2 & -11.32 \times 10^{10} b \\ -11.32 \times 10^{10} b & 11.88e \times 10^{10} b^2 + 5.94 \times 10^{10} - 5307v^2 \end{bmatrix} = 0$$

$$v = 2719.95 \text{ m/s}$$

and

$$b_1 = 0.561 - 0.401i,\ b_3 = -0.561 - 0.401i,$$
$$\alpha_1 = 1,\ \alpha_3 = -0.321 - 0.947i$$

Note that in Examples 1 and 4, b vales are found to be pure negative imaginary numbers, but here these two b values are complex numbers with negative imaginary part.

Based on the discussions we had on the Rayleigh waves for isotropic materials and for ZnO and CdS, we pick the real part of u_x and the imaginary part of u_z to capture the magnitudes of the particle displacements as follows:

$$u_x = 0.844e^{-2.524z/\lambda}\cos(3.527z/\lambda) - 1.178e^{-2.524z/\lambda}\sin(3.527z/\lambda)$$
$$u_z = e^{-2.524z/\lambda}\cos(3.527z/\lambda) + 1.396e^{-2.524z/\lambda}\sin(3.527z/\lambda)$$

For Si, the governing matrix, phase velocity of the Rayleigh waves, the b and α values, and the magnitude of the displacements are

$$\det\begin{bmatrix} 7.95\times 10^{10}b^2 + 16.56\times 10^{10} - 2329v^2 & -14.34\times 10^{10}b \\ -14.34\times 10^{10}b & 16.56e\times 10^{10}b^2 + 7.95\times 10^{10} - 2329v^2 \end{bmatrix} = 0$$

$$v = 4916.36\ \text{m/s}$$

$$b_1 = 0.481 - 0.456i,\ b_3 = -0.481 - 0.456i,\ \alpha_1 = 1,\ \alpha_3 = -0.439 - 0.898i$$

and

$$u_x = 0.815e^{-2.864z/\lambda}\cos(3.021z/\lambda) - 1.307e^{-2.864z/\lambda}\sin(3.021z/\lambda)$$
$$u_z = e^{-2.864z/\lambda}\cos(3.021z/\lambda) + 1.602e^{-2.864z/\lambda}\sin(3.021z/\lambda)$$

Note that the two b values are complex numbers also for Si crystal.

For TiO_2, the governing matrix, phase velocity of the Rayleigh waves, the b and α values, and the magnitude of the displacements are

$$\det\begin{bmatrix} 19.4\times10^{10}b^2+27.3\times10^{10}-4250v^2 & -37.0\times10^{10}b \\ -37.0\times10^{10}b & 27.3\times10^{10}b^2+19.4\times10^{10}-4250v^2 \end{bmatrix}=0$$

$$v = 4238.16 \text{ m/s}$$

$$b_1 = -0.784 - 0.213i,\ b_3 = 0.785 - 0.213i,\ \alpha_1 = 1,\ \alpha_3 = -0.057 + 0.998i$$

and

$$u_x = 0.958e^{-1.339z/\lambda}\cos(4.929z/\lambda) - 1.014e^{-1.339z/\lambda}\sin(4.929z/\lambda)$$
$$u_z = e^{-1.339z/\lambda}\cos(4.929z/\lambda) + 1.059e^{-1.339z/\lambda}\sin(4.929z/\lambda)$$

Again, the two b values are complex numbers for TiO_2 crystal.

Judging from the expressions of these displacement fields, it is apparent that the displacements are not only dictated by exponential decaying functions but also sinusoidal functions, suggesting that the decaying patterns of the displacements possess sinusoidal type of oscillation. The oscillating character in the wave magnitude comes from the fact that the b values for these cases are all complex numbers rather than pure imaginary numbers.

Figure 6.15 shows the sketches of the displacements for all three cases. For a better visual comparison we also include the case of isotropic material discussed in Section 6.2. Clearly, as A increases, the degree of oscillation in the displacements increases. For example, for the isotropic case, $A = 1$, both the u_x and u_z fields show no oscillation, but as A increases to $A = 1.49$ (for Si) and $A = 1.81$ (for GaAs), both displacement fields show slight oscillation with displacement values going across the zero line, and when $A = 4$ (for TiO_2), both displacement fields exhibit strong oscillation jumping back and forth a few times from positive to negative values before they die out with depth.

6.6.3 Piezoelectric Stiffened Bleustein–Gulyaev Waves

As we observed in Example 4, for SAWs traveling in YX-cut ZnO, their governing $[\Gamma]$ matrix can be reduced to two sub-matrices, one

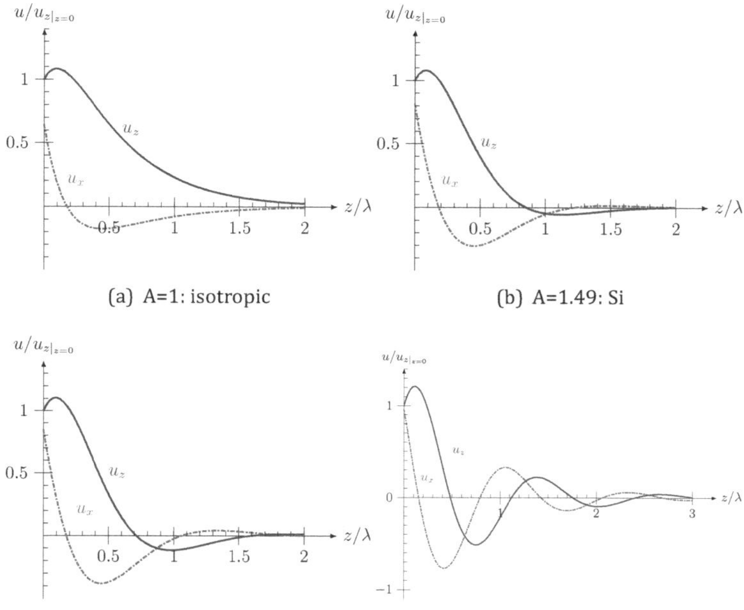

(a) A=1: isotropic (b) A=1.49: Si

(c) A=1.83: GaAs (d) A=4: TiO_2

Figure 6.15 Relative magnitude of displacements under the influence of isotropy factor in various isotropic and anisotropic materials.

governs the non-piezoelectric Rayleigh waves by Eq. (6.21) and the other governs a piezoelectric stiffened transverse wave by Eq. (6.22). This type of piezoelectric stiffened surface transverse wave exists simply because of the piezoelectric coupling effect. It is commonly known as the Bleustein–Gulyaev wave to honor their discovery of this type wave. In the case study provided below we will solve its governing equation to find the wave velocity and examine the displacement and electric fields induced by the wave.

Example 6: Solve the governing Eq. (6.22) for SAWs traveling in the YX-cut ZnO crystal.

Answer: This partial wave is piezoelectric stiffened. Referring again to the six-fold symmetry case listed in Section 6.5.1, with substitution of the known material constants from Section 4.9, we write the following governing equation for the piezoelectric stiffened transverse wave:

$$\begin{bmatrix} 4.25\times 10^{10}b^2 - 5676v^2 + 4.25\times 10^{10} & -0.48(b^2+1) \\ -0.48(b^2+1) & -8.55\times 10^{-11}(b^2+1) \end{bmatrix} \begin{Bmatrix} {}^{\circ}u_y \\ {}^{\circ}\phi \end{Bmatrix} = 0$$

Apparently, this wave polarizes in the y direction, a TH wave, formed by the piezoelectric coupling effect. By setting the following determinant to zero

$$\det \begin{bmatrix} 4.25\times 10^{10}b^2 - 5676v^2 + 4.25\times 10^{10} & -0.48(b^2+1) \\ -0.48(b^2+1) & -8.55\times 10^{-11}(b^2+1) \end{bmatrix} = 0$$

we find two roots with negative imaginary part that satisfy the wave decaying requirement as depth (z) increases

$$b_2 = -1.0i,\ \ b_4 = -\sqrt{\rho v^2/(c_{44} + e_{x5}^2/\epsilon_{xx}) - 1} = -\sqrt{1.26\times 10^{-7}v^2 - 1}$$

Knowing these two b expressions, we find their corresponding null polarization vectors as

$$\begin{Bmatrix} {}^{\circ}u_y^2 \\ {}^{\circ}\phi^2 \end{Bmatrix} = \begin{Bmatrix} 0 \\ 1 \end{Bmatrix} \quad \text{and} \quad \begin{Bmatrix} {}^{\circ}u_y^4 \\ {}^{\circ}\phi^4 \end{Bmatrix} = \begin{Bmatrix} -1.78\times 10^{-10} \\ 1 \end{Bmatrix}$$

By regarding $b_1 = b_3 = 0$ and $\alpha_1 = \alpha_3 = 0$ in Eq. (6.15), we express the displacement and potential fields as

$$\begin{Bmatrix} u_x \\ u_y \\ u_z \end{Bmatrix} = \begin{Bmatrix} 0 \\ {}^{\circ}u_y^2\alpha_2 e^{-ikb_2z} + {}^{\circ}u_y^4\alpha_4 e^{-ikb_4z} \\ 0 \end{Bmatrix} e^{ik(vt-x)}$$

and

$$\{\phi\} = ({}^{\circ}\phi^2\alpha_2 e^{-ikb_2z} + {}^{\circ}\phi^4\alpha_4 e^{-ikb_4z})e^{ik(vt-x)}$$

By referring to Eq. (6.16), we write

$$\{T\} = \begin{Bmatrix} \sigma_x \\ \sigma_y \\ \sigma_z \\ \tau_{yz} \\ \tau_{xz} \\ \tau_{xy} \end{Bmatrix} = -ik[c_{xyz}]\left\{ \begin{bmatrix} 1 & 0 & 0 \\ 0 & 0 & 0 \\ 0 & 0 & b_2 \\ 0 & b_2 & 0 \\ b_2 & 0 & 1 \\ 0 & 1 & 0 \end{bmatrix} \begin{Bmatrix} 0 \\ \circ u_y^2 \alpha_2 e^{-ikb_2 z} \\ 0 \end{Bmatrix} \right.$$

$$\left. + \begin{bmatrix} 1 & 0 & 0 \\ 0 & 0 & 0 \\ 0 & 0 & b_4 \\ 0 & b_4 & 0 \\ b_4 & 0 & 1 \\ 0 & 1 & 0 \end{bmatrix} \begin{Bmatrix} 0 \\ \circ u_y^4 \alpha_4 e^{-ikb_4 z} \\ 0 \end{Bmatrix} \right\} e^{ik(vt-x)}$$

$$- ik[e_{xyz}]^T \left\{ \begin{Bmatrix} 1 \\ 0 \\ b_2 \end{Bmatrix} \circ \phi^2 \alpha_2 e^{-ikb_2 z} + \begin{Bmatrix} 1 \\ 0 \\ b_4 \end{Bmatrix} \circ \phi^4 \alpha_4 e^{-ikb_4 z} \right\} e^{ik(vt-x)}$$

and Eq. (6.17), we have

$$\{D\} = \begin{Bmatrix} D_x \\ D_y \\ D_z \end{Bmatrix} = -ik[e_{xyz}]\left\{ \begin{bmatrix} 1 & 0 & 0 \\ 0 & 0 & 0 \\ 0 & 0 & b_2 \\ 0 & b_2 & 0 \\ b_2 & 0 & 1 \\ 0 & 1 & 0 \end{bmatrix} \begin{Bmatrix} 0 \\ \circ u_y^2 \alpha_2 e^{-ikb_2 z} \\ 0 \end{Bmatrix} \right.$$

$$\left. + \begin{bmatrix} 1 & 0 & 0 \\ 0 & 0 & 0 \\ 0 & 0 & b_4 \\ 0 & b_4 & 0 \\ b_4 & 0 & 1 \\ 0 & 1 & 0 \end{bmatrix} \begin{Bmatrix} 0 \\ \circ u_y^4 \alpha_4 e^{-ikb_4 z} \\ 0 \end{Bmatrix} \right\} e^{ik(vt-x)}$$

$$+ ik[\epsilon_{xyz}]^T \left\{ \begin{Bmatrix} 1 \\ 0 \\ b_2 \end{Bmatrix} \circ \phi^2 \alpha_2 e^{-ikb_2 z} + \begin{Bmatrix} 1 \\ 0 \\ b_4 \end{Bmatrix} \circ \phi^4 \alpha_4 e^{-ikb_4 z} \right\} e^{ik(vt-x)}$$

From these expressions we pick out and determine the two relevant terms

$$\tau_{xz} = -0.48ic_2 - 8.05c_4\sqrt{1.26 \times 10^{-7}v^2 - 1}$$

and

$$D_z = -8.55ic_2 \times 10^{-11}$$

From the ϕ expression along with the null vectors, we know that the potential at the surface can be expressed as

$$\phi|_{z=0} = c_2 + c_4$$

To consider metallized surface with zero-stresses, we have the following boundary conditions:

$$\tau_{xz}|_{z=0} = 0, \ \phi|_{z=0} = 0$$

After applying these boundary conditions we obtain two boundary-satisfying equations and with further variable separation we set the following determinant of the boundary-satisfying matrix to zero, as

$$\det\begin{bmatrix} -0.48i & -1.04\sqrt{7.50 \times 10^{-6}v^2 - 59.73} \\ 1 & 1 \end{bmatrix} = 0$$

Solving it for phase velocity we find

$$v = 2816.76 \text{ m/s}$$

Knowing this phase velocity, we further find

$$b_2 = -i, b_4 = -0.0596i, \alpha_2 = 1, \alpha_4 = -1$$

By substituting these values into the displacement and potential expressions we obtain

$$\begin{Bmatrix} u_x \\ u_y \\ u_z \end{Bmatrix} = \begin{Bmatrix} 0 \\ 0.178 \times 10^{-9}e^{-0.374z/\lambda} \\ 0 \end{Bmatrix} e^{ik(vt-x)}$$

and

$$\phi = e^{-6.28z/\lambda} - e^{-0.375z/\lambda}$$

Figure 6.16(a) sketches the obtained displacement and potential induced by this piezoelectric stiffened TH wave in YX-cut ZnO with metallized surface.

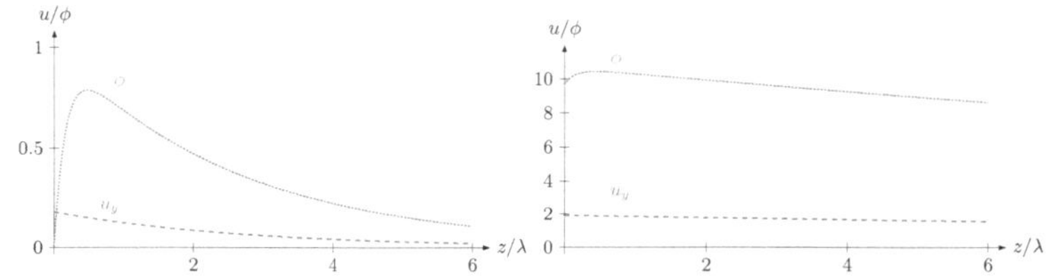

(a) Metallized surface (v = 2816.67 m/s) (b) Free-floating surface (v = 2821.73 m/s)

Figure 6.16 Displacement and potential generated by the TH wave in YX-cut ZnO with metallized surface and free-floating surface.

To consider an electrically free-floating surface with zero-stresses, we have the following boundary conditions:

$$\tau_{xz}|_{z=0} = 0, \;\; D_z|_{z=0} = \epsilon_{\text{vacuum}} k\phi|_{z=0}$$

These two boundary conditions will lead us to two equations and after further variable separation as we did before we set the determinant of the boundary-satisfying matrix to zero, as

$$\det \begin{bmatrix} -0.48i & -1.04\sqrt{7.50 \times 10^{-6}v^2 - 59.73} \\ 9.44 & 0.88 \end{bmatrix} = 0$$

to solve for phase velocity. Doing so we find

$$v = 2821.73 \text{ m/s}$$

Knowing v, we further determine

$$b_2 = -i, \; b_4 = -0.0055951i, \; \alpha_2 = 1, \; \alpha_4 = -10.657$$

By substituting all the obtained values into the displacement and potential expressions we obtain

$$\begin{Bmatrix} u_x \\ u_y \\ u_z \end{Bmatrix} = \begin{Bmatrix} 0 \\ 0.189 \times 10^{-9} e^{-0.035z/\lambda} \\ 0 \end{Bmatrix} e^{ik(vt-x)}$$

and

$$\phi = e^{-6.28z/\lambda} - 10.657e^{-0.035z/\lambda}$$

Figure 6.16(b) sketches the obtained displacement and potential induced by this piezoelectric stiffened TH wave in YX-cut ZnO with electrically free-floating surface. Note that since determining the actual displacement and potential values would require more information, in both cases, the displacements are just normalized by a factor of 10^{-9} and the potential is plotted as is. Doing so, the reader can have a relative sense of how a grounded metallization surface will affect the particle displacements and potential generation in the material when compared with an electrically free-floating surface. It is apparent that the difference in the two cases and fairly drastic in that the magnitudes of both the displacements and potential are significantly reduced in the metallized case.

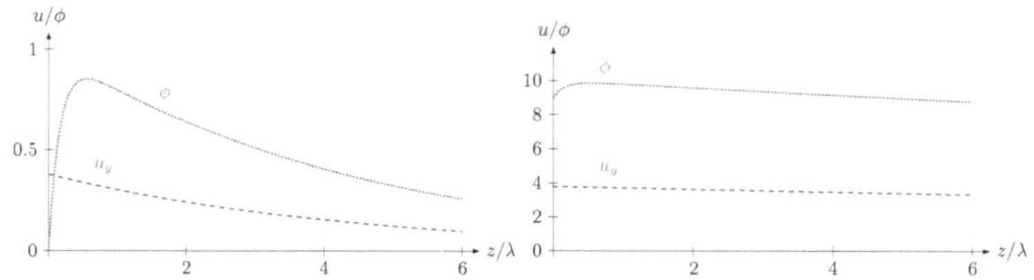

(a) Metallized surface (v = 1788.59 m/s) (b) Free-floating surface (v = 1789.72 m/s)

Figure 6.17 Displacement and potential generated by the TH wave in YX-cut CdS with metallized surface and free-floating surface.

Like ZnO, CdS possesses similar characteristics. Figure 6.17 shows the magnitudes of the particle displacement and potential generated by the piezoelectric stiffened TH wave in YX-cut CdS with metallized and free-floating surface, respectively.

6.6.4 Piezoelectric Stiffened Rayleigh Waves

When piezoelectric crystals possess mirror-plane symmetry and when the *xz* plane of a crystal cut is parallel with the mirror plane, the governing Eq. (6.23) will decouple into piezoelectric stiffened Rayleigh waves governed by

$$\begin{bmatrix} \Gamma_{11}-\rho v^2 & \Gamma_{13} & \gamma_1 \\ \Gamma_{13} & \Gamma_{33}-\rho v^2 & \gamma_3 \\ \gamma_1 & \gamma_3 & -\epsilon_4 \end{bmatrix} \begin{Bmatrix} {}^\circ u_x \\ {}^\circ u_z \\ {}^\circ \phi \end{Bmatrix} = 0$$

and a non-piezoelectric TH wave governed by

$$[\Gamma_{22}-\rho v^2]\{{}^\circ u_y\} = 0$$

The YZ cut ZY selections of lithium niobate ($LiNbO_3$) are two of one such cases.

Example 7: Examine the piezoelectric stiffened Rayleigh waves and the non-piezoelectric TH wave traveling in the ZY-cut $LiNbO_3$ crystal.

Answer: From the discussions in Section 6.5.2, we know that the governing matrix equation for SAWs in ZY-cut $LiNbO_3$ will decouple into two separate sets of waves, one set is the piezoelectric stiffened Rayleigh waves governed by Eq. (6.24) and the other set is a non-piezoelectric bulk TH wave governed by Eq. (6.25).

For the piezoelectric stiffened Rayleigh waves, referring to the single mirror plane (trigonal 3m, YZ) case, with substitution of the material constants given in Section 4.9, we write the governing equation as follows:

$$\begin{bmatrix} 24.5 + 6b^2 - 4.7 \times 10^{-7} v^2 & 0.9b^2 - 13.5b & (3.7b^2 + 1.3) \times 10^{-10} \\ 0.9b^2 - 13.5b & 6 + 20.3b^2 - 1.8b - 4.7 \times 10^{-7} v^2 & -(3.9b + 2.5b^2) \times 10^{-10} \\ (3.7b^2 + 1.3) \times 10^{-10} & -(3.9b + 2.5b^2) \times 10^{-10} & -(3.89b^2 + 2.57) \times 10^{-20} \end{bmatrix} \begin{Bmatrix} {}^{\circ}u_x \\ {}^{\circ}u_z \\ {}^{\circ}\phi \end{Bmatrix} = 0$$

By setting the determinant of the above matrix to zero, we can find three b expressions and three null polarization vectors in terms of phase velocity (v). With substitution of these b and null vector expressions into Eq. (6.15), we write the displacement fields and potential as

$$\begin{Bmatrix} u_x \\ u_y \\ u_z \end{Bmatrix} = \begin{Bmatrix} u_x^1 \alpha_1 e^{-ikb_1 z} + {}^{\circ}u_x^3 \alpha_3 e^{-ikb_3 z} + {}^{\circ}u_x^4 \alpha_4 e^{-ikb_4 z} \\ 0 \\ {}^{\circ}u_z^1 \alpha_1 e^{-ikb_1 z} + {}^{\circ}u_z^3 \alpha_3 e^{-ikb_3 z} + {}^{\circ}u_z^4 \alpha_4 e^{-ikb_4 z} \end{Bmatrix} e^{ik(vt-x)}$$

$$\{\phi\} = [{}^{\circ}\phi^1 \alpha_1 e^{-ikb_1 z} + {}^{\circ}\phi^3 \alpha_3 e^{-ikb_3 z} + {}^{\circ}\phi^4 \alpha_4 e^{-ikb_4 z}] e^{ik(vt-x)}$$

By Eq. (6.16), we can determine the stresses as

$$\{T\} = \begin{Bmatrix} \sigma_x \\ \sigma_y \\ \sigma_z \\ \tau_{yz} \\ \tau_{xz} \\ \tau_{xy} \end{Bmatrix} = -ik[c_{xyz}]\left\{ \begin{bmatrix} 1 & 0 & 0 \\ 0 & 0 & 0 \\ 0 & 0 & b_1 \\ 0 & b_1 & 0 \\ b_1 & 0 & 1 \\ 0 & 1 & 0 \end{bmatrix} \begin{Bmatrix} {}^{\circ}u_x^1 \alpha_1 e^{-ikb_1 z} \\ 0 \\ {}^{\circ}u_z^1 \alpha_1 e^{-ikb_1 z} \end{Bmatrix} \right.$$

$$+ \begin{bmatrix} 1 & 0 & 0 \\ 0 & 0 & 0 \\ 0 & 0 & b_3 \\ 0 & b_3 & 0 \\ b_3 & 0 & 1 \\ 0 & 1 & 0 \end{bmatrix} \begin{Bmatrix} {}^{\circ}u_x^3 \alpha_3 e^{-ikb_3 z} \\ 0 \\ {}^{\circ}u_z^3 \alpha_3 e^{-ikb_3 z} \end{Bmatrix} + \left. \begin{bmatrix} 1 & 0 & 0 \\ 0 & 0 & 0 \\ 0 & 0 & b_4 \\ 0 & b_4 & 0 \\ b_4 & 0 & 1 \\ 0 & 1 & 0 \end{bmatrix} \begin{Bmatrix} {}^{\circ}u_x^4 \alpha_4 e^{-ikb_4 z} \\ 0 \\ {}^{\circ}u_z^4 \alpha_4 e^{-ikb_4 z} \end{Bmatrix} \right\} e^{ik(vt-x)}$$

$$- ik[e_{xyz}]^T \left\{ \begin{Bmatrix} 1 \\ 0 \\ b_1 \end{Bmatrix} {}^{\circ}\phi^1 \alpha_1 e^{-ikb_1 z} + \begin{Bmatrix} 1 \\ 0 \\ b_3 \end{Bmatrix} {}^{\circ}\phi^3 \alpha_3 e^{-ikb_3 z} \right.$$

$$\left. + \begin{Bmatrix} 1 \\ 0 \\ b_4 \end{Bmatrix} {}^{\circ}\phi^4 \alpha_4 e^{-ikb_4 z} \right\} e^{ik(vt-x)}$$

and by Eq. (6.17), the electrical displacements

$$\{D\} = \begin{Bmatrix} D_x \\ D_y \\ D_z \end{Bmatrix} = -ik[e_{xyz}]\left\{ \begin{bmatrix} 1 & 0 & 0 \\ 0 & 0 & 0 \\ 0 & 0 & b_1 \\ 0 & b_1 & 0 \\ b_1 & 0 & 1 \\ 0 & 1 & 0 \end{bmatrix} \begin{Bmatrix} {}^{\circ}u_x^1 \alpha_1 e^{-ikb_1 z} \\ 0 \\ {}^{\circ}u_z^1 \alpha_1 e^{-ikb_1 z} \end{Bmatrix} \right.$$

$$+ \begin{bmatrix} 1 & 0 & 0 \\ 0 & 0 & 0 \\ 0 & 0 & b_3 \\ 0 & b_3 & 0 \\ b_3 & 0 & 1 \\ 0 & 1 & 0 \end{bmatrix} \begin{Bmatrix} {}^{\circ}u_x^3 \alpha_3 e^{-ikb_3 z} \\ 0 \\ {}^{\circ}u_z^3 \alpha_3 e^{-ikb_3 z} \end{Bmatrix} + \left. \begin{bmatrix} 1 & 0 & 0 \\ 0 & 0 & 0 \\ 0 & 0 & b_4 \\ 0 & b_4 & 0 \\ b_4 & 0 & 1 \\ 0 & 1 & 0 \end{bmatrix} \begin{Bmatrix} {}^{\circ}u_x^4 \alpha_4 e^{-ikb_4 z} \\ 0 \\ {}^{\circ}u_z^4 \alpha_4 e^{-ikb_4 z} \end{Bmatrix} \right\} e^{ik(vt-x)}$$

$$+ ik[\epsilon_{xyz}]^T \left\{ \begin{Bmatrix} 1 \\ 0 \\ b_1 \end{Bmatrix} {}^{\circ}\phi^1 \alpha_1 e^{-ikb_1 z} + \begin{Bmatrix} 1 \\ 0 \\ b_3 \end{Bmatrix} {}^{\circ}\phi^3 \alpha_3 e^{-ikb_3 z} \right.$$

$$\left. + \begin{Bmatrix} 1 \\ 0 \\ b_4 \end{Bmatrix} {}^{\circ}\phi^4 \alpha_4 e^{-ikb_4 z} \right\} e^{ik(vt-x)}$$

For metallized surface, after plugging in the three obtained b expressions and applying the boundary conditions given in Eq. (6.19), we can set up a 3×3 boundary-satisfying matrix equation. By letting the determinant of this boundary-satisfying matrix to zero we find the phase velocity of the waves as

$$v = 3403.72 \text{ m/s}$$

With that we determine

$$b_1 = -0.40 - 0.78i,\ b_3 = 0.07 - 0.18i,\ b_4 = 0.38 - 1.05i$$

and

$$\alpha_1 = 1,\ \alpha_3 = -0.48 - 0.18i,\ \alpha_4 = 0.59 + 1.34i$$

After plugging these determined values into the displacement and stress expressions, we obtain the displacement fields, the potential, and stresses induced by piezoelectric stiffened Rayleigh waves in YZ-cut $LiNbO_3$ with metallized surface. Figure 6.18(a) shows the magnitudes of the displacements and potential as a function depth for the matallized case and Figure 6.18(b) shows the relevant stresses.

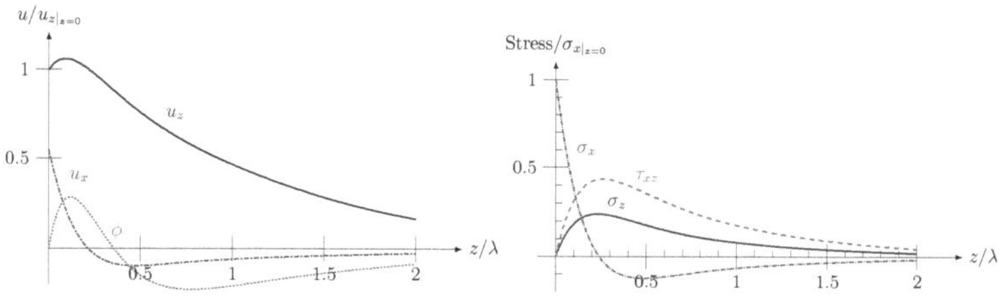

(a) Displacements and potential (b) Stresses

Figure 6.18 Displacements and potential and stresses in YZ-cut $LiNbO_3$ (v = 3403.72 m/s) with metallized surface.

For free-floating surface, with the three obtained b expressions and the boundary conditions given in Eq. (6.18), we can also set up a 3×3 boundary-satisfying matrix equation. Letting the determinant of this boundary-satisfying matrix to zero we find the phase velocity as

$$v = 3487.76 \text{ m/s}$$

along with

$$b_1 = -0.39 - 0.77i, \; b_3 = 0.06 - 0.12i, \; b_4 = 0.38 - 1.04i$$

and

$$\alpha_1 = 1, \; \alpha_3 = -0.22 - 0.08i, \; \alpha_4 = 0.03 + 1.25i$$

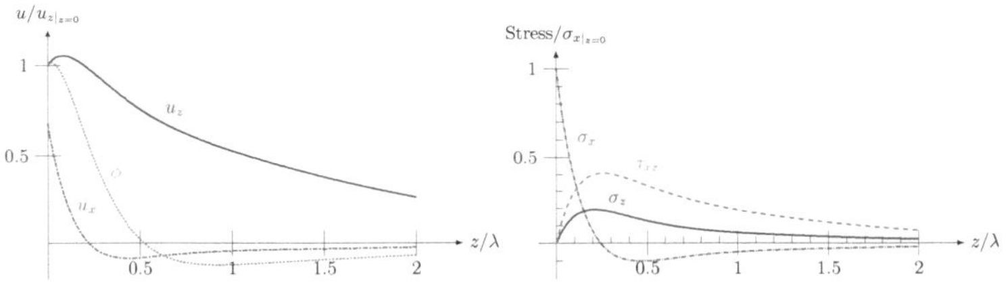

(a) Displacements and potential (b) Stresses

Figure 6.19 Displacements and potential and stresses in YZ-cut $LiNbO_3$ (v = 3487.76 m/s) with free-floating surface.

With these values, we obtain the displacement fields, the potential, and stresses induced by piezoelectric stiffened Rayleigh waves in Y-cut Z-propagation $LiNbO_3$ with a free-floating surface. Figure 6.19(a) shows the magnitudes of the displacements and potential as a function depth for the free-floating surface case and Figure 6.19(b) shows the relevant stresses. By comparing the cases with metallized surface and a free-floating surface, we can see a major difference in the potential profiles: the potential is grounded (zero) at the surface for the metallized case and it floats high in the other case. Moreover, the u_z field decays a bit faster in the metallized case than in the free-floating case, while the difference in the u_x field is less obvious. For stresses, the profile of σ_x appears to be similar, but the peak values of σ_z and τ_{xz} are slightly higher in the metallized case than the free-floating case.

Next we examine the non-piezoelectric TH wave. With the given material property constants, the governing equation for this wave is expressed

$$(7.5b^2 + 1.8b + 6) \times 10^{10} - 4700v^2)\{^\circ u_y\} = 0$$

Like in the isotropic material case, this wave has one b value and with only one component ($^\circ u_y^2$) in its null polarization vector; hence, we express its wave field as

$$\begin{Bmatrix} u_x \\ u_y \\ u_z \end{Bmatrix} = \begin{Bmatrix} 0 \\ ^\circ u_y^2 \alpha_2 e^{-ikb_2 z} \\ 0 \end{Bmatrix} e^{ik(vt-x)}$$

in which α_2 is a constant. Applying the zero-stress condition, $\tau_{yz} = 0$, we arrive at

$$b = 0$$

This means that this decoupled surface TH wave in $LiNbO_3$ is actually a bulk T(||) wave. With $b = 0$, we find its phase velocity from the governing equation as follows:

$$v = \sqrt{\frac{c_{44}}{\rho}} = \sqrt{\frac{6 \times 10^{10}}{4700}} = 3572.95 \text{ m/s}$$

6.7 Fully Coupled SAWs in Piezoelectric Solids

We have now seen that SAWs traveling in piezoelectric crystals can be decoupled into non-piezoelectric Rayleigh waves and piezoelectric TH wave, or piezoelectric stiffened Rayleigh waves and non-piezoelectric bulk TH wave, depending on the selections of materials and cut orientations. For instances, examples for the former situation include cases of ZnO and CdS in YX-cut selections, and examples for the latter situation include the case of $LiNbO_3$ in YZ-cut selection. In practical applications of piezoelectric crystals, however, if we are limited to these specific cut orientations, we will not be able to unleash the hidden power of numerous piezoelectric materials. Overcoming this challenge requires us to deal with fully coupled SAWs in various piezoelectric crystal materials.

The theoretical foundation for finding fully coupled surface acoustic waves has been provided in Section 6.3. In such a case, we need to consider all four components, namely, u_x, u_y, u_z and ϕ representing the L wave, the TH wave and the TV wave, respectively, and the corresponding potential field. Due to the mathematical complexity caused by the four components coupled together, solving the governing equations for finding the phase velocity of SAWs is challenging. A practical approach is to devise a numerical programming routine to search for the phase velocity of the SAWs numerically. Typically for a given cut angle and material, by setting the determinant of a 4×4 boundary-satisfying matrix equation to zero after substitutions of the resolved displacement and electric field expressions obtained from solving the governing equation given in (6.14), the search begins by starting with an initial estimate value for the phase velocity. Since the phase velocity of SAWs is typically less than that of the slowest bulk wave in the material, this phase velocity can be used as the initial value for the search. In an iterative searching process with decremental estimate value in which the decrement is determined by the precision needs and computation power, a velocity value that makes the determinant zero will be regarded as the phase velocity. In this section, we will present some of the results obtained using such a searching method for the phase velocity of surface acoustic waves in various common crystal cuts. The angles shown in the results provided in the subsequent subsections are the rotation angles illustrated in each of the six cuts shown in Figure 6.20. With a known rotation angle, the corresponding Euler angles of a specific crystal cut can be determined using the expression given in each of the six sub-figures, where a crystal slab is positioned in reference to both the fixed XYZ coordinates and the rotated *xyz* coordinates. Moreover, recalling the discussions in Section 5.3, SAWs behave the same way as BAWs in that the phase velocity remains the same if we rotate either +180° or -180° from a preselected orientation. Because of this, only results from 0° to 180° are provided. For angles beyond this range, we can either add or subtract 180° to determine.

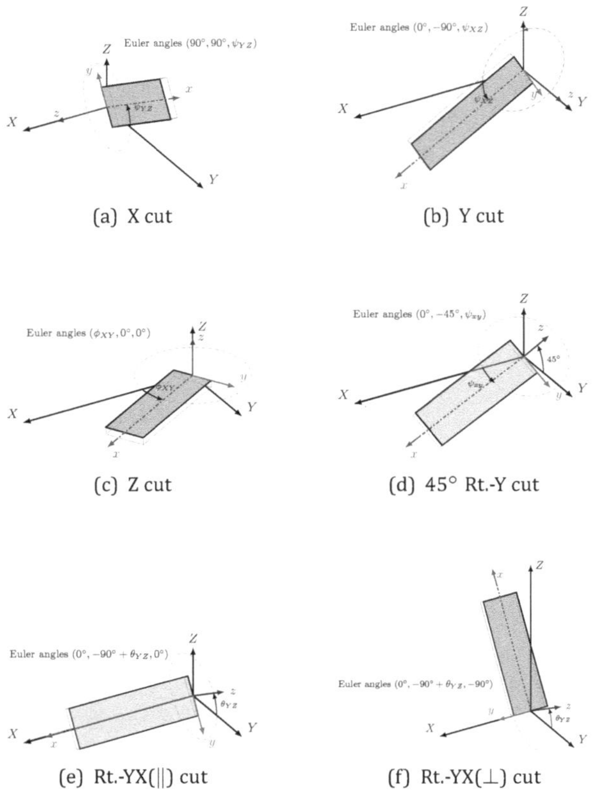

Figure 6.20 Six common crystal cuts with rotation angles illustrated.

6.7.1 Phase Velocity of SAWs in $LiTaO_3$

Figure 6.21 shows six panels of results for the phase velocity of SAWs found in trigonal Class 3m crystal, lithium tantalate ($LiTaO_3$), in the six common cuts illustrated in Figure 6.20, namely, the X, Y, Z, 45° Rt.-Y, Rt.-YX(||) and Rt.-YX(⊥) cuts. In these results, both cases of metallized surface and free-floating surface are provided. The obvious difference in each cut between the two cases show a strong piezoelectric effect in $LiTaO_3$. Note that in the Rt.-YX(||) cut (Figure 6.21(e)) the cut-off feature in the curve is due to the interference from the slower transverse bulk wave. Referring to Figure 5.22, we can estimate from both the XY and XZ sections

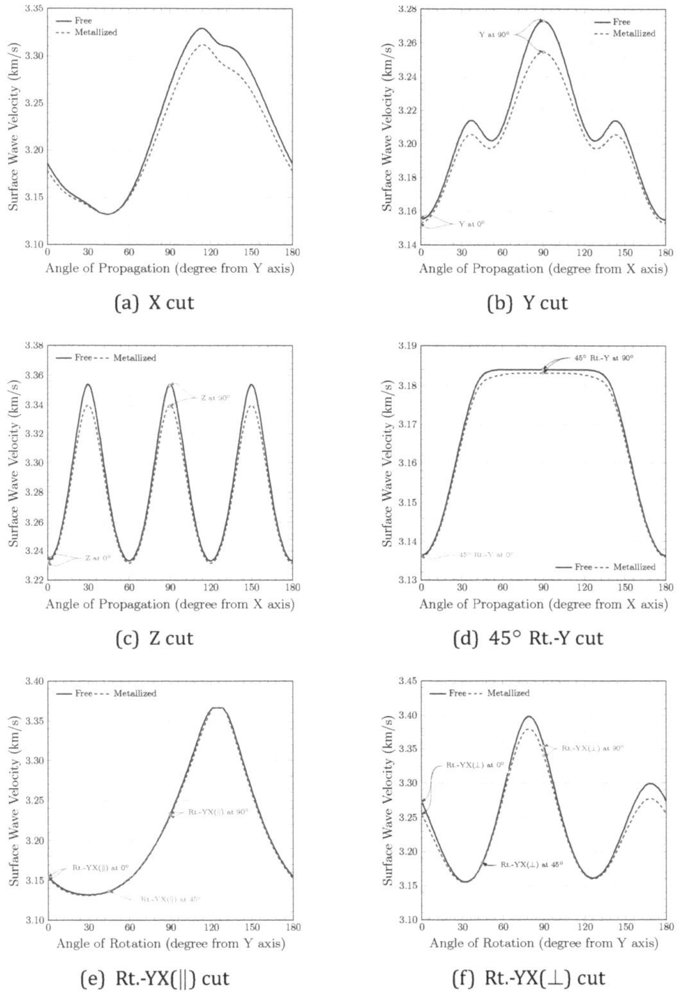

Figure 6.21 Phase velocity of SAWs propagating in $LiTaO_3$ in six common crystal cuts.

that the phase velocity of the slowest T wave propagating in the X direction is at this velocity.

By referring to the discussions in Section 6.4, the same crystal orientations in different cuts are marked for matching comparison in these figures. For example, the phase velocity of the Rt.-YX(||) cut at 0°, 45° and 90° in Figure 6.21(e) matches exactly the Y cut at 0° in Figure 6.21(b), the 45° Rt.-Y cut at 0° in Figure 6.21(d) and the Z cut at 0° in Figure 6.21(c), respectively. Figure 6.22 provides an illustrative look of the corresponding crystal orientations under these cuts: Figures 6.22(a) and 6.22(b), 6.22(c) and 6.22(d), and 6.22(e) and 6.22(f), respectively, show side-by-side the identical crystal orientations in each of these three cuts.

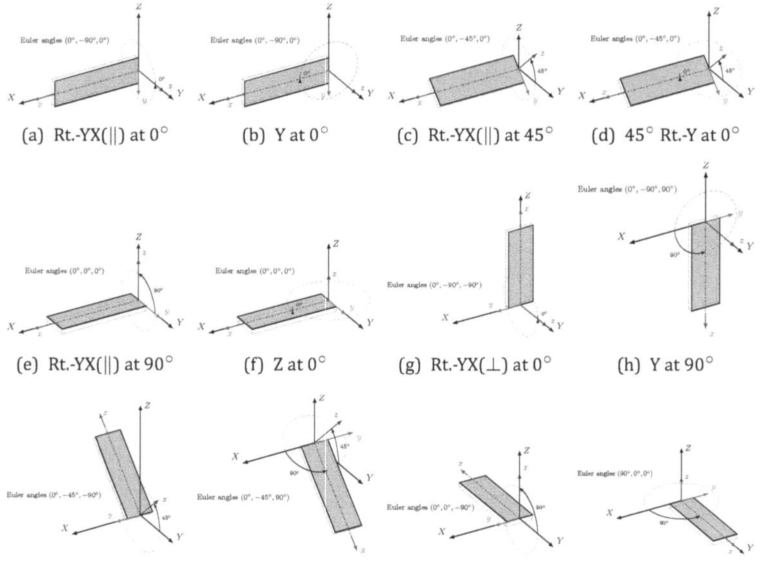

(a) Rt.-YX(||) at 0° (b) Y at 0° (c) Rt.-YX(||) at 45° (d) 45° Rt.-Y at 0°
(e) Rt.-YX(||) at 90° (f) Z at 0° (g) Rt.-YX(⊥) at 0° (h) Y at 90°
(i) Rt.-YX(⊥) at 45° (j) 45° Rt.-Y at 90° (k) Rt.-YX(⊥) at 90° (l) Z at 90°

Figure 6.22 Six pairs of same crystal orientations obtained from different cuts.

Similarly, the phase velocity of the Rt.-YX(⊥) cut at 0°, 45° and 90° in Figure 6.21(f) matches exactly the Y cut at 90° in Figure 6.21(b), the 45° Rt.-Y cut at 90° in Figure 6.21(d) and the Z cut at 90° in Figure 6.21(c), respectively. The side-by-side illustrations provided in Figures 6.22(g) and 6.22(h), 6.22(i) and 6.22(j), and 6.22(k) and 6.22(l), respectively, show the crystal orientations in these cuts. Unlike in the previous three cases where identical crystal orientations and wave propagation directions are seen, here although the crystal orientations are the same, the directions of the wave propagation differ by 180°. Recalling the discussion that the phase velocity values repeat themselves after a rotation by either 180° clockwise (−180°) or counterclockwise (+180°), we know that once a crystal cut is selected, it makes no difference whether the x direction or the $-x$ direction is used as the wave propagation direction, though some elements of the material properties may differ by either positive or negative sign (see Examples 8 and 9 in Chapter 3).

6.7.2 Phase Velocity of SAWs in $LiNbO_3$

Now we take a look of one of the most commonly used piezoelectric materials: trigonal Class 3m $LiNbO_3$. Figures 6.23, 6.24 and 6.25 show the phase velocity of SAWs in the X, Y, Z, cuts, respectively. In these results, both cases of metallized surface and free-floating surface are provided. Comparing with $LiTaO_3$, the differences between the metallized and free-floating surface cases in $LiNbO_3$ are much higher. This means that $LiNbO_3$ possesses even stronger piezoelectric effect than $LiTaO_3$, confirming what we observed for BAWs in these materials. The results in these three cuts seem to agree well with the results initially reported by Campbell and Jones (*IEEE Transactions of Sonics and Ultrasonics*, Vol. SU-15, No. 4, 1968).

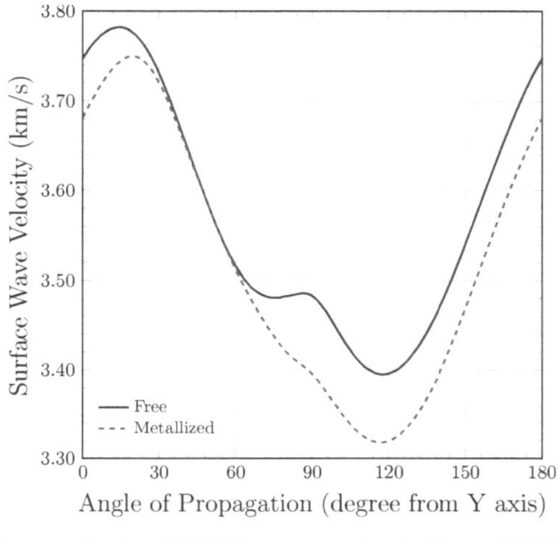

Figure 6.23 Phase velocity of SAWs propagating in X-cut $LiNbO_3$.

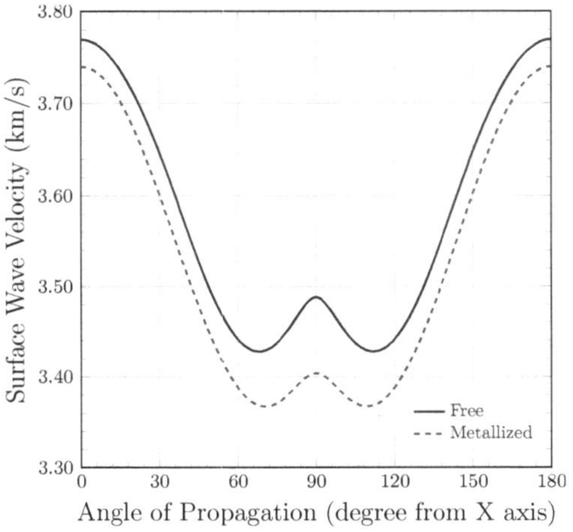

Figure 6.24 Phase velocity of SAWs propagating in Y-cut $LiNbO_3$.

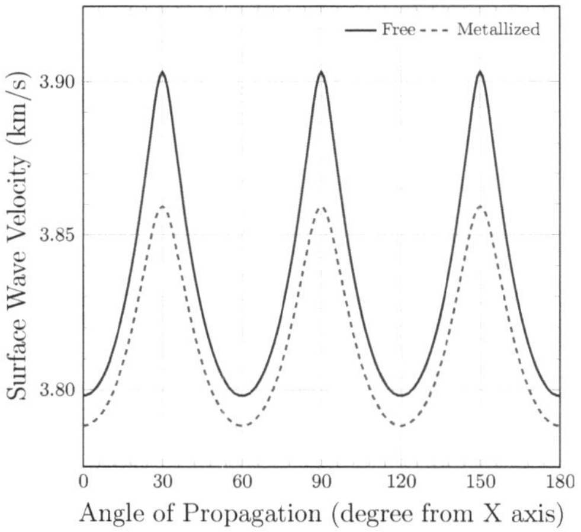

Figure 6.25 Phase velocity of SAWs propagating in Z-cut $LiNbO_3$.

Figure 6.26 shows the phase velocity results for SAWs in 45° Rt.-Y cut, Figure 6.27 for the Rt.-YX(||) cut, and Figure 6.28 the Rt.-YX($\perp$) propagation. Apparently, the difference between the metallized and free-floating cases in the Rt.-YX(||) cut in Figure 6.27 is very large at around 131.5°, suggesting that in this cut angle, $LiNbO_3$ will have a very high piezoelectric coupling effect.

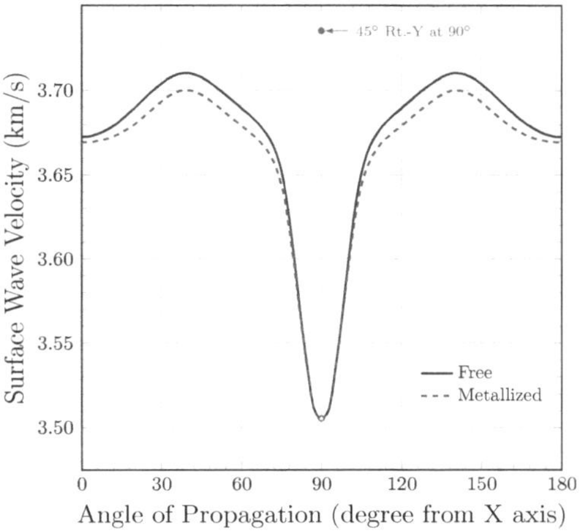

Figure 6.26 Phase velocity of SAWs propagating in 45° Rt.-Y cut $LiNbO_3$.

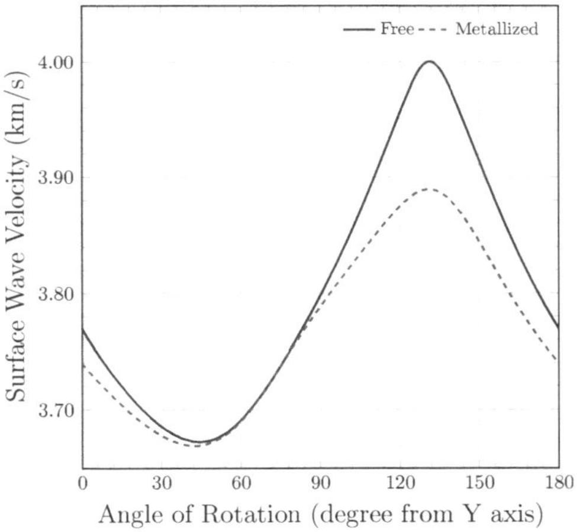

Figure 6.27 Phase velocity of SAWs propagating in Rt.-YX(||) cut $LiNbO_3$.

In a close examination of Figures 6.26 and 6.28, we note a discrepancy in the result for the 45° Rt.-Y cut at 90° in Figure 6.26 and that for the Rt.-YX(⊥) at 45° in Figure 6.28 which was not reported in the work of Campbell and Jones. Per the discussion for $LiTaO_3$ crystal in the proceeding section and the illustrations in Figure 6.22, these two orientations are of the same cut; hence, the same value for phase velocity is expected. However, the phase velocity value shown in Figure 6.26 (marked by a hollow dot at the bottom of the valley of curves) does not agree with that shown in Figure 6.28 (marked by a solid dot). It is worth noting that the phase velocity value marked by the hollow dot at the bottom of the valley in Figure 6.26 is not the result from the numerical search. It is marked based on the smooth transition of the phase velocity curve.

Recalling the discussion in Section 6.5.2, we know that the YZ plane is a mirror symmetric plane of $LiNbO_3$. This means that in this particular cut, SAWs are decoupled into independent sub-waves. With Figures 6.22(i) and 6.22(j) in mind, we can see that of all the rotation angles shown in Figure 6.26, the 90° case indeed has its *xz*-plane coincide with the YZ mirror plane. Because of this, waves are no longer fully coupled like in all other cases and they consist of two decoupled independent sub waves, namely, stiffened Rayleigh waves and bulk T(||) wave as discussed in Section 6.6.4. In Figure 6.28, however, since every single cut has its *xz* plane coincide with the XZ plane, all the results are obtained based on the decoupled SAWs. For this reason, a solid dot is provided in Figure 6.26 to mark the velocity for the decoupled stiffened Rayleigh waves for the case of 45° Rt-Y cut at 90°. Interestingly, the Y-cut situation shown in Figure 6.24 possesses the same issue for the 90° case, but the results based on the smooth transition of the phase velocity curves seem to match exactly those obtained from the decoupled SAWs (see Section 6.6.4).

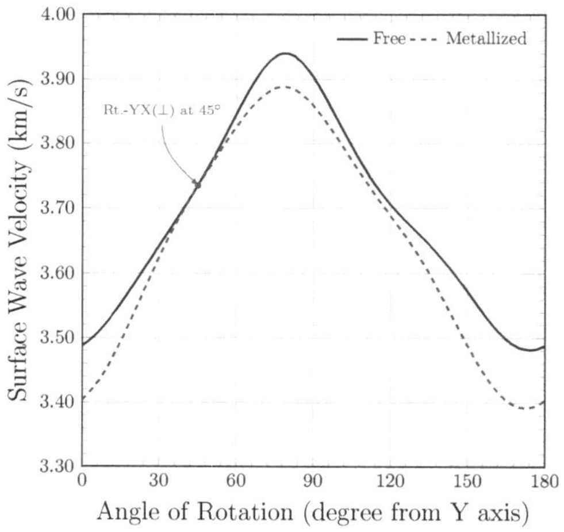

Figure 6.28 Phase velocity of SAWs propagating in Rt.-YX(⊥) cut $LiNbO_3$.

6.7.3 Phase Velocity of SAWs in Langasite

Figure 6.29 shows the phase velocity of SAWs in 45° Rt.-Y cut (Figure 6.29(a)) and Z cut (Figure 6.29(b)) of trigonal Class 23 crystal Langasite ($La_3Ga_5SiO_{14}$, or LGS). In these results, both cases of metallized surface and free-floating surface are provided. Clearly, the differences between the metallized and free-floating surface cases is barely discernible, suggesting that LGS is not a strong piezoelectric material in comparison with $LiTaO_3$ and $LiNbO_3$.

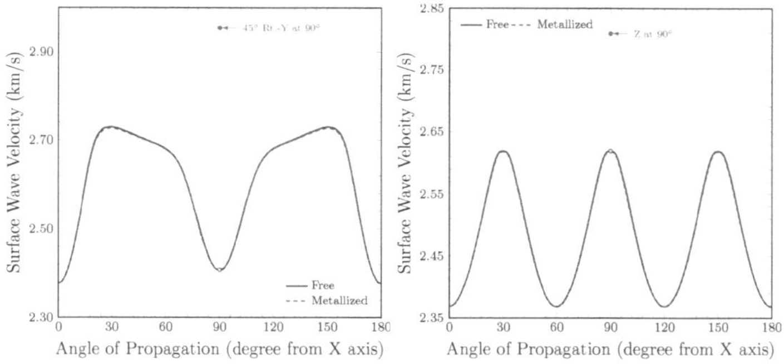

(a) 45° Rt.-Y cut (b) Z cut

Figure 6.29 Phase velocity of SAWs propagating in LGS in Y and Z cuts.

Because of the two-fold symmetry in LGS crystal, per the discussions in Section 6.5.1, we know that in the Rt.-YX(⊥) cut, SAWs are decoupled into non-piezoelectric Rayleigh waves and piezoelectric stiffened TH wave. Figure 6.30 shows the results obtained by solving the decoupled Rayleigh waves (Figure 6.30(a)) and the TH wave (Figure 6.30(b)) in the Rt.-YX(⊥) cut LGS.

In these results, we also note two discrepancies in phase velocity: one between the Rt.-YX(⊥) cut at 45° (in Figure 6.30(a)) and the 45° Rt.-Y cut at 90° (in Figure 6.29(a)), and the other between the Rt.-YX(⊥) cut at 90° (in Figure 6.30(a)) and in the Z cut at 90° (in Figure 6.29(b)). The reason for this is that in both cases the *xz*-plane of the two crystal cuts is perpendicular to the axis of two-fold symmetry, namely, the X axis of the trigonal Class 23 langasite. Referring to the discussions in Section 6.5.1, we know that in these

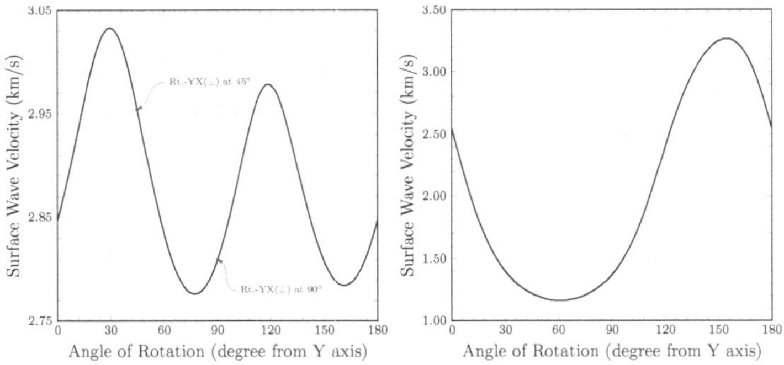

(a) Decoupled Rayleigh waves (b) Decoupled TH wave

Figure 6.30 Phase velocity of SAWs propagating in LGS in Rt.-YX(⊥) cut.

two cuts SAWs are decoupled into non-piezoelectric Rayleigh waves and piezoelectric stiffened TH wave. Since the results shown in Figure 6.30(a) are for non-piezoelectric Rayleigh waves, solid dots representing the respective decoupled results for the 45° Rt.-Y cut at 90° and the Z cut at 90° are added in Figures 6.29(a) and 6.29(b), respectively.

6.7.4 Phase Velocity of SAWs in SiO_2

Figure 6.31 shows the results of phase velocity for SAWs in silicon oxide (SiO_2) in six common cuts: X cut in Figure 6.31(a), Y cut in Figure 6.31(b), Z cut in Figure 6.31(c), 45° Rt.-Y cut in Figure 6.31(d), Rt.-YX(||) cut in Figure 6.31(e) and Rt.-YX(⊥) cut in Figure 6.31(f).

Of all these results, both cases of metallized surface and free-floating surface are provided. Like in LGS, the differences between the metallized and free-floating surface cases is barely discernible, indicating that SiO_2 is not a material with strong piezoelectric coupling effect. Moreover, in SiO_2, we also note a discrepancy in phase velocity, i.e., between the Rt.-YX(⊥) cut at 90° and the Z cut at 90°. The reason for this discrepancy is the same as for the two Langasite cases in Section 6.7.3 because SiO_2 is also a trigonal Class 23 crystal. We therefore use a solid dot in Figure 6.31(c) to mark the wave velocity for the decoupled non-piezoelectric Rayleigh waves. Note that in the Rt.-YX(||)-cut case (Figure 6.31(e)) we see a similar

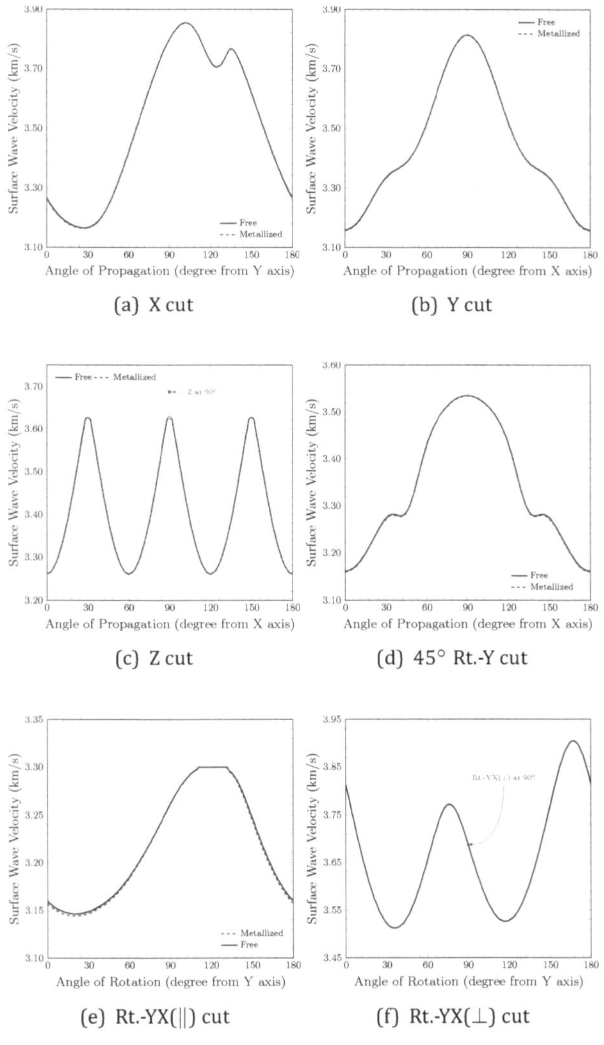

Figure 6.31 Phase velocity of SAWs propagating in SiO_2. Note that in (f) the waves are decoupled Rayleigh waves.

feature as in Figure 6.21(e) for $LiTaO_3$, that is a cut-off top due to the interference from the slower transverse bulk wave. Referring to Figure 5.20, we can estimate from both the XY and XZ sections that the phase velocity of the slowest T wave propagating in the X direction is indeed at this velocity.

6.7.5 Electromechanical Coupling Coefficient K for SAWs

In Section 5.4.3 we have defined the *electromechanical coupling coefficient K* for BAWs. Because the polarization and propagation of SAWs are in a different manner, we need to introduce a different way to quantify this coupling coefficient for SAWs. As we have seen in several previous sections, the phase velocity of SAWs can be adjusted by either using a metal film to electrically ground the surface or letting the surface free floating electrically. The difference in the phase velocity of the two cases with either metallized surface or free-floating surface is often used to quantify the *electromechanical coupling coefficient K* for SAWs in the following expression:

$$K_{SAW} = \sqrt{2(1 - \frac{v_{\mathrm{metallized}}}{v_{\mathrm{free}}})}$$

where $v_{\mathrm{metallized}}$ and v_{free} are the phase velocity of SAWS with metallized surface and free-floating surface, respectively. Very often, we use its squared expression as given below:

$$K_{SAW}^2 = [2(1 - \frac{v_{\mathrm{metallized}}}{v_{\mathrm{free}}})] \times 100\% \qquad (6.27)$$

By applying Eq. (6.27) to the velocity data for $LiNbO_3$ shown in Section 6.7.2 and plotting the K_{SAW}^2 values as a function of propagation angle from 0° to 180°, we obtain the K_{SAW}^2 curves for $LiNbO_3$ in the X, Y Z, and 45° Rt.-Y cuts as shown in Figure 6.32.

Of these four cases, the X and Y cuts possess a higher degree of piezoelectric coupling for SAWs than the other cut cases, with peak coupling effect reaching approximately 5.2% near 93° in the X cut and approximately 4.8% at 90° in the Y cut.

Figure 6.33 shows the K_{SAW}^2 values for the Rt.-YX(||) and Rt.-YX(⊥) cuts. Apparently, both cuts possess high piezoelectric coupling effect. For instance, in the case of Rt.-YX(||), K_{SAW}^2 reaches almost 5.6% at an angle of approximately 131.5° and in the case of Rt.-YX(⊥) propagation, K_{SAW}^2 reaches about 5.4% at an angle of approximately 163°. As we have noted in Section 5.4.3, K_{BAW}^2 values for $LiNbO_3$ are in the range of 0–30%, or close to 40% in some cuts, but here we see that K_{SAW}^2 values are in the 0–5.6% range. This difference provides another means to note the difference between BAWs and SAWs, as

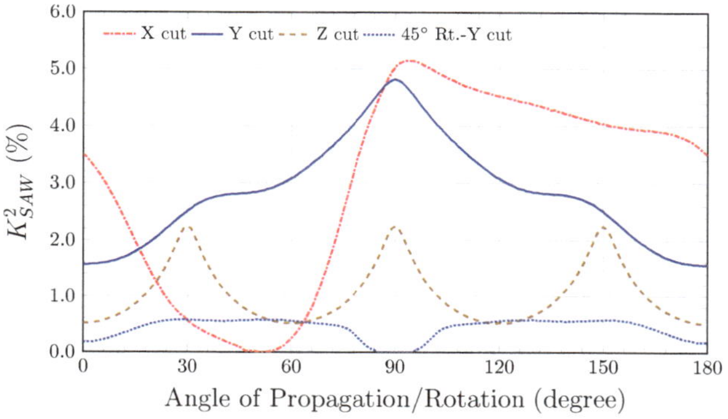

Figure 6.32 Electromechanical coupling coefficient of SAWs in $LiNbO_3$.

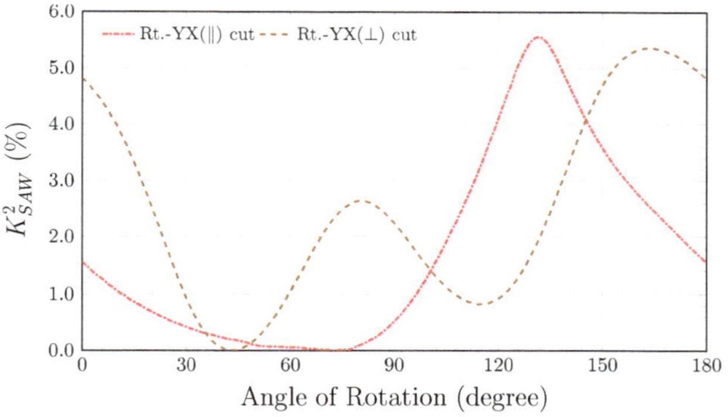

Figure 6.33 Electromechanical coupling coefficient of SAWs in $LiNbO_3$.

well as the difference in quantifying the K^2 values for BAWs and SAWs.

In the case of rotated Rt.-YX(||) cut, the 131.5° orientation angle has the highest piezoelectric coupling effect for SAW device applications. In practice, however, the 128° Rt.-YX(||) cut is commonly used for many $LiNbO_3$ SAW devices. This practice is believed to reduce the unwanted excitation of a bulk wave as confirmed experimentally. One piece of supporting evidence for this argument can be found by referring to the discussions in Section 5.4.3: From Figure 5.18(b), we can see that the electromechanical coupling coefficient for the bulk L wave in the YZ section of $LiNbO_3$

reaches a peak value at the same cut angle (i.e., 131.5°), so deviating away from this angle, even slightly, would logically help reduce the excitation of the unwanted bulk L wave.

Figure 6.34 shows the K^2_{SAW} results $LiTaO_3$ in the X, Y Z, and 45° Rt.-Y cuts, and Figure 6.35 the results for the Rt.-YX(||) and Rt.-YX(⊥) cuts, both with velocity data from Section 6.7.1. Obviously, the piezoelectric coupling effect of $LiTaO_3$ is not as higher as $LiNbO_3$, but it is still relatively high when compared with many other piezoelectric materials. Of the cuts shown here, the X, Y, Z, and Rt.-YX(⊥) cuts possess higher piezoelectric coupling effect than the other cuts, with peak coupling effect reaching approximately 1.4% at around 138° in the X cut, 1.15% at 90° in the Y cut, and approximately

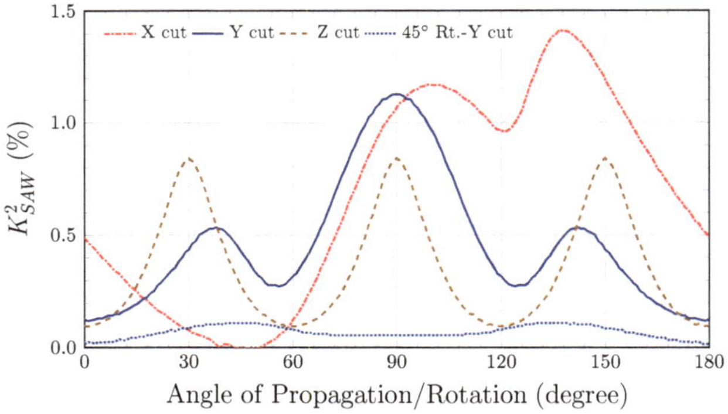

Figure 6.34 Electromechanical coupling coefficient of SAWs in $LiTaO_3$.

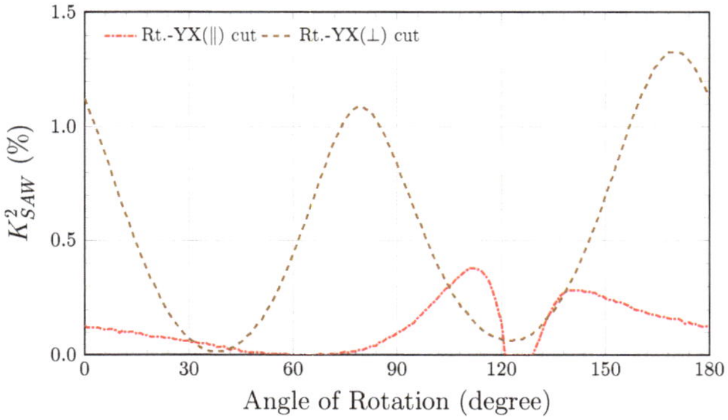

Figure 6.35 Electromechanical coupling coefficient of SAWs in $LiTaO_3$.

1.33% at around 168° in the Rt.-YX(⊥) cut. So, for applications where converting electrical energy to mechanical one, or vice versa, is of importance, these cuts with high electromechanical coupling coefficient would be preferred choices.

Figure 6.36 shows the K^2_{SAW} curves for three selected cuts in LGS from the wave velocity data presented in Section 6.7.3, and Figure 6.37 shows the K^2_{SAW} curves for five selected cuts in SiO_2 from the wave velocity data presented in Section 6.7.4. Comparing with $LiNbO_3$ and $LiTaO_3$, LGS and SiO_2 possess much lower piezoelectric coupling effect, with the electromechanical coupling coefficient (K^2_{SAW}) varying below 0.4% for LGS and below 0.2% for SiO_2. This is consistent with what we have observed for BAWs in these materials. For the Y-cut LGS, K^2_{SAW} value higher than 0.2% can be reached when

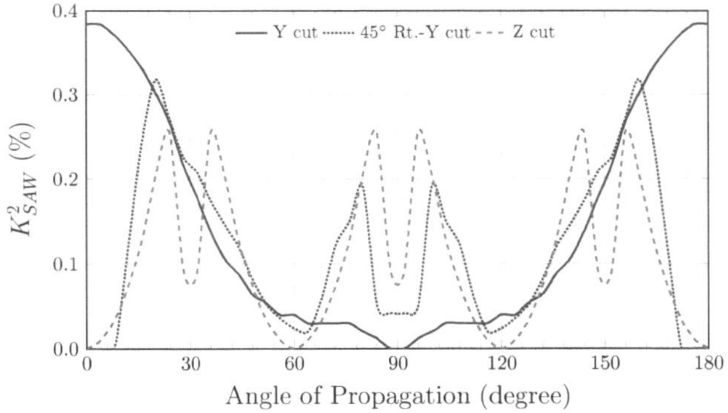

Figure 6.36 Electromechanical coupling coefficient of SAWs in LGS.

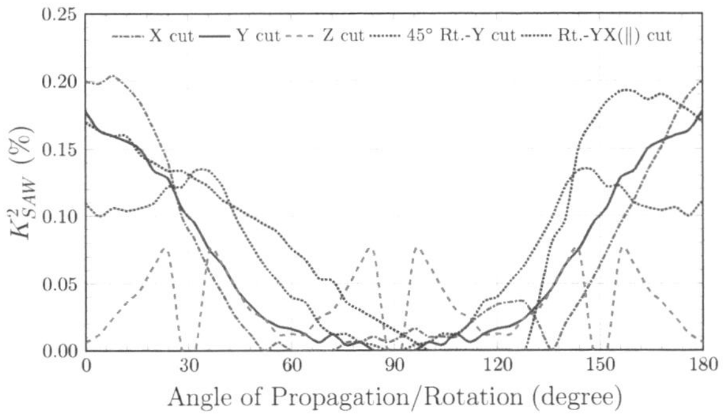

Figure 6.37 Electromechanical coupling coefficient of SAWs in SiO_2.

the propagation direction is within 0° to 30° (or 150° to 180°, and so on), and for the 45° Rt.-Y-cut LGS, K^2_{SAW} value higher than 0.2% can be reached when the propagation direction is within 16° to 33° (or 147° to 164°, and so on).

6.7.6 Poynting Angle for SAWs

As we have discussed in Section 5.3.4, the Poynting vector captures the directional energy flux of BAWs. To expand this concept to SAWs, we can use the Poynting angle (θ_{Poynting}), that is, the angle between the a wave's power flow direction, as defined by the direction of its energy velocity, and its propagation direction, to assess any deviation of acoustic energy of SAWs away from the propagation direction within the surface along which the SAWs travel.

Figure 6.38 shows the variation of Poynting angle as a function of propagation direction for the X and Y cuts $LiNbO_3$ with either free floating or metallized surface. Apparently, there are several orientations in which the Poynting angle is zero. Note that an orientation with zero Poynting angle is a direction in which the acoustic power flow coincides with the wave propagation direction. For the X-cut case (Figure 6.38(a)), within the range of propagation directions shown, there are four orientations in the free-floating surface situation where the Poynting angle is zero (at approximately 20°, 75°, 87° and 118°) and two orientations in the metallized-surface situation, where the Poynting angle is zero (at approximately 14° and 117°). For the Y-cut case (Figure 6.38(b)), there are five orientations at which the Poynting angle is zero: at 0°, 68°, 90°, 112° and 180° for the free-surface case, and at 0°, 71°, 90°, 110° and 180° for the metallized-surface case.

Figure 6.39 shows the shows the variation of Poynting angle for two more cuts in $LiNbO_3$. For the Rt.-YX(||)-cut case (Figure 6.39(a)), there are two orientations at which the Poynting angle is zero: at 44° and 131.5° for the free-surface case, and at 43.5° and 131.5° for the metallized-surface case. For the Rt.-YX(⊥)-cut case (Figure 6.39(b)) there are also two orientations at which the Poynting angle is zero: at 79° and 354° for the free-surface case, and at 78° and 354° for the metallized-surface case..

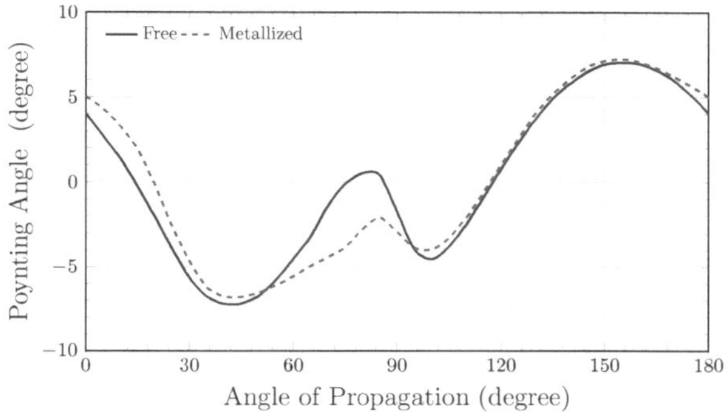

(a) X cut

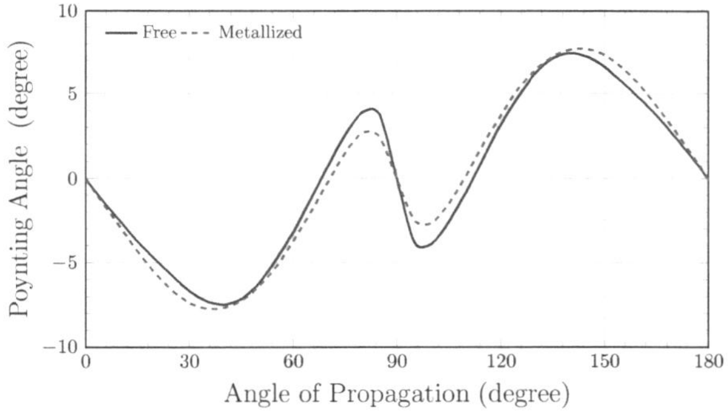

(b) Y cut

Figure 6.38 Poynting angle of SAWs in $LiNbO_3$ in X and Y cuts.

Figure 6.40 shows the variation of Poynting angle for the X and Rt.-YX(||) cuts SiO_2. In both cases, the difference between free-surface and metallized-surface situations is hardly seen. Moreover, the range in which the Poynting angle varies in the Rt.-YX(||) case is more than three times larger than in the X-cut case. For the X-cut case (Figure 6.40(a)), there are four orientations at which the Poynting angle is zero: at approximately 27°, 102°, 124° and 135° for both the free-surface and metallized-surface situations. In the Rt.-YX(||)-cut case (Figure 6.40(b)) there is only one orientation, at approximately 20°, where the Poynting angle is zero. Note that the flat region between 110° and 132° is due to bulk wave interference.

Figure 6.41 shows the variation of Poynting angle for the X and Rt.-YX(⊥) cuts $LiTaO_3$. Within the range of propagation directions

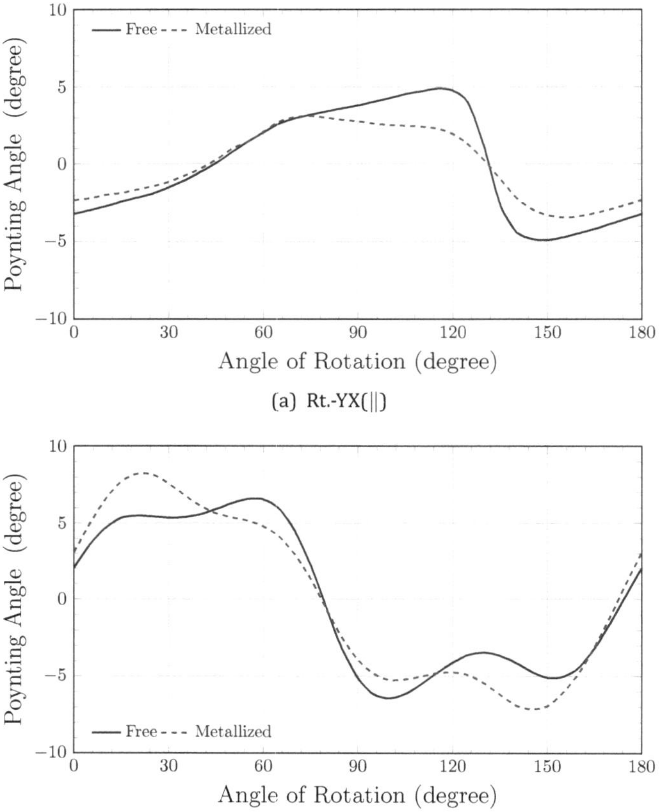

(a) Rt.-YX(||)

(b) Rt.-YX(⊥)

Figure 6.39 Poynting angle of SAWs in $LiNbO_3$ in Rt.-YX(||) and Rt.-YX(⊥) cuts.

shown, the X-cut case has only one orientation at which the Poynting angle is zero which is at approximately 49° (note that the flat region between 122° and 130° is due to bulk wave interference, like the situation in SiO_2). The Rt.-YX(⊥)-cut case has three orientations at which the Poynting angle is zero, that is at approximately 28°, 77.5° and 126°.

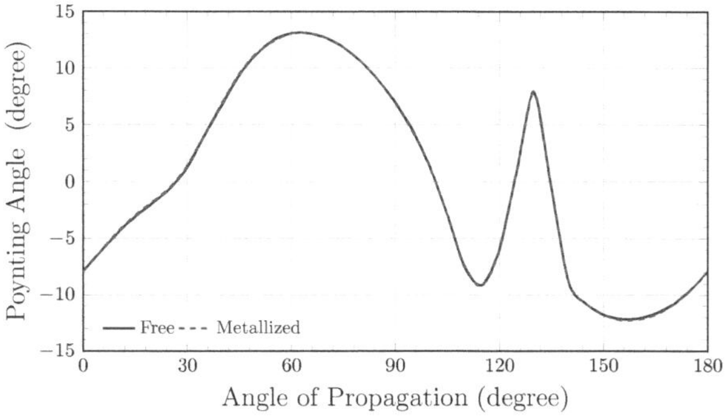

(a) X cut

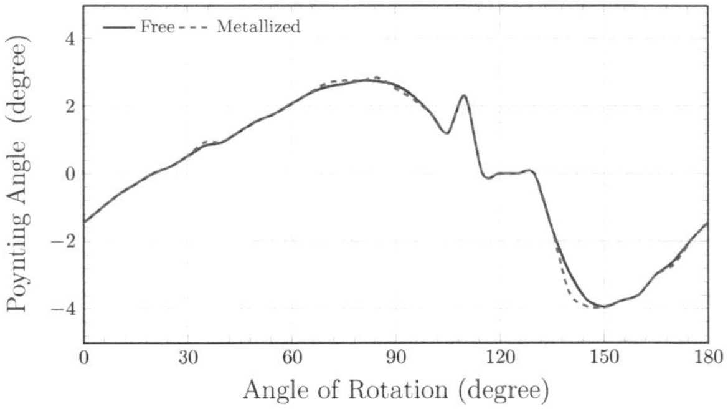

(b) Rt.-YX(||) cut

Figure 6.40 Poynting angle of SAWs in SiO_2 in the X and Rt.-YX(||) cuts.

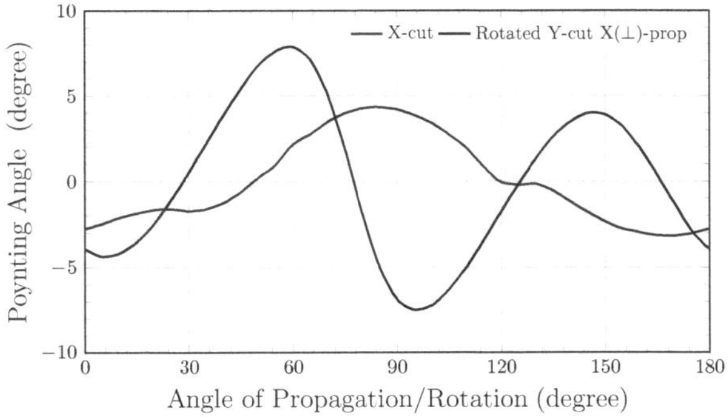

Figure 6.41 Poynting angle of SAWs in $LiTaO_3$.

6.8 Study Problems

1. Follow the steps in Example 1 to solve the velocity for the Rayleigh waves in an isotropic elastic material with c_{11} = 25.3×10^{10} N/m^2, $c_{44} = 3.7 \times 10^{10}$ N/m^2, and $\rho = 3623$ kg/m^3. Also, find the velocity for the decoupled bulk transverse wave.
2. At the end of Example 4 and Example 6, results for hexagonal 6mm cadmium sulfide are also given. Follow the steps used in these example cases to provide detailed solutions for both the Rayleigh waves and Bleustein–Gulyaev wave traveling in the XY-cut crystal.
3. For trigonal 3m Lithium tantalate, find the wave velocity for Rayleigh waves traveling in the YZ direction and the corresponding K_{SAW}^2 value.
4. For orthorhombic 2mm barium sodium niobate, use Eq. (6.14) along with the material properties given in Section 4.9 to establish the governing equation for SAWs traveling in the ZX direction.
5. For tetragonal 4mm barium titanate, find the wave velocity for Rayleigh waves traveling in the ZX direction and the corresponding K_{SAW}^2 value.

Chapter 7

Temperature Effect

In this chapter, we will discuss the effect of temperature on piezoelectric material properties and acoustic wave propagation. Specifically, we will learn the concept of temperature coefficient of velocity and temperature coefficient of delay and evaluate their values for various piezoelectric materials. Knowing these temperature coefficients of a piezoelectric crystal is important. When an acoustic device is intended for applications in a temperature fluctuating environment, materials exhibiting zero temperature coefficient of delay would be ideal.

7.1 Temperature-Dependent Behavior

For many temperature-sensitive applications, it is desirable to use crystals with cuts possessing better temperature compensated behavior. A quantitative measure of such temperature compensated behavior is temperature coefficient of delay (TCD).

Bulk and Surface Acoustic Waves: Fundamentals, Devices, and Applications
Guigen Zhang

ISBN 978-981-4877-73-2 (Hardcover), 978-1-003-25662-5 (eBook)
www.jennystanford.com

7.1.1 Temperature Coefficient of Delay

Temperature coefficient of delay is a parameter often used to quantify the time delay behavior of acoustic waves caused by a change in temperature. It is characterized by the following expression:

$$\mathrm{TCD} = \frac{1}{\tau}\frac{\partial \tau}{\partial T}|_{T_0} \tag{7.1}$$

in which, τ is the time it takes for a surface wave with a phase velocity of v_s (note that a subscript s is used to denote surface wave velocity) to travel a distance of l ($\tau = l/v_s$), T is temperature, and T_0 is a specific point in temperature at which TCD is evaluated.

Considering both τ and l as a function of temperature (T), we can expand the above expression as follows:

$$\begin{aligned}\mathrm{TCD} = \frac{1}{\tau}\frac{\partial \tau}{\partial T}|_{T_0} &= \frac{v_s}{l}\frac{\partial}{\partial T}(\frac{l}{v_s})|_{T_0} = \frac{v_s}{l}\left[\frac{1}{v_s}\frac{\partial l}{\partial T} - \frac{l}{v_s^2}\frac{\partial v_s}{\partial T}\right]_{T_0} \\ &= \frac{\partial l}{l\partial T}|_{T_0} - \frac{1}{v_s}\frac{\partial v_s}{\partial T}|_{T_0} = \alpha(T_0) - \frac{1}{v_s}\frac{\partial v_s}{\partial T}|_{T_0}\end{aligned} \tag{7.2}$$

where $\alpha(T)$ is linear coefficient of thermal expansion.

7.1.2 Temperature Coefficient of Velocity

If we replace the time term τ with the velocity term v_s in Eq. (7.1), we have the parameter commonly known as the temperature coefficient of velocity (TCV).

$$\mathrm{TCV} = \frac{1}{v_s}\frac{\partial v_s}{\partial T}|_{T_0} \tag{7.3}$$

From Eq. 7.2, it is clear that the sum of TCD and TCV equals to the thermal coefficient of expansion through the following relationship:

$$TCD + TCV = \alpha(T)$$

7.1.3 Approximate Values of TCD and TCV

Although v_s is a function of T, an analytical expression for this function is typically unknown. Thus, TCV and TCD are usually determined by using the following approximate expression:

$$\begin{aligned} \text{TCV} &= \frac{1}{v_s(T_0)} \frac{v_s(T_0 + \Delta T) - v_s(T_0 - \Delta T)}{2\Delta T} \\ \text{TCD} &= \alpha(T_0) - \frac{1}{v_s(T_0)} \frac{v_s(T_0 + \Delta T) - v_s(T_0 - \Delta T)}{2\Delta T} \end{aligned} \tag{7.4}$$

For example, if $T_0 = 25°\text{C}$ and $\Delta = 10°\text{C}$, we have

$$\text{TCD} = \alpha(25°C) - \text{TCV} = \alpha(25°C) - \frac{1}{v_s(25°C)} \frac{v_s(35°C) - v_s(15°C)}{20°C}$$

This means that to determine the approximate values for TCD and TCV at 25°C, we need to know the phase velocity at 15°C, 25°C and 35°C, respectively, along with the coefficient of thermal expansion at 25°C. Since piezoelectric crystals are anisotropic materials and phase velocity is propagation-direction dependent, now with the consideration of temperature effect, we can see that phase velocity of SAWs is temperature dependent as well.

7.2 Temperature-Dependent Material Properties

7.2.1 Temperature-Dependent $[c]$, $[e]$ and $[\epsilon]$ Constants

To determine the temperature-dependent phase velocity of SAWs in crystal materials, we need to know the temperature-dependent material properties. To do that, we first apply the concept of Taylor series expansion. Using Taylor series, we can express function $f(x)$ in terms of its value at a point when $x = a$, as

$$f(x) = f(a) + \frac{f'(a)}{1!}(x-a) + \frac{f''(a)}{2!}(x-a)^2 + \frac{f'''(a)}{3!}(x-a)^3 + \cdots$$

This can be rearranged as

$$f(x) = f(a)\left[1 + \frac{f'(a)}{1!f(a)}(x-a) + \frac{f''(a)}{2!f(a)}(x-a)^2 + \frac{f'''(a)}{3!f(a)}(x-a)^3 + \cdots\right]$$

If we regard all the fraction terms as coefficients, we can then write

$$f(x) = f(a)\left[1 + \alpha_1(x-a) + \alpha_2(x-a)^2 + \alpha_3(x-a)^3 + \cdots\right]$$

Now by applying this expression to each of the constants in the property matrices of a given piezoelectric material including the stiffness matrix $[c]$, piezoelectric stress constant matrix $[e]$, and permittivity matrix $[\epsilon]$, we have the following temperature-dependent expressions for these respective material properties:

$$\begin{aligned}
c_{ij}(T) &= c_{ij}(T_0)\left[1 + \alpha_{ij}^{c_1}(T-T_0) + \alpha_{ij}^{c_2}(T-T_0)^2 + \cdots\right];\ i,j = 1,2,\ldots,6\\
e_{ij}(T) &= e_{ij}(T_0)\left[1 + \alpha_{ij}^{e_1}(T-T_0) + \alpha_{ij}^{e_2}(T-T_0)^2 + \cdots\right];\ i = x,y,z;\ j = 1,2,\ldots,6\\
\epsilon_{ij}(T) &= \epsilon_{ij}(T_0)\left[1 + \alpha_{ij}^{\epsilon_1}(T-T_0) + \alpha_{ij}^{\epsilon_2}(T-T_0)^2 + \cdots\right];\ i,j = x,y,z
\end{aligned} \tag{7.5}$$

where T_0 is a reference temperature, $c_{ij}(T_0)$, $e_{ij}(T_0)$ and $\epsilon_{ij}(T_0)$ are the respective constants in the $[c]$, $[e]$ and $[\epsilon]$ matrices at temperature T_0, $\alpha_{ij}^{c_1}$ and $\alpha_{ij}^{c_2}$ are the first- and second-order thermal coefficients of the respective $[c]$ constants, $\alpha_{ij}^{e_1}$ and $\alpha_{ij}^{e_2}$ are the first- and second-order thermal coefficients of the respective $[e]$ constants, and $\alpha_{ij}^{\epsilon_1}$ and $\alpha_{ij}^{\epsilon_2}$ are the first- and second-order thermal coefficients of the respective $[\epsilon]$ constants. Typically, the first-order coefficients are available for most of the piezoelectric materials. When higher-order coefficients are not available, we just simply ignore the higher-order terms and truncate the expressions.

7.2.2 Temperature-Dependent Thermal Coefficient of Expansion

For the temperature-dependent coefficient of thermal expansion at T_0, by definition we have

$$\alpha_{ii}(T) = \frac{1}{l_{ii}}\frac{\partial l_{ii}(T)}{\partial T}|_{T_0}, \quad i = x, y, z \tag{7.6}$$

By Taylor series expansion, $l_{ii}(T)$ can also be expressed as

$$l_{ii}(T) = l_{ii}(T_0)\left[1 + \alpha_{ii}^{T_1}(T - T_0) + \alpha_{ii}^{T_2}(T - T_0)^2 + \cdots\right];\ i = x, y, z$$

in which, $\alpha_{ii}^{T_1}$ and $\alpha_{ii}^{T_2}$ are the first- and second-order coefficients of thermal expansion, respectively. By applying this expression to Eq. (7.6), we have

$$\alpha_{ii}(T) = \alpha_{ii}^{T_1} + 2\alpha_{ii}^{T_2}(T - T_0) + \cdots;\ i = x, y, z \tag{7.7}$$

To capture the orthogonal nature of the coefficient of thermal expansion for an anisotropic material, we further express it in a matrix form as

$$[\alpha(T)] = \begin{bmatrix} \alpha_{xx} & 0 & 0 \\ 0 & \alpha_{yy} & 0 \\ 0 & 0 & \alpha_{zz} \end{bmatrix}$$
$$= \begin{bmatrix} \alpha_{xx}^{T_1} + 2\alpha_{xx}^{T_2}(T - T_0) + \cdots & 0 & 0 \\ 0 & \alpha_{yy}^{T_1} + 2\alpha_{yy}^{T_2}(T - T_0) + \cdots & 0 \\ 0 & 0 & \alpha_{zz}^{T_1} + 2\alpha_{zz}^{T_2}(T - T_0) + \cdots \end{bmatrix}$$

For most piezoelectric materials, $\alpha_{xx} = \alpha_{yy}$ due to certain crystal symmetry as discussed in Chapter 4.

Because the coefficient of thermal expansion is a secondary variable, its transformation by a rotation matrix $[a]$ can be determined by

$$[\alpha'(T)] = [a][\alpha(T)][a]^T$$

7.2.3 Temperature-Dependent Density

The temperature-dependent density can be expressed in a similar way using Taylor series as follows:

$$\rho(T) = \rho(T_0)\left[1 + \alpha_\rho^1(T - T_0) + \alpha_\rho^2(T - T_0)^2 + \cdots\right] \tag{7.8}$$

where α_ρ^1 and α_ρ^2 are the first- and second-order thermal coefficients of density, respectively.

In the event that these coefficients are not readily available, we can turn to the coefficients of thermal expansion to determine the temperature-dependent density. By definition, the density (ρ) of a material measures the mass per unit volume, which can be expressed mathematically as

$$\rho = \frac{m}{l_{xx}l_{yy}l_{zz}} = m\left(l_{xx}^{-1}l_{yy}^{-1}l_{zz}^{-1}\right)$$

where m is the mass, l_{xx}, l_{yy} and l_{zz} are respective length of the unit volume in the three orthogonal directions. By the law of conservation of mass, m is temperature independent, and all other parameters including ρ, l_{xx}, l_{yy} and l_{zz} are regarded as temperature dependent, or as a function of T. Taking the first derivative of ρ with respect to T, we have

$$\begin{aligned}\frac{\partial\rho(T)}{\partial T} &= m\frac{\partial}{\partial T}\left(l_{xx}^{-1}l_{yy}^{-1}l_{zz}^{-1}\right)\\ &= -m\left[l_{yy}^{-1}l_{zz}^{-1}\frac{\partial l_{xx}^{-1}}{\partial T} + l_{xx}^{-1}l_{zz}^{-1}\frac{\partial l_{yy}^{-1}}{\partial T} + l_{xx}^{-1}l_{yy}^{-1}\frac{\partial l_{zz}^{-1}}{\partial T}\right]\end{aligned}$$

With some simplification along with using ρ_0 to represent the density at T_0, we arrive at

$$\begin{aligned}\frac{\partial\rho(T)}{\partial T} &= -ml_{xx}^{-1}l_{yy}^{-1}l_{zz}^{-1}\left[\frac{\partial l_{xx}^{-1}}{l_{xx}^{-1}\partial T} + \frac{\partial l_{yy}^{-1}}{l_{yy}^{-1}\partial T} + \frac{\partial l_{zz}^{-1}}{l_{zz}^{-1}\partial T}\right]\\ &= -\rho_0\left[\frac{\partial l_{xx}^{-1}}{l_{xx}^{-1}\partial T} + \frac{\partial l_{yy}^{-1}}{l_{yy}^{-1}\partial T} + \frac{\partial l_{zz}^{-1}}{l_{zz}^{-1}\partial T}\right]\end{aligned}$$

By definition, the three terms in the brackets are coefficients of thermal expansion, α_{xx}, α_{yy} and α_{zz}, in the three orthogonal directions, respectively. This leads us to

$$\frac{\partial \rho(T)}{\rho_0 \partial T} = -(\alpha_{xx} + \alpha_{yy} + \alpha_{zz}) \tag{7.9}$$

By applying the method of separation of variables, we can solve this differential equation and obtain

$$\rho(T) = \rho_0 e^{-(\alpha_{xx}+\alpha_{yy}+\alpha_{zz})(T-T_0)} \tag{7.10}$$

Alternatively, letting $\partial\rho(T) = \rho(T) - \rho_0$ and $\partial T = T - T_0$, we can express Eq. (7.9) in the following approximate form:

$$\rho(T) = \rho_0 \left[1 - (\alpha_{xx} + \alpha_{yy} + \alpha_{zz})(T - T_0)\right] \tag{7.11}$$

With all these material property constants expressed as functions of temperature, we can use the same procedures discussed in Chapter 6 to determine the phase velocity of SAWs in a selected material for a given temperature and propagation direction.

7.3 Case Studies of Temperature Effect on SAWs

7.3.1 TCD for Langasite

Let us now examine a few quantitative examples. The first case we will discuss is for Class 32 trigonal crystal Langasite (LGS). As we have seen in Figure 6.36, LGS exhibits much lower electromechanical coupling coefficient than that of $LiNbO_3$ and $LiTaO_3$, with K^2_{SAW} values varying below 0.4%. However, when it comes to temperature compensated performance, we will soon see that LGS possesses superior TCD to $LiNbO_3$ and $LiTaO_3$.

By using the thermal coefficients listed in Table 7.1 and letting T_0 = 25°C, we first search for SAW velocity for various selected crystal cuts at three different temperatures, namely, at 15°C, 25°C and 35°C, respectively. Then by substituting the obtained velocity values into Eq. (7.4) we determine the approximate values for the

Table 7.1 Temperature-dependent properties and coefficients for LGS

Variables	Properties (at 25°C)	Temperature coefficient	
	c_{ij} (10^{10} N/m^2)	α_{ij}^{c1} (10^{-4}/°C)	α_{ij}^{c2} (10^{-7}/°C^2)
c_{11}	18.89	-1.77	-2.39
c_{12}	10.41	-3.2	-3.78
c_{13}	10.15	-0.169	13.7
c_{14}	1.442	3.0	1.94
c_{33}	26.83	-0.0919	-2.36
c_{44}	5.33	-0.6	-1.57
	e_{ij} (C/m^2)	α_{ij}^{e1} (10^{-4}/°C)	α_{ij}^{e2} (10^{-7}/°C^2)
e_{x1}	-0.437	6.13	11.4
e_{x4}	0.104	13.57	41.2
	ϵ_{ij} (10^{-11} F/m)	$\alpha_{ij}^{\epsilon1}$ (10^{-4}/°C)	$\alpha_{ij}^{\epsilon2}$ (10^{-7}/°C^2)
ϵ_{xx}, ϵ_{yy}	16.87	1.9	4.69
ϵ_{zz}	45.87	-7.61	11.6
		α_{ij}^{T1} (10^{-6}/°C)	α_{ij}^{T2} (10^{-9}/°C^2)
α_{xx}, α_{yy}		5.08	–
α_{zz}		3.49	–

TCD as a function of wave propagation direction. It is worth noting that the effect of temperature on SAW phase velocity is on the scale of part-per-million (ppm), it is necessary to have a good precision, ideally to the 0.001 m/s, for the velocity values. In the event that such a precision is not economical to obtain because doing so requires a much longer searching time, reasonable approximation is still attainable by applying a multiple-point running average (essentially a digital lower-pass filter) to remove the noise in the data. This is the reason that all the TCD data shown in this chapter are plotted with the original noised data as well as the smoothed data.

Figure 7.1 shows the obtained TCD curves for the Y-cut LGS at 25°C with free floating and metallized surface. Apparently, the TCD curves for both the free floating and metallized surface cases intersect with the line of TCD = 0 several times, reflecting the fact that the Y-cut LGS possesses several orientations at which the TCD

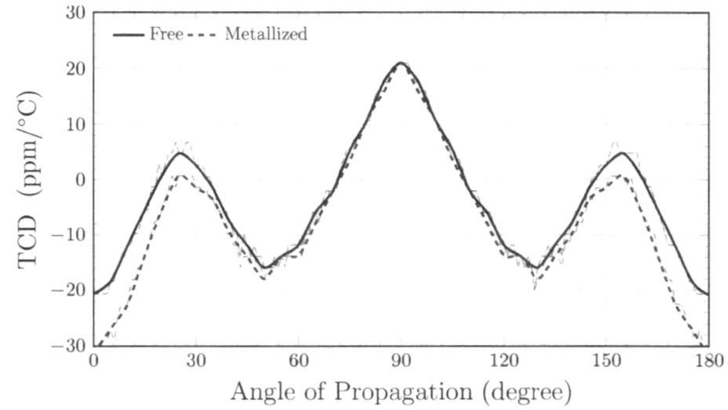

Figure 7.1 Temperature coefficient of delay in the Y-cut LGS.

is zero. Note that a wave propagation direction having Zero TCD is a direction at which changing temperature will have a diminished effect on wave propagation. This will be ideal for temperature sensitive applications. For example, for the free-floating case, TCD is zero at approximately 23°, 36°, 70°, 110°, 144° and 157°. To provide a visual comparison, Figure 7.2 shows the variation of SAW velocity with wave propagation direction in the Y-cut LGS at 25°C with free floating and metallized surface. Although the difference in the wave phase velocity between the free-floating and metalized cases is hardly discernible (see Figure 7.2), the difference in the TCD is much obvious (see Figure 7.1).

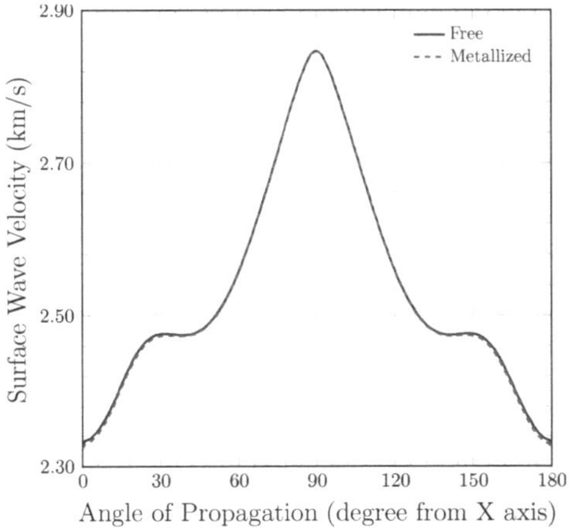

Figure 7.2 SAW velocity in the Y-cut LGS.

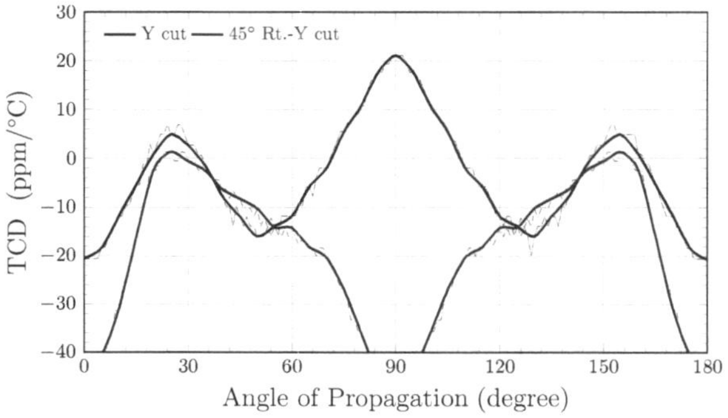

Figure 7.3 Temperature coefficient of delay in Y- and Rt.-Y-cut LGS.

Figure 7.3 shows a side-by-side comparison between the TCD curves for the Y-cut and 45° Rt.-Y-cut in LGS with free floating surface. Although the two curves differ notably, the curve for the 45° rotated Y-cut case also intersects with the zero line.

7.3.2 TCD for Quartz

Figure 7.4 shows the TCD curve for the Rt.-YX(||) cut of SiO_2 whose temperature-dependent coefficients are listed in Table 7.2.

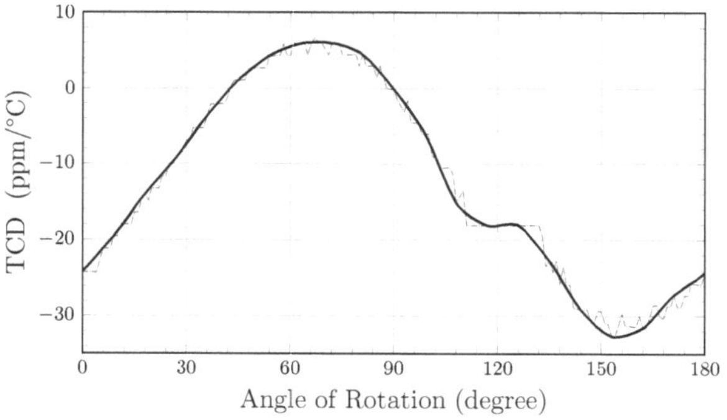

Figure 7.4 Temperature coefficient of delay in the Rt.-YX||-cut SiO_2.

For SiO_2 (quartz) the series of Rt.-YX(||) cuts contains a few commonly used crystal cuts such as the AT cut at 35.25°, CT cut at 38° and ST cut at 42.75° (see Chapter 9 for more discussions).

Table 7.2 Temperature-dependent properties and coefficients for quartz

Variables	Properties (at 25°C) c_{ij} (10^{10} N/m^2)	Temperature coefficient $\alpha_{ij}^{c_1}$ (10^{-4}/°C)	$\alpha_{ij}^{c_2}$ (10^{-7}/°C^2)	
c_{11}	8.67	−0.443	-	
c_{12}	0.70	−26.9	-	
c_{13}	1.19	−5.5	-	
c_{14}	−1.79	1.17	-	
c_{33}	10.72	−1.6	-	
c_{44}	5.79	−1.754	-	
		$\alpha_{ij}^{T_1}$ (10^{-6}/°C)	$\alpha_{ij}^{T_2}$ (10^{-9}/°C^2)	$\alpha_{ij}^{T_3}$ (10^{-12}/°C^3)
α_{xx}, α_{yy}		13.65	10.02	−19.21
α_{zz}		7.5	8.0	−10.44
	ρ (Kg/m^3)	α_{ρ}^{1} (10^{-6}/°C)	α_{ρ}^{2} (10^{-9}/°C^2)	
ρ	2650	−34.92	−15.9	

In a close examination of the TCD curve in Figure 7.4, we can see that the TCD curve passes across zero at around 42.75°. This is the angle of the ST cut, where ST stands for steady temperature. This fact confirms that ST-cut quartz will experience the least temperature effect around room temperature of 25°C. From the TCD curve, we can see that the TCD value of the 90° orientation in the Rt.-YX(||) cut is also zero. Figure 7.5 shows the variation of TCD in 42.75° Rt.-Y-cut

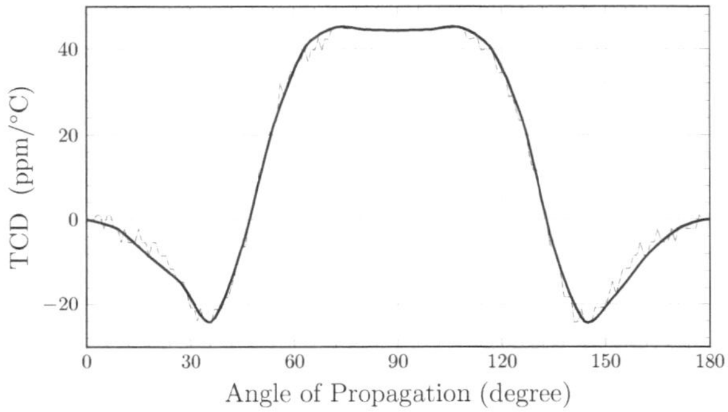

Figure 7.5 Temperature coefficient of delay in 42.75° Rt.-Y-cut SiO_2.

SiO_2. Apparently, in this cut view, the 0° case is the ST cut at which TCD is zero. Aside from that, there are two more orientations, i.e., at approximately 47.25° and 132.25°, in which TCD is also zero.

Figure 7.6 shows the TCD curve for the X-cut SiO_2 and Figure 7.7 shows that of the Y cut. Apparently, the TCD values for these cases also possess orientations with zero TCD, making SiO_2 an attractive crystal material for SAWs devices for temperature sensitive application even though its electromechanical coupling effect is relatively low.

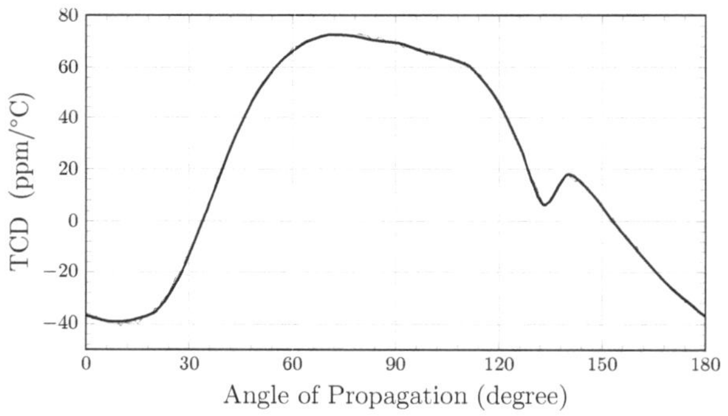

Figure 7.6 Temperature coefficient of delay in the X-cut SiO_2.

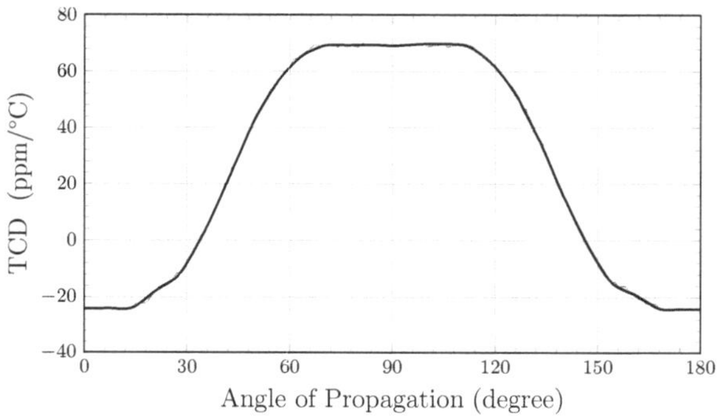

Figure 7.7 Temperature coefficient of delay in the Y-cut SiO_2.

7.3.3 TCD for Lithium Niobate

Table 7.3 lists thermal coefficients for $LiNbO_3$ and Figure 7.8 shows the TCD curves for the X cut, Rt.-YX(||)-cut and Rt.-YX(⊥)-cut $LiNbO_3$, plotted together.

Table 7.3 Temperature-dependent properties and coefficients for $LiNbO_3$

Variables	Properties (at 25°C) c_{ij} (10^{10} N/m^2)	Temperature coefficient α_{ij}^{c1} (10^{-4}/°C)	α_{ij}^{c2} (10^{-7}/°C^2)
c_{11}	20.3	−1.74	–
c_{12}	5.3	−2.52	–
c_{13}	7.5	−1.59	–
c_{14}	0.9	−2.14	–
c_{33}	24.5	−1.53	–
c_{44}	6.0	−2.04	–
	e_{ij} (C/m^2)	α_{ij}^{e1} (10^{-4}/°C)	α_{ij}^{e2} (10^{-7}/°C^2)
e_{x5}	3.7	1.47	–
e_{y2}	2.5	0.79	–
e_{z1}	0.2	2.21	–
e_{z3}	1.3	8.87	–
	ϵ_{ij} (10^{-11} F/m)	$\alpha_{ij}^{\epsilon 1}$ (10^{-4}/°C)	$\alpha_{ij}^{\epsilon 2}$ (10^{-7}/°C^2)
ϵ_{xx}, ϵ_{yy}	38.9	3.23	-
ϵ_{zz}	25.7	6.27	–
		α_{ij}^{T1} (10^{-6}/°C)	α_{ij}^{T2} (10^{-9}/°C^2)
α_{xx}, α_{yy}		15.4	5.3
α_{zz}		7.6	−7.7
	ρ (Kg/m^3)	α_{ρ}^{1} (10^{-6}/°C)	α_{ρ}^{2} (10^{-9}/°C^2)
ρ	4700	−38.3	–

Comparing these TCD curves with those for SiO_2, the TCD values for $LiNbO_3$ are much larger and they are far away from zero. Specifically, the TCD values for these cases vary between approximately 70 and 96 ppm/°C. This fact suggests that $LiNbO_3$ is not as good as LGS or SiO_2 when it comes to applications in a temperature varying environment due to high TCD values.

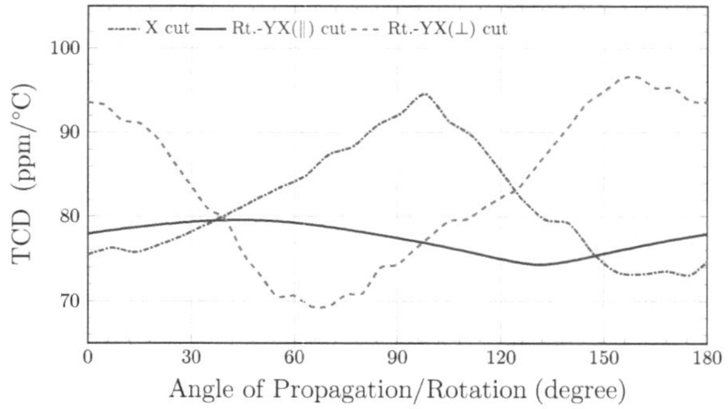

Figure 7.8 Temperature coefficient of delay in three different cuts of $LiNbO_3$.

7.3.4 TCD for Lithium Tantalate

Table 7.4 lists thermal coefficients for $LiTaO_3$ and Figure 7.9 shows the TCD curves for the X-cut and Rt.-YX(⊥)-cut $LiTaO_3$. The TCD values for these cases, though not close to zero, are generally lower than those for $LiNbO_3$. The lowest TCD for $LiTaO_3$ in these two cuts is at about 22.5 ppm/°C for the X cut at approximately 112° angle and about 30 ppm/°C for the Rt.-YX(⊥) cut at about 168°.

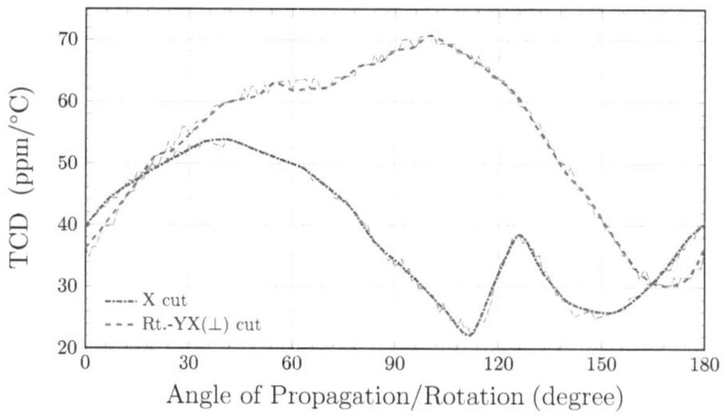

Figure 7.9 Temperature coefficient of delay of $LiTaO_3$.

Table 7.4 Temperature-dependent properties and coefficients for $LiTaO_3$

Variables	Properties (at 25°C) c_{ij} (10^{10} N/m^2)	Temperature coefficient α_{ij}^{c1} (10^{-4}/°C)	α_{ij}^{c2} (10^{-7}/°C^2)
c_{11}	22.96	−1.03	0.77
c_{12}	4.4	−3.41	−1.18
c_{13}	8.12	−0.5	6.0
c_{14}	−1.04	6.67	16.7
c_{33}	27.98	−0.96	−3.21
c_{44}	9.68	−0.43	1.67
	e_{ij}(C/m^2)	α_{ij}^{e1} (10^{-4}/°C)	α_{ij}^{e2} (10^{-7}/°C^2)
e_{x5}	2.72	−1.32	−7.17
e_{y2}	1.67	−0.6	−6.28
e_{z1}	-0.38	0.87	51.8
e_{z3}	1.09	1.54	1.41
	ϵ_{ij} (10^{-11} F/m)	$\alpha_{ij}^{\epsilon 1}$ (10^{-4}/°C)	$\alpha_{ij}^{\epsilon 2}$ (10^{-7}/°C^2)
$\epsilon_{xx}, \epsilon_{yy}$	37.7	3.29	4.28
ϵ_{zz}	37.9	11.6	78.0
		$\alpha_{ij}^{T_1}$ (10^{-6}/°C)	$\alpha_{ij}^{T_2}$ (10^{-9}/°C^2)
α_{xx}, α_{yy}		16.1	7.0
α_{zz}		4.1	−10
	ρ (Kg/m^3)	α_{ρ}^{1} (10^{-6}/°C)	α_{ρ}^{2} (10^{-9}/°C^2)
ρ	7450	−36.3	-

7.4 Study Problems

1. Find the TCD value for Langasite in a Y cut crystal at 90°.
2. Find the TCD value for SiO_2 in a X cut crystal at 45°.

Chapter 8

Acoustic Waves in Thin Layers

Aside from the Rayleigh-type elliptically polarized surface waves, the discovery of the transverse horizontal (TH) (which sometimes is also referred to as shear horizontal, SH) type Love waves in a thin layer provided us with the crucial insight into understanding the presence of the horizontal movement perpendicular to the direction of wave propagation. In this chapter, we will first take a close look of the mathematic foundation that Love established in the 1910s, and examine the formation of Love waves in various modes and their corresponding phase velocity. We will then discuss other types of SAWs traveling in thin layers including the Lamb waves.

8.1 Love Waves

In Section 6.5, we have encountered the Bleustein–Gulyaev type TH surface waves in certain crystals like the YX cut ZnO and CdS. In these cases, the occurrence of the Bleustein–Gulyaev type TH waves at the surface is the result of piezoelectric coupling. The Love type TH waves are different in that their existence requires a wave–guide surface layer to confine the wave energy near the surface, regardless whether there exists a piezoelectric coupling effect or not.

Bulk and Surface Acoustic Waves: Fundamentals, Devices, and Applications
Guigen Zhang

ISBN 978-981-4877-73-2 (Hardcover), 978-1-003-25662-5 (eBook)
www.jennystanford.com

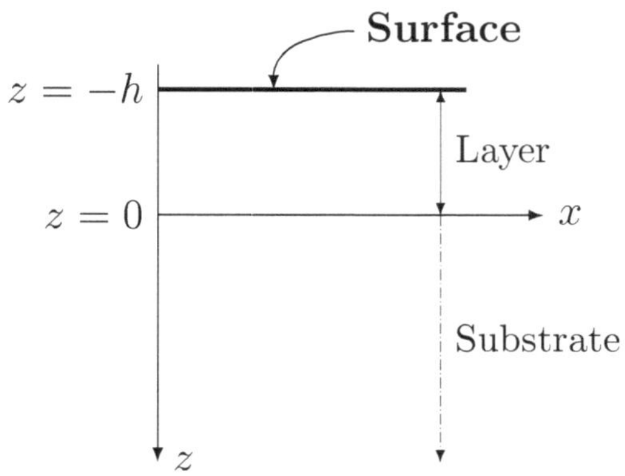

Figure 8.1 A section view of a thin layer material bonded to a thick substrate material.

In other words, the waves traveling in a thin layer bonded to a thick layer (or substrate) are known as Love waves. The generation and confinement of Love waves in such a thin layer require that the materials selected for the layer and the substrate must satisfy the condition that the bulk TH wave generated in the substrate travels faster than that of the bulk TH wave in the thin layer.

In this section, we will show how the governing equation for Love waves is solved. Figure 8.1 illustrates a section view of a thin layer material bonded to a thick bulk (or substrate) material. Considering meeting the surface boundary condition of maximum displacement at $z = -h$ and the continuity conditions at the interface of the layer and substrate at $z = 0$, along with the known trigonometric relationship $\cos(z) = [\exp(iz) + \exp(-iz)]/2$, we express the horizontal particle displacement, u_y, as proportional to

$$u_y \propto \begin{cases} \cos[kb(z+h)]e^{ik(vt-x)} & -h \leq z \leq 0 \\ e^{-ikbz}e^{ik(vt-x)} & z \geq 0 \end{cases}$$

With this expression, by taking the same steps as discussed in Sections 6.2.1 and 6.2.4 (also, the reader may refer to the discussions in Section 8.3), we note that the same governing equation applies to the TH wave in the thin layer, namely,

$$\left[c_{44}b^2 + c_{44} - \rho v^2\right]\{^\circ u_y\} = 0$$

Using the same concepts discussed in Section 6.2.4, we know that this wave has one b value and one component $({}^{\circ}u_y^2)$ in its null polarization vector, which will be regarded as ${}^{\circ}u_y^2 = 1$; hence, we express its wave field as

$$u_y = \begin{cases} \alpha_2^l \cos[kb_2^l(z+h)]e^{ik(vt-x)} & -h \le z \le 0 \\ \alpha_2^s e^{-ikb_2^s z} e^{ik(vt-x)} & z \ge 0 \end{cases} \tag{8.1}$$

in which α_2^l and α_2^s are constants, and b_2^l and b_2^s are the b values for the layer and substrate materials, respectively. Here again, since the selection of the displacement fields given in Eq. (8.1) not only satisfies the wave governing equation but also assures that the boundary conditions at the top surface ($z = -h$) will be met. Thus we only need to consider the two continuity conditions at the interface at $z = 0$, namely

$$u_z|_{z=0+} = u_z|_{z=0-} \quad \text{and} \quad \tau_{yz}|_{z=0+} = \tau_{yz}|_{z=0-} \tag{8.2}$$

Example 1: Use a thin layer of a material having a thickness of 2λ, $c_{44} = 3.12\times10^{10}$ N/m^2, and $\rho = 2203$ kg/m^3 bonded to a substrate material with $c_{44} = 7.95\times10^{10}$ N/m^2, and $\rho = 2239$ kg/m^3 to illustrate how Love wave propagate, determine their phase velocity, and sketch the displacement fields.

Answer: The selection of the layer and substrate materials provides a phase velocity for the bulk TH wave of $v_{TH} = 3763.2$ m/s in the layer and of $v_{TH} = 5842.5$ m/s in the substrate, thus meeting the requirement for the generation and confinement of TH wave travels in the layer. By substituting the given material constants into

$$\det\left[c_{44}b^2 + c_{44} - \rho v^2\right] = 0$$

for both the layer and substrate materials, we obtain the following:

$$b_2^l = -\sqrt{\rho v^2/c_{44} - 1} = -\sqrt{7.061\times10^{-8}v^2 - 1}$$

$$b_2^s = -\sqrt{\rho v^2/c_{44} - 1} = -\sqrt{2.930\times10^{-8}v^2 - 1}$$

With the b values, we obtain the following displacements and stresses at the interface ($z = 0$):

$$u_z|_{z=0+} = \alpha_2^l \cos(kh\sqrt{7.061 \times 10^{-8} v^2 - 1})$$

$$u_z|_{z=0-} = \alpha_2^s$$

$$\sigma_z|_{z=0+} = 3.12 \times 10^{10} \times \alpha_2^l k \sin[kh(7.061 \times 10^{-8} v^2 - 1)]$$

$$\sigma_z|_{z=0-} = -7.95i \times 10^{10} \times \alpha_2^s k\sqrt{2.930 \times 10^{-8} v^2 - 1}$$

By substituting these expressions into the boundary conditions given in Eq. (8.2), and after variable separation by α_2^l and α_2^s, we obtain a 2×2 matrix equation. By setting the determinant of the matrix to zero, we find that the resulting characteristic equation has multiple roots. This reflects the fact that Love waves are dispersive waves made of a series of sub-waves traveling at different phase velocities. The lowest phase velocity case represents the first-mode Love wave, which is

$$v = 3790.34 \text{ m/s}$$

With this velocity value, we further find its corresponding parameters

$$b_2^l = -0.120, \ b_2^s = -0.761i, \ \alpha_2^l = 1, \ \alpha_2^s = 0.062$$

With these known values along with $k = 2\pi/\lambda$ we obtain the following relative amplitude of the particle displacements in the layer and the substrate for the first-mode Love wave:

$$u_y = \begin{cases} \cos(0.754z/\lambda + 1.509) & -2\lambda \le z \le 0 \\ 0.062 \exp(-4.782z/\lambda) & z \le 0 \end{cases}$$

The next several roots represent the second and third modes of Love waves, and so on. For the second mode, its phase velocity is

v_2 = 3999.5 m/s, and its relative amplitudes of the particle displacements in the layer and the substrate are

$$u_y = \begin{cases} \cos(2.261z/\lambda + 4.521) & -2\lambda \le z \le 0 \\ -0.190\exp(-4.580z/\lambda) & z \le 0 \end{cases}$$

The third mode has a phase velocity of v_3 = 4384.6 m/s, and its relative amplitudes of the particle displacements in the layer and the substrate are

$$u_y = \begin{cases} \cos(3.756z/\lambda + 7.513) & -2\lambda \le z \le 0 \\ 0.335\exp(-4.153z/\lambda) & z \le 0 \end{cases}$$

The fourth mode has a phase velocity of v_4 = 4895.7 m/s, and its relative amplitudes of the particle displacements in the layer and the substrate are

$$u_y = \begin{cases} \cos(5.228z/\lambda + 10.456) & -2\lambda \le z \le 0 \\ -0.513\exp(-3.429z/\lambda) & z \le 0 \end{cases}$$

The fifth mode has a phase velocity of v_4 = 5472.7 m/s, and its relative amplitudes of the particle displacements in the layer and the substrate are

$$u_y = \begin{cases} \cos(6.634z/\lambda + 13.268) & -2\lambda \le z \le 0 \\ 0.764\exp(-2.200z/\lambda) & z \le 0 \end{cases}$$

Apparently, the phase velocity in all these modes remains greater than that of the bulk TH wave in the layer (3763.2 m/s) and less than that in the substrate (5842.5 m/s). This suggests that the phase velocity for all modes of Love waves remains within a range bonded at the lower end by the phase velocity of the bulk TH wave in the layer and at the upper end by the phase velocity of the bulk TH wave in the substrate.

Figure 8.2 shows the relative amplitudes of the particle displacements in the layer and the substrate for the first- through

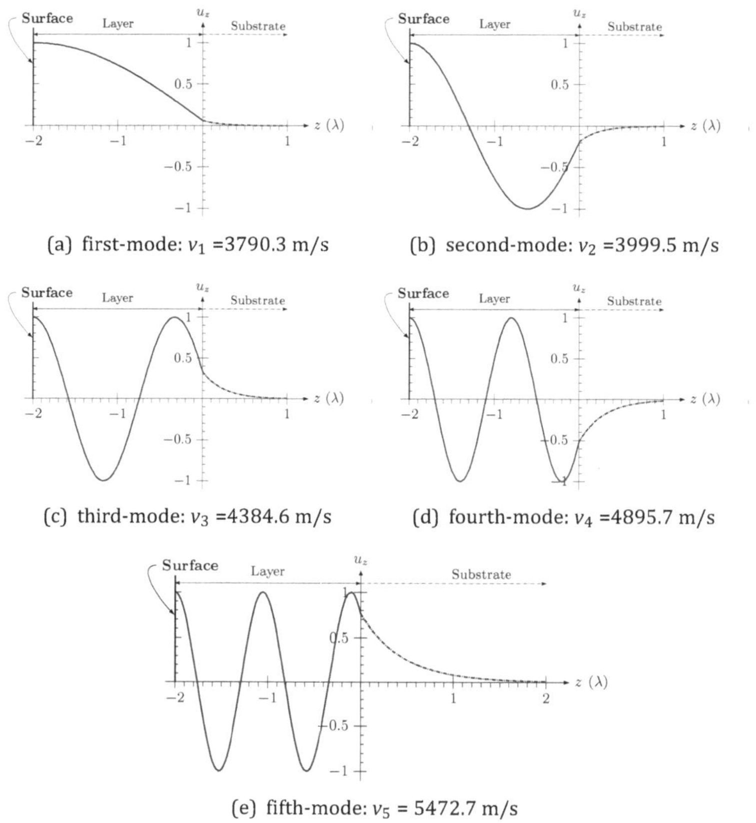

(a) first-mode: v_1 =3790.3 m/s

(b) second-mode: v_2 =3999.5 m/s

(c) third-mode: v_3 =4384.6 m/s

(d) fourth-mode: v_4 =4895.7 m/s

(e) fifth-mode: v_5 = 5472.7 m/s

Figure 8.2 Relative amplitude of particle displacement caused by Love waves in a thin layer with a thickness $h = 2\lambda$ bonded to a thick substrate in the first five modes.

fifth-mode Love waves, respectively. In all cases, the amplitude of the Love wave is at its peak, or maximum, at the surface, and it dies down gradually in the substrate as depth increases. Moreover, except in the case of the first mode, in all other modes the particle displacement oscillates in the layer with vibration amplitude changes sign from positive to negative at least once. Aside from the increase in wave phase velocity with the order of the modes, the number of oscillation peaks in the displacement amplitude in the layer also increases. For instance, there is one peak in the first mode (Figure 8.2(a)), two peaks in the second mode (Figure 8.2(b)), three peaks in the third mode (Figure 8.2(c)), and so on. Figure 8.2(e) shows the results of the fifth mode, in which the particle displacement oscillates five times in the layer with three positive peaks and two negative peaks before it decays out gradually with depth in the substrate. In comparison with the piezoelectric induced Bleustein–Gulyaev type TH surface waves (see Section 6.5) in which the amplitude of particle displacement dies down gradually as depth increases in a simple decaying pattern as shown in Figures 6.16 and 6.17, the profile of the particle displacement of the Love-type TH surface waves, particularly those of higher modes, is totally different and much more complicated.

8.2 Effect of Layer Thickness on Love Waves

Aside from the influence of materials, Love waves are affected also by the thickness of the layer. Table 8.1 lists the phase velocity of the first through fifth modes of Love waves traveling in a thin layer with two different thicknesses. Apparently, for the same material, the thinner the layer the faster the wave travels in each of the five modes. Please note that all these phase velocity values are within the range between v_{TH} = 3763.2 m/s (in the layer) and v_{TH} = 5842.5 m/s (in the substrate).

Table 8.1 Phase velocity of Love waves (m/s)

Thickness	1st	2nd	3rd	4th	5th
$h = 1.75\lambda$	3798.2	4065.3	4546.4	5163.2	5789.0
$h = 2.00\lambda$	3790.3	3999.5	4384.6	4895.7	5472.7

Figure 8.3(a) shows the relative amplitudes of the particle displacements of the fifth-mode Love wave traveling in a layer with a thickness h = 1.75λ bonded to a thick substrate, and Figure 8.3(b) shows the relative amplitudes of the particle displacements of the

third-mode Love wave traveling in a layer with a thickness $h = 1.88\lambda$ bonded to a thick substrate. Figure 8.4 shows the relative amplitudes of the particle displacements of the first-mode Love wave traveling in a layer with a thickness $h = 3.5\lambda$ bonded to a thick substrate. Comparing with the other two first-mode cases listed in Table 8.1, this case has the lowest phase velocity due to its thickest layer, again confirming that the thinner the layer the faster the Love wave travels.

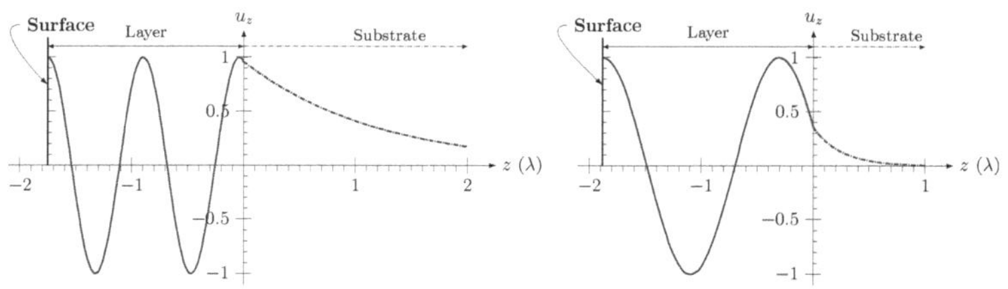

(a) h = 1.75λ, fifth mode, $v_5 = 5789.0$ m/s (b) h = 1.88λ, third mode, $v_3 = 4455.3$ m/s

Figure 8.3 Particle displacements by Love wave in a thin layer bonded to a thick substrate.

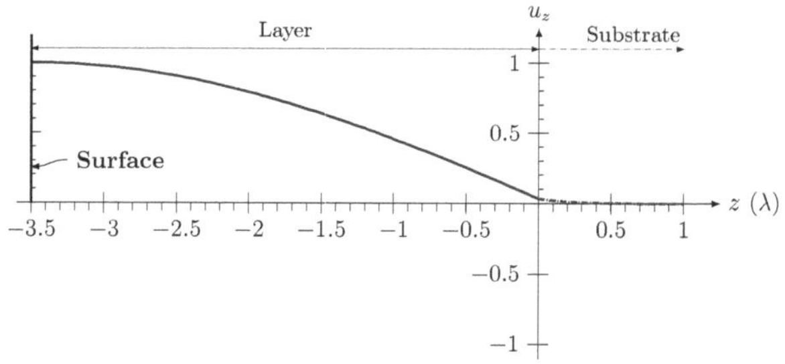

Figure 8.4 Particle displacement caused by the 1st mode of Love wave ($v_1 = 3772.5$ m/s) in a thin layer with a thickness $h = 3.5\lambda$ bonded to a thick substrate.

8.3 Lamb Waves

We have learned that Rayleigh waves are waves generated at the surface of a semi-infinite body, i.e., an infinitely large body bonded by a large flat plane. In 1916, Horace Lamb (1849–1934) extended the concept of Rayleigh waves to a plate structure, which can be regarded as an infinitely large body bonded by two parallel flat planes or surfaces placed very close (on the order of a fraction of to a few wavelengths) to each other. In this section, we will examine the generation and propagation of acoustic waves in a very thin plate.

Figure 8.5 illustrates a section view of a thin plate in which Lamb waves will be generated which will then propagate along the x direction. For the consideration of particle displacements at both the upper ($z = h$) and lower ($z = -h$) surfaces of the plate, we express the particle displacement fields as proportional to

$$\{u\} \propto \left[e^{-ikbz} \pm e^{ikbz}\right] e^{ik(vt-x)} \tag{8.3}$$

thus the particle displacements take the forms of

$$\begin{Bmatrix} u_x \\ u_y \\ u_z \end{Bmatrix} = \begin{Bmatrix} {}^{\circ}u_x \\ {}^{\circ}u_y \\ {}^{\circ}u_z \end{Bmatrix} \left[e^{-ikbz} \pm e^{ikbz}\right] e^{ik(vt-x)}$$

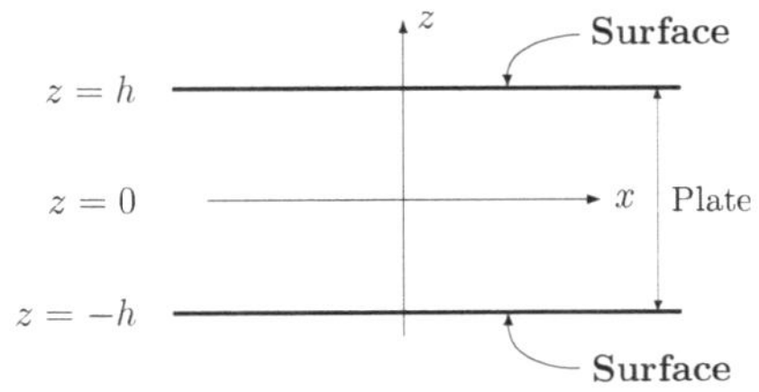

Figure 8.5 A section view of a thin plate in which Lamb waves will generate and propagate.

From this expression we note that taking a partial derivative of these displacement fields with respect to either x, y or z is equivalent to multiplying a constant factor as listed below

$$\frac{\partial}{\partial x} = -ik, \quad \frac{\partial}{\partial y} = 0, \quad \frac{\partial}{\partial z} = \pm ikb = -ikb', \quad \frac{\partial^2}{\partial t^2} = -k^2 v^2$$

where $b' = \mp b$. By taking the same steps as discussed in Section

6.2.1, we obtain the following governing equation for an isotropic material:

$$\begin{bmatrix} c_{11}+c_{44}b'^2-\rho v^2 & 0 & (c_{11}-c_{44})b' \\ 0 & c_{44}b'^2+c_{44}-\rho v^2 & 0 \\ (c_{11}-c_{44})b' & 0 & c_{11}b'^2+c_{44}-\rho v^2 \end{bmatrix} \begin{Bmatrix} {}^\circ u_x \\ {}^\circ u_y \\ {}^\circ u_z \end{Bmatrix} = 0$$

After decoupling the two independent groups of waves, we have

$$\begin{bmatrix} c_{11}+c_{44}b'^2-\rho v^2 & (c_{11}-c_{44})b' \\ (c_{11}-c_{44})b' & c_{11}b'^2+c_{44}-\rho v^2 \end{bmatrix} \begin{Bmatrix} {}^\circ u_x \\ {}^\circ u_z \end{Bmatrix} = 0 \qquad (8.4)$$

for governing the coupled L and TV waves, and

$$\left[c_{44}b'^2+c_{44}-\rho v^2\right]\left\{{}^\circ u_y\right\} = 0 \qquad (8.5)$$

for governing the TH wave. Note that with b' replacing b, Eq. (8.4) is essentially the same as that for the Rayleigh waves, and Eq. (8.5) the same as that for the transverse wave in Section 6.2.1. Solutions to these governing equations can be found by setting the determinant of their characteristic matrix to zero, namely,

$$\det\begin{bmatrix} c_{11}+c_{44}b'^2-\rho v^2 & (c_{11}-c_{44})b' \\ (c_{11}-c_{44})b' & c_{11}b'^2+c_{44}-\rho v^2 \end{bmatrix} = 0,$$
$$\text{and } \det\left[c_{44}b'^2+c_{44}-\rho v^2\right] = 0$$

After term expansion, it becomes clear that all b' terms appear in b'^2. Since $b' = \mp b$, the above equations are exactly the same as

$$\det\begin{bmatrix} c_{11}+c_{44}b^2-\rho v^2 & (c_{11}-c_{44})b \\ (c_{11}-c_{44})b & c_{11}b^2+c_{44}-\rho v^2 \end{bmatrix} = 0,$$
$$\text{and } \det\left[c_{44}b^2+c_{44}-\rho v^2\right] = 0$$

This means that even with the consideration of the displacement fields given in Eq. (8.3), the waves' characteristic equations remain the same.

8.4 Lamb Waves of the Zeroth Order

The coupled L and TV waves in a thin plate with a thickness on the order of a fraction of to a few wavelengths are known as Lamb waves. In this section, we will solve the following characteristic equation with particle displacements in the form given in Eq. (8.3):

$$\det\begin{bmatrix} c_{11}+c_{44}b^2-\rho v^2 & (c_{11}-c_{44})b \\ (c_{11}-c_{44})b & c_{11}b^2+c_{44}-\rho v^2 \end{bmatrix}=0 \tag{8.6}$$

Apparently, Eq. (8.6) provides two roots as solutions for b in terms of wave phase velocity (v), b_1 and b_3, and for each b we can find a corresponding null polarization vector that satisfies Eq. (8.6)

$$\begin{Bmatrix} {}^{\circ}u_x^1 \\ {}^{\circ}u_z^1 \end{Bmatrix}, \quad \begin{Bmatrix} {}^{\circ}u_x^3 \\ {}^{\circ}u_z^3 \end{Bmatrix}$$

8.4.1 Lamb Waves with Asymmetric Flexure-Mode Displacements

The asymmetric flexure mode describes a situation where the particle displacements in the upper surface and lower surface resemble a flexure or bending type of deformation. This type of deformation can be captured by the following displacement fields:

$$\begin{Bmatrix} u_x \\ u_y \\ u_z \end{Bmatrix} = \begin{Bmatrix} {}^{\circ}u_x^1\alpha_1\sinh(-ikb_1z)+{}^{\circ}u_x^3\alpha_3\sinh(-ikb_3z) \\ 0 \\ {}^{\circ}u_z^1\alpha_1\cosh(-ikb_1z)+{}^{\circ}u_z^3\alpha_3\cosh(-ikb_3z) \end{Bmatrix} e^{ik(vt-x)} \tag{8.7}$$

where α_1 and α_3 are constants that adjust the contributions of the two partial waves towards each of the two components of Lamb waves, and they will need to be solved along with the phase velocity of the waves, v.

When the plate is mechanically free from any constraints, both the upper and lower surfaces ($z = \pm h$) have the following zero-stress boundary conditions:

$$\sigma_z|_{z=\pm h} = \tau_{xz}|_{z=\pm h} = 0$$

Since the selection of the displacement fields given in Eq. (8.7) not only satisfies the wave governing equation (i.e., Eq. (8.6)) but also assures that the boundary conditions at both the upper and lower surfaces will be met simultaneously, albeit in an asymmetric manner. Thus we only need to consider the two conditions at either $z = h$ or $z = -h$.

With Eq. (8.7) we write the stress vector as

$$\{T\} = \begin{Bmatrix} \sigma_x \\ \sigma_y \\ \sigma_z \\ \tau_{yz} \\ \tau_{xz} \\ \tau_{xy} \end{Bmatrix} = -ik[c]\left\{ \begin{bmatrix} 1 & 0 & 0 \\ 0 & 0 & 0 \\ 0 & 0 & b_1 \\ 0 & b_1 & 0 \\ b_1 & 0 & 1 \\ 0 & 1 & 0 \end{bmatrix} \begin{Bmatrix} {}^{\circ}u_x^1 \alpha_1 \cosh(-ikb_1 z) \\ 0 \\ {}^{\circ}u_z^1 \alpha_1 \sinh(-ikb_1 z) \end{Bmatrix} \right.$$

$$\left. + \begin{bmatrix} 1 & 0 & 0 \\ 0 & 0 & 0 \\ 0 & 0 & b_3 \\ 0 & b_3 & 0 \\ b_3 & 0 & 1 \\ 0 & 1 & 0 \end{bmatrix} \begin{Bmatrix} {}^{\circ}u_x^3 \alpha_3 \cosh(-ikb_3 z) \\ 0 \\ {}^{\circ}u_z^3 \alpha_3 \sinh(-ikb_3 z) \end{Bmatrix} \right\} e^{ik(vt-x)} \tag{8.8}$$

After applying the two zero-stress conditions at $z = h$, we will obtain two equations in terms of α_1 and α_3 as a function of phase velocity v. With further variable separation by α_1 and α_3 and setting the determinant of the resulting 2×2 matrix to zero, we can solve for the phase velocity of the Lamb waves.

Example 2: Use an isotropic elastic plate having a thickness of $2h$ with $kh = 3$, $c_{11} = 10.7\times10^{10}$ N/m^2, $c_{44} = 2.8\times10^{10}$ N/m^2, and $\rho = 2700$ kg/m^3 to show how asymmetric flexure-mode Lamb waves form and propagate, determine their phase velocity, and sketch the displacement fields.

Answer: By substituting the given material constants into Eq. (8.6), we have

$$\det \begin{bmatrix} 2.8b^2 + 10.7 - 2.7 \times 10^{-7}v^2 & 7.9b \\ 7.9b & 10.7b^2 + 2.8 - 2.7 \times 10^{-7}v^2 \end{bmatrix} = 0$$

Since this is exactly the same equation as that solved in Section 6.2.1, we just relist the two roots here

$$b_1 = -\sqrt{\rho v^2/c_{11} - 1} = -\sqrt{2.523 \times 10^{-8}v^2 - 1}$$

$$b_3 = -\sqrt{\rho v^2/c_{44} - 1} = -\sqrt{9.643 \times 10^{-8}v^2 - 1}$$

along with the two corresponding null polarization vectors

$$\begin{Bmatrix} {}^{\circ}u_x^1 \\ {}^{\circ}u_z^1 \end{Bmatrix} = \begin{Bmatrix} 1 \\ \sqrt{\rho v^2/c_{11} - 1} \end{Bmatrix} = \begin{Bmatrix} 1 \\ \sqrt{2.523 \times 10^{-8}v^2 - 1} \end{Bmatrix}$$

$$\begin{Bmatrix} {}^{\circ}u_x^3 \\ {}^{\circ}u_z^3 \end{Bmatrix} = \begin{Bmatrix} 1 \\ \dfrac{-1}{\sqrt{\rho v^2/c_{44} - 1}} \end{Bmatrix} = \begin{Bmatrix} 1 \\ \dfrac{-1}{\sqrt{9.643 \times 10^{-8}v^2 - 1}} \end{Bmatrix}$$

By substituting these b values and null polarization vectors into stress expressions given in Eq. (8.8), along with letting $z = h$ for surface stresses, we can find σ_z and τ_{xz} in terms of α_1, α_3 and v. By applying the two zero-stress boundary conditions of $\sigma_{z|_{z=h}} = \tau_{xz|_{z=h}} = 0$, and writing them in a 2×2 matrix equation through variable separation by α_1 and α_3, we obtain the characteristic equation after setting the determinant of the 2×2 matrix to zero. The solutions of the obtained characteristic equation gives the values for the phase velocity (v) of the Lamb waves, of which the smallest root is

$$v = 2851.8 \text{ m/s}$$

Knowing this v value, we can find the following:

$$b_1 = -0.892i,\ b_3 = -0.465i;\ {}^{\circ}u_x^1 = 1,\ {}^{\circ}u_z^1 = -0.892i,\ {}^{\circ}u_x^3 = 1,$$
$${}^{\circ}u_z^3 = -2.153i$$

and

$$\alpha_3 = -2.321\alpha_1$$

By plugging these obtained values into Eq. (8.7), we can find the displacement fields. After normalizing the field magnitude with u_z at the surface (i.e., $u_{z|z=h}$), thus with $k = 2\pi/\lambda$, we express the displacement fields as follows:

$$\{u\} = \begin{Bmatrix} u_x \\ u_y \\ u_z \end{Bmatrix} = \begin{Bmatrix} 0.238\sinh(0.892kz) - 0.554\sinh(0.464kz) \\ 0 \\ (1.193\cosh(0.464kz) - 0.213\cosh(0.892kz))\, i \end{Bmatrix} e^{ik(vt-x)}$$

Apparently, like Rayleigh waves, Lamb waves also possess the L and TV wave components. The two waves exhibit a phase shift of $\pi/2$, indicating that the particle displacements of Lamb waves are also elliptic in nature. Figure 8.6 shows the relative magnitude of the two displacements plotted as a function of plate thickness from the upper surface ($h = 3/k$) to the lower surface ($h = -3/k$). As expected from the flexure-mode displacement expressions, u_x is an odd function and u_z an even function of z, thus at the two surfaces ($h = \pm 3/\text{k}$), u_x has the same magnitude with opposite signs and u_z has the same magnitude with the same sign.

Figure 8.7 sketches an illustration of the particle displacements induced by flexure-mode Lamb waves in a xz-view, with the deformed meshes sketched in a way that both the x and z displacements are in the right proportion to the respective displacement fields.

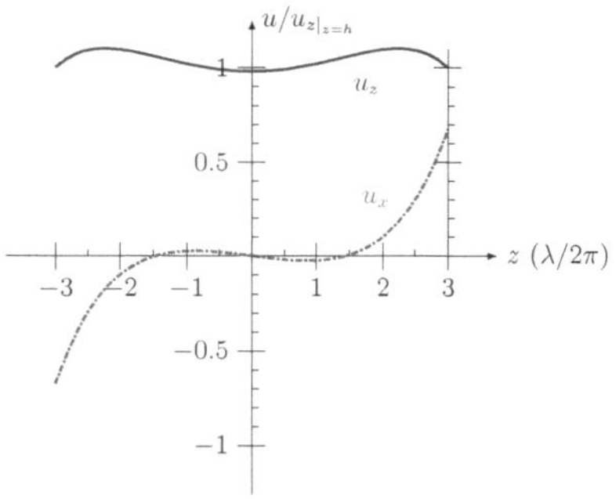

Figure 8.6 Relative amplitude of particle displacements of flexure-mode Lamb waves in a thin isotropic plate with a thickness of $2h$, where $kh = 3$ or $h = 3\lambda/2\pi$ (v = 2851.88 m/s).

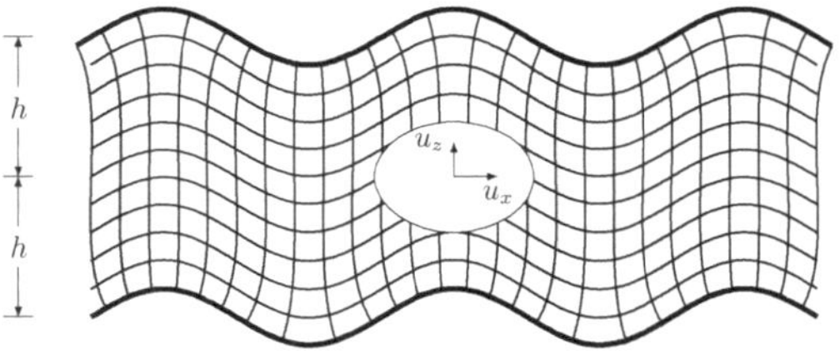

Figure 8.7 Illustration of particle displacements of asymmetric flexure-mode Lamb waves in a thin isotropic plate with a thickness of $2h$.

8.4.2 Lamb Waves with Symmetric Dilation-Mode Displacements

The symmetric dilation mode describes a situation where the particle displacements in the upper surface and lower surface resemble a dilation type of deformation, and this type of deformation can be captured by the following displacement fields:

$$\begin{Bmatrix} u_x \\ u_y \\ u_z \end{Bmatrix} = \begin{Bmatrix} {}^\circ u_x^1 \alpha_1 \cosh(-ikb_1 z) + {}^\circ u_x^3 \alpha_3 \cosh(-ikb_3 z) \\ 0 \\ {}^\circ u_z^1 \alpha_1 \sinh(-ikb_1 z) + {}^\circ u_z^3 \alpha_3 \sinh(-ikb_3 z) \end{Bmatrix} e^{ik(vt-x)} \tag{8.9}$$

Again, the selection of such displacement fields not only satisfies the wave governing equation (i.e., Eq. (8.6)) but also assures that the boundary conditions at both the upper and lower surfaces will be met simultaneously in a symmetric manner. Thus for a plate with zero-stress at the two surfaces ($z = \pm h$), we can use the following zero-stress boundary conditions:

$$\sigma_z|_{z=\pm h} = \tau_{xz}|_{z=\pm h} = 0$$

to solve for the phase velocity and displacement fields for this type of Lamb waves. In this case, the stress vector is as follows for the particle displacements given in Eq. (8.9):

$$\{T\} = \begin{Bmatrix} \sigma_x \\ \sigma_y \\ \sigma_z \\ \tau_{yz} \\ \tau_{xz} \\ \tau_{xy} \end{Bmatrix} = -ik[c]\left\{ \begin{bmatrix} 1 & 0 & 0 \\ 0 & 0 & 0 \\ 0 & 0 & b_1 \\ 0 & b_1 & 0 \\ b_1 & 0 & 1 \\ 0 & 1 & 0 \end{bmatrix} \begin{Bmatrix} {}^\circ u_x^1 \alpha_1 \sinh(-ikb_1 z) \\ 0 \\ {}^\circ u_z^1 \alpha_1 \cosh(-ikb_1 z) \end{Bmatrix} \right.$$

$$\left. + \begin{bmatrix} 1 & 0 & 0 \\ 0 & 0 & 0 \\ 0 & 0 & b_3 \\ 0 & b_3 & 0 \\ b_3 & 0 & 1 \\ 0 & 1 & 0 \end{bmatrix} \begin{Bmatrix} {}^\circ u_x^3 \alpha_3 \sinh(-ikb_3 z) \\ 0 \\ {}^\circ u_z^3 \alpha_3 \cosh(-ikb_3 z) \end{Bmatrix} \right\} e^{ik(vt-x)} \tag{8.10}$$

Example 3: Use the same isotropic elastic plate given in Example 2 to show how symmetric dilation-mode Lamb waves propagate, determine their phase velocity, and sketch the displacement fields.

Answer: By taking the same steps as for the asymmetric flexure-mode case, we obtain the smallest phase velocity

$$v = 3336.9 \text{ m/s}$$

along with its corresponding parameters including

$$b_1 = -0.848i, \ b_3 = -0.272i$$

and

$$^\circ u_x^1 = 1,\ ^\circ u_z^1 = -0.848i,\ ^\circ u_x^3 = 1,\ ^\circ u_z^3 = 3.682,\ \alpha_3 = -4.323\alpha_1$$

By plugging all these obtained values into Eq. (8.9), we find the displacement fields as follows:

$$\{u\} = \begin{Bmatrix} u_x \\ u_y \\ u_z \end{Bmatrix} = \begin{Bmatrix} 0.161\cosh(0.848kz) - 0.695\cos(0.272kz) \\ 0 \\ (2.560\sin(0272kz) - 0.136\sinh(0.848kz))\, i \end{Bmatrix} e^{ik(vt-x)}$$

Figure 8.8 shows the relative magnitude of the two displacements plotted as a function of plate thickness. As expected, u_x is an even function and u_z an odd function of z, thus at the two surfaces ($h = \pm 3/\mathrm{k}$), u_x has the same magnitude with the same sign, and u_z has the same magnitude with opposite signs.

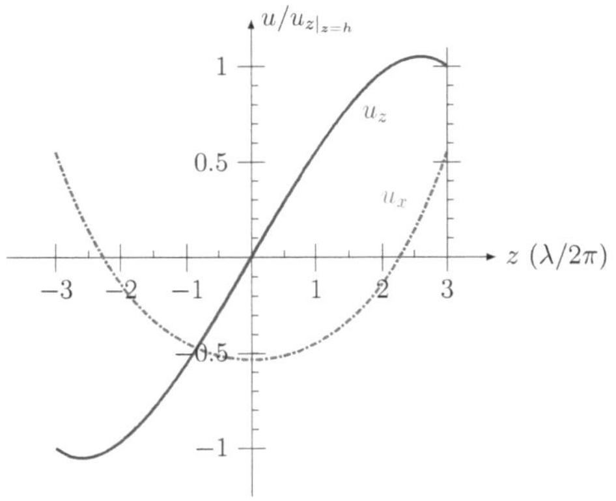

Figure 8.8 Relative amplitude of particle displacements of dilation-mode Lamb waves in a thin isotropic plate with a thickness of $2h$, where $kh = 3$ or $h = 3\lambda/2\pi$ (v = 3336.9 m/s).

Figure 8.9 sketches an illustration of the particle displacements induced by the dilation-mode Lamb waves in a xz-view, with the deformed meshes sketched in a way that both the x and z displacements are in the right proportion to the respective displacement fields.

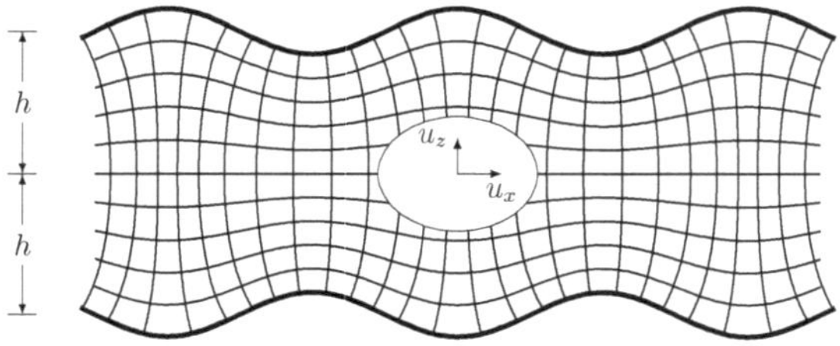

Figure 8.9 Illustration of particle displacements of symmetric dilation-mode Lamb waves in a thin isotropic plate with a thickness of 2h.

8.5 Lamb Waves of Higher Orders

The flexure and dilation modes of Lamb waves discussed and illustrated in Section 8.4 are known as the zeroth-order Lamb waves, because these waves are associated with the smallest roots of the characteristic equation in each of the asymmetric flexure-mode and symmetric dilation-mode cases. In fact, the characteristic equations in both cases have multiple roots. This indicates that Lamb waves are dispersive waves made of a series of sub-waves traveling at different phase velocities. To understand this concept better, let us take a close look of some of these roots and their corresponding wave phase velocity and particle displacements using a slightly thick plate ($kh = 5$) made of the same isotropic material.

Figure 8.10 shows the relative amplitude of particle displacements for the zeroth through third orders of Lamb waves in both the flexure and dilation modes. In general, the wave phase velocity increases as the order increases, and the phase velocity in the symmetric dilation mode is typically higher than that in the asymmetric flexure mode. Table 8.2 lists the phase velocity values for these different orders and modes of Lamb waves for the case with $kh = 5$. In all cases, the wave velocity for the symmetric dilation mode is typically higher than that for the asymmetric flexure mode.

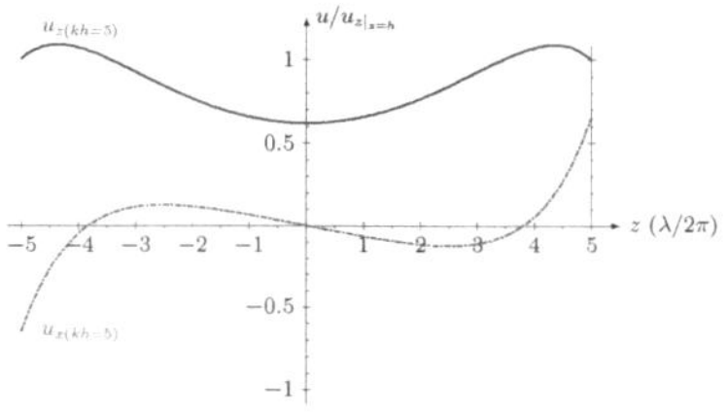

(a) zeroth-orderflexure-mode

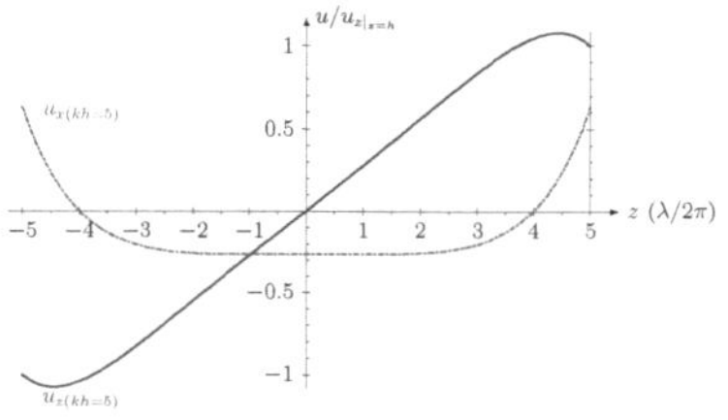

(b) zeroth-orderdilation-mode

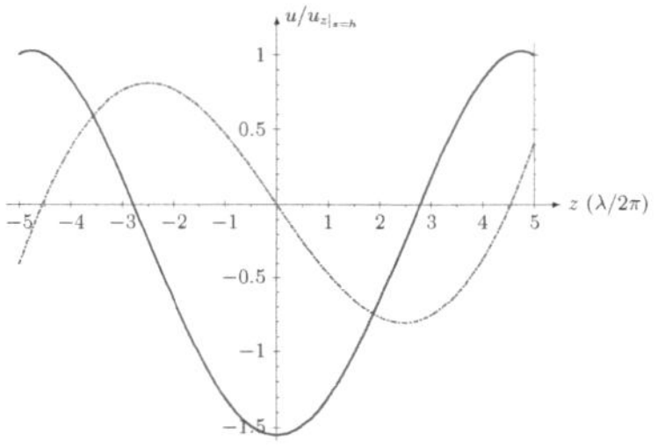

(c) first-orderflexure-mode

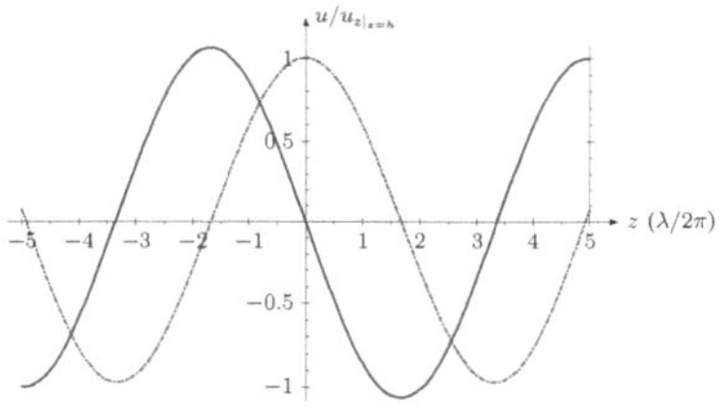

(d) first-orderdilation-mode

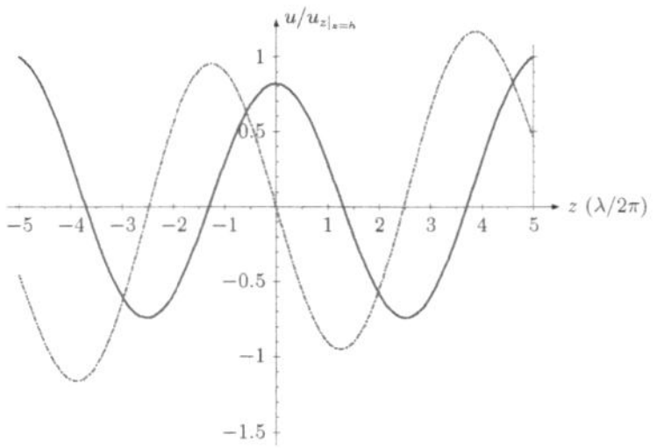

(e) second-orderflexure-mode

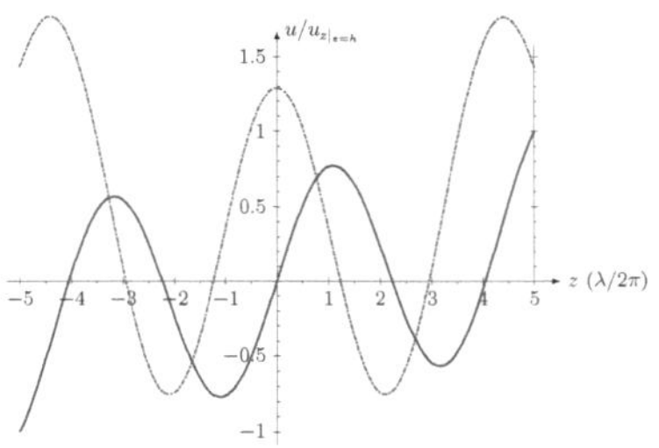

(f) second-orderdilation-mode

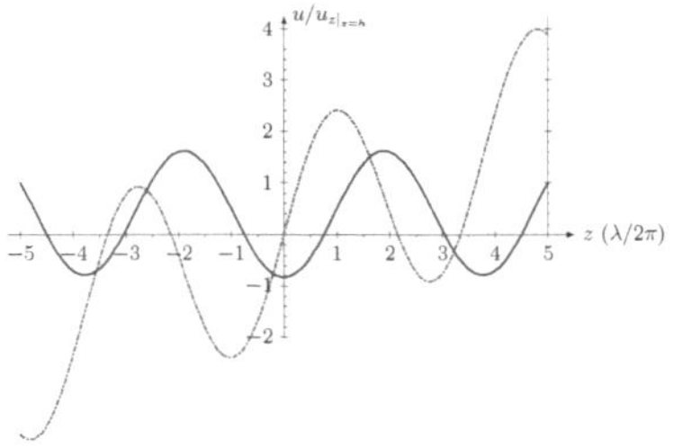

(g) third-orderflexure-mode

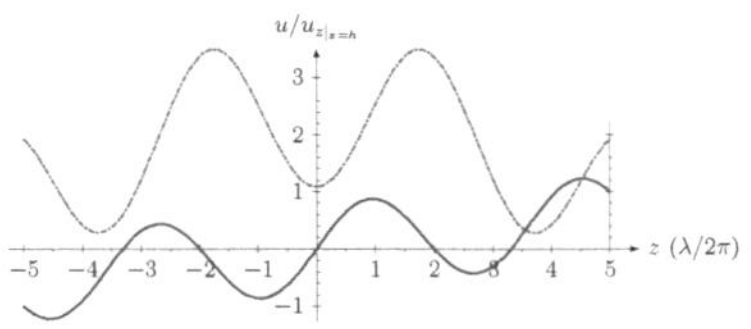

(h) third-order dilation-mode

Figure 8.10 Relative amplitude of particle displacements caused by Lamb waves in a thin plate with $kh = 5$ in the flexure mode and the dilation mode of the first four modes.

Table 8.2 Phase velocity of Lamb waves in a plate with $kh = 5$

Order	Asymmetric flexure mode (m/s)	Symmetric dilation mode (m/s)
0th	2962.1	3050.3
1st	3727.2	4423.0
2nd	5122.9	5720.9
3rd	6244.1	6461.3

8.6 Effect of Plate Thickness on Lamb Waves

The thickness of a plate affects not only the profile of the particle displacements but also the wave velocity. Figure 8.11 shows the profiles of particle displacements of the zeroth-order Lamb waves in plates with $kh = 3$ and $kh = 5$ plotted together.

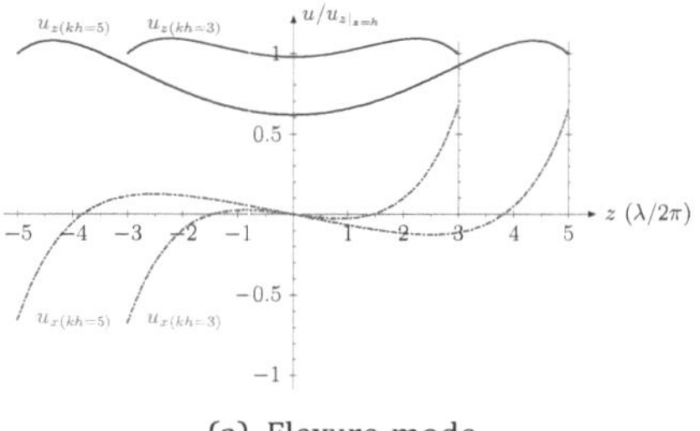

(a) Flexure-mode

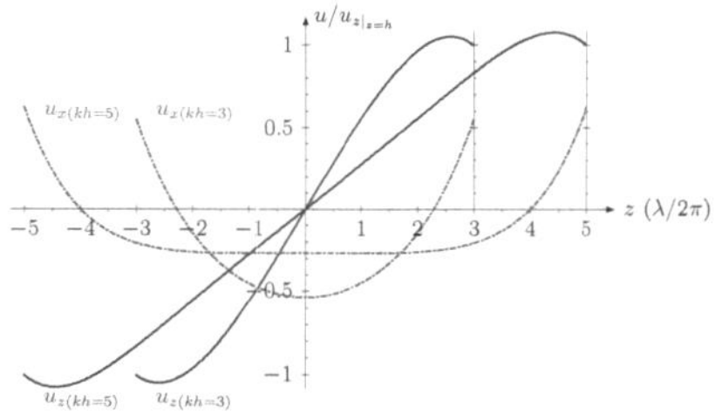

(b) Dilation-mode

Figure 8.11 Particle displacements in flexure-mode and dilation-mode Lamb waves.

Though the effect of plate thickness on wave velocity is better visualized from dispersion curves, a general trend can be discerned from the two cases discussed here. Table 8.3 lists the phase velocity values for different orders and modes of Lamb waves for the case with $kh = 3$. Comparing with the values listed in Table 8.2, we can see that for the zeroth-order Lamb wave, the phase velocity for the asymmetric flexure mode is lower for the $kh = 3$ plate than for the $kh = 5$ plate, and that for the symmetric dilation mode is higher for the $kh = 3$ plate than for the $kh = 5$ plate. For higher orders Lamb waves, the wave velocity for the $kh = 3$ plate is higher than that for the $kh = 5$ plate in both the asymmetric flexure mode and symmetric dilation mode across-the-board.

Table 8.3 Phase velocity of Lamb waves in a plate with $kh = 3$

Order	Asymmetric flexure mode (m/s)	Symmetric dilation mode (m/s)
0th	2851.9	3336.9
1st	4660.5	5606.4
2nd	6295.2	6717.5
3rd	6427.9	7913.2

8.7 Rayleigh Waves in Thin Layers

Finally, let us put the concepts of Rayleigh waves and Love waves together and examine a situation in which Rayleigh waves travel in a thin layer bonded to a thick substrate.

In this situation, the wave equation governing the SAWs in both the thin layer and thick substrate is solved with a continuity condition in displacements imposed at the interface between the layer and the substrate. Figure 8.12 shows the relative magnitudes of the particle displacements caused by the Rayleigh waves (v = 3408.8 m/s) in a thin layer of two different thicknesses bonded to a thick substrate. Note that the dashed lines in the substrate represent a hypothetic situation where the substrate is made of the same material at the layer material. Apparently, although continuity in both displacements is ensured, a change in the gradient of the

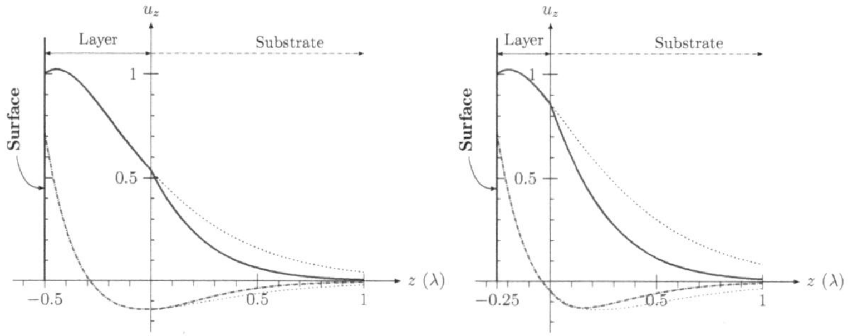

(a) $h = 0.5\lambda$ (b) $h = 0.25\lambda$

Figure 8.12 Particle displacement caused by Rayleigh wave ($v = 3408.8$ m/s) in a thin layer bonded to a thick substrate of two different thicknesses.

displacement fields is observed at the interface between the layer and substrate materials.

8.8 Study Problems

1. For a thin layer having a thickness of 2.5λ and $c_{44} = 5.35\times10^{10}$ N/m^2, and ρ = 3560 kg/m^3 bonded to a substrate material with $c_{44} = 27.55\times10^{10}$ N/m^2, and ρ = 4675 kg/m^3, find the velocity of the first mode Love wave.
2. For Study Problem 1, find the wave velocity for the second and third modes of Love waves.
3. By increasing the thickness of the think layer to 3.5λ, redo Problem 1 and Problem 2.
4. Use an isotropic elastic plate having a thickness of $2h$ with $kh = 4$, $c_{11} = 15.4\times10^{10}$ N/m^2, $c_{44} = 4.5\times10^{10}$ N/m^2, and ρ = 3540 kg/m^3 to find the phase velocity of the asymmetric flexure-mode Lamb waves.
5. Find the phase velocity of the symmetric dilation-mode Lamb waves for Study Problem 4.

Chapter 9

Applications and Devices

Bulk acoustic wave (BAW)- and surface acoustic wave (SAW)-based devices have been used for countless applications including oscillators, resonators, filters, and sensors, to just name a few. These devices have been utilized mainly for signal generation and transmission, time keeping, signal processing, as well as for various different types of actuators, detectors, sensors and biosensors. In this chapter, we will put the fundamental knowledge learned in previous chapters to practical use for the analysis of some common BAW and SAW devices using piezoelectric crystals of optimal properties and cuts. We will do so through case studies of analyzing a few BAW and SAW devices including resonators, filters and sensors. These case studies will be performed using a finite element modeling approach to not only examine the effect of various crystals in different cuts on the frequency responses, resonant frequencies and insertion loss of the respective devices but also at the same time provide some visualization of wave propagation either through the bulk part of a crystal substrate or along the surface of a device, as well as wave leakage into the bulk part of a piezoelectric substrate.

Bulk and Surface Acoustic Waves: Fundamentals, Devices, and Applications
Guigen Zhang

ISBN 978-981-4877-73-2 (Hardcover), 978-1-003-25662-5 (eBook)
www.jennystanford.com

9.1 BAW Devices

The most common type of BAW devices is that of resonators. One example of BAW resonators takes the shape of a simple circular disc. Another commonly known resonator is the tuning-fork resonator. As a resonator, aside from its commonly intended use for time keeping, Quartz Crystal Microbalance (QCM) is often used as a sensor for weighing the mass or estimating the thickness of a deposited film. Quartz (SiO_2) is probably the most abundant and cost-effective crystal material available. Because of this, quartz is also the most common material used for BAW resonators. Both the disc and tuning-fork types of resonators are made of quartz crystals in countless quantity. In the following sections, we will examine a disc and a tuning-fork resonator made of quartz crystals in various different cuts.

9.1.1 Disc Resonators

We begin with a case study in which a disc resonator is modeled using finite element (FE) modeling software COMSOL. A cautionary note: In using COMSOL to produce reasonably acceptable results for all the case studies, it seems necessary to code in the rotation matrices presented in Chapter 3 for the transformation of matrices of the stiffness, piezoelectric constant and permittivity. When investigating the effect of temperature on the performances of acoustic devices, it is then necessary to consider temperature-dependent material constants as we have discussed in Chapter 7.

To simplify matters we consider the resonator as having a circular disc shape with a diameter of 850 μm and a thickness of 330 μm. Figure 9.1 shows several views of the FE model.

For electromechanical excitation and BAWs generation in the resonator, electrodes are created on the upper and lower surfaces of the disc. In a top view (see Figure 9.1(a)), the positive electric terminal is highlighted in green, and in a bottom view (Figure 9.1(b)), the negative terminal is highlighted. In the modeling analysis, the positive terminal is supplied with an alternating potential signal with an amplitude of 10 V (Figure 9.1(c)) and the negative terminal is

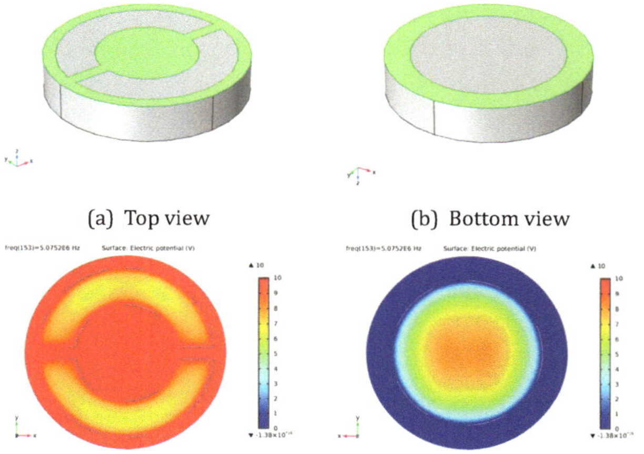

(a) Top view (b) Bottom view

(c) 10 V applied at the top electrode (d) Ground at the bottom electrode)

Figure 9.1 Illustration of a quartz resonator model developed for finite element analysis.

grounded (Figure 9.1(d)). For resonator materials, we use quartz crystal with several different Rt.-Y cuts listed in Table 9.1.

Table 9.1 Quartz crystals used for disc resonators

Cases	Cuts	Euler angles (ϕ, θ, ψ)
1	AT cut	$(0°, -54.75°, 0°)$
2	CT cut	$(0°, -52°, 0°)$
3	ST cut	$(0°, -47.25°, 0°)$
4	DT cut	$(0°, -142°, 0°)$
5	BT cut	$(0°, -139°, 0°)$

Figures 9.2 and 9.3 show the frequency response curves obtained from frequency sweep analyses of various resonators made of quartz with cuts specified in the legends. Apparently, disc resonators made of quartz in these different cuts have different resonant frequencies, defined as the frequency at which a peak vibration amplitude is reached. For example, the resonant frequency

is approximately 5.0757 MHz, 5.0850 MHz, and 5.1037 MHz for resonators made of quartz in the AT, CT and ST cuts, respectively as shown in Figure 9.2, and it is 5.6341 MHz and 5.6462 MHz for those in the DT and BT cuts respectively as shown in Figure 9.3.

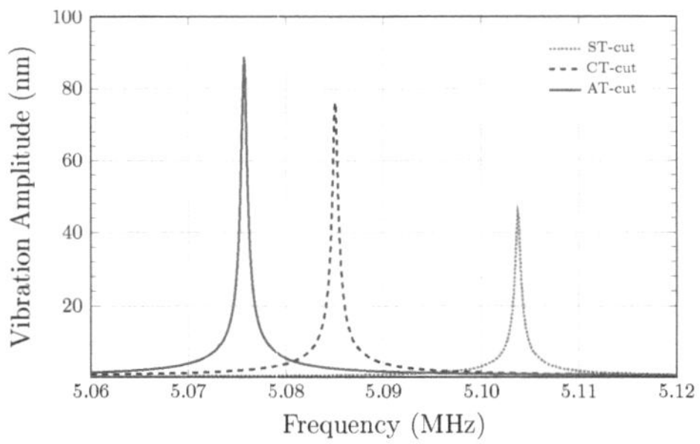

Figure 9.2 Frequency responses of a disc resonator made of quartz in ST, CT and AT cuts.

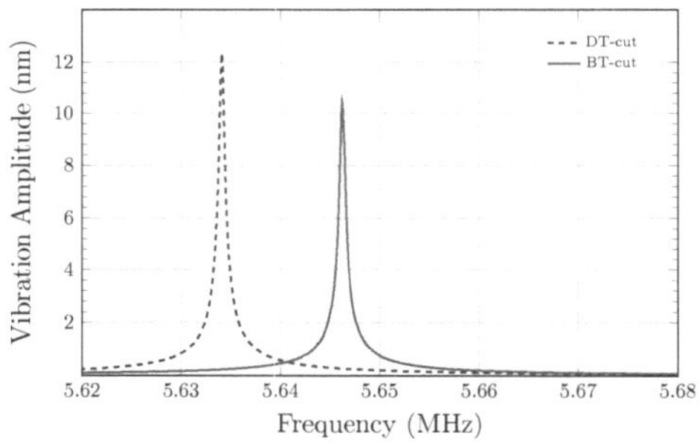

Figure 9.3 Frequency responses of a disc resonator made of quartz in BT and DT cuts.

By taking advantage of the computational modeling approach, we can visualize the deformation (albeit on an exaggerated scale) generated in the quartz resonators by the transverse shearing BAWs. Figure 9.4 shows some typical displacement results generated in the resonators made of quartz in AT and BT cuts. An interesting observation can be made from a close examination of the results. Of the two cut cases presented here, the AT and BT cuts quartz can be used either with the x direction (as defined in Chapter 6, and as shown in these figures) parallel with the X axis of the crystal, or

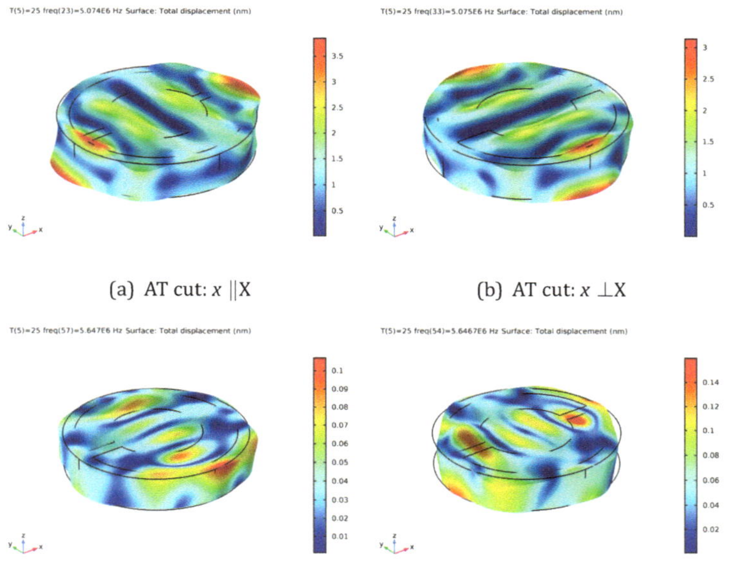

(a) AT cut: x ||X (b) AT cut: x ⊥X

(c) BT cut: x ||X (d) BT cut: x ⊥X

Figure 9.4 Particle displacements in disc resonators made of AT-cut and BT-cut quartz.

with the x perpendicular to the X axis. Using the cut convention we discussed in Chapter 6, in the former case, we refer the crystal as a Rt.-YX(||) cut, and in the latter case, we call it a Rt.-YX(⊥) cut.

Apparently, with AT-cut quartz, when the x axis is parallel with the X axis of the crystal (see Figure 9.4(a)), the ripples of the generated wave appear to be aligned perpendicular to the X axis, and when the x axis is perpendicular to the X axis (see Figure 9.4(b)), the generated ripples are aligned in parallel with the X axis. With BT-cut quartz, the situation seems opposite: When the x axis is parallel with the X axis (see Figure 9.4(c)), the generated ripples are aligned in parallel with X axis, and when the x axis is perpendicular to the X axis (see Figure 9.4(d)), the generated ripples are aligned perpendicular to the X axis. These results provide a direct and yet interesting way for us to picture the anisotropic nature of quartz crystal.

9.1.2 Temperature Influence on Quartz Disc Resonators

By considering the temperature effect discussed in Chapter 7 along with Eqs. (7.5), (7.7) and (7.11) to deal with temperature-induced changes in quartz material properties, we further evaluate thermal influence on the quartz disc resonators made of AT cut and BT cut quartz. To do that, we consider some selected temperature points from −40°C to 90°C and determine the frequency response curves of the disc resonators made of AT and BT cuts quartz, respectively.

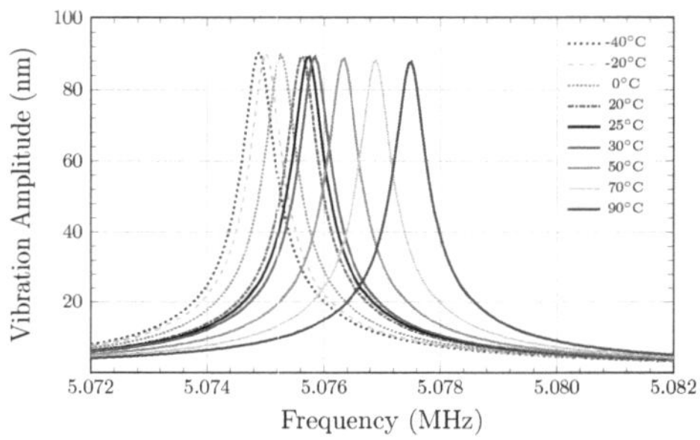

Figure 9.5 Temperature effect on the frequency response of a disc resonator made of AT-cut quartz.

Figure 9.5 shows the frequency response of the AT-cut resonator and Figure 9.6 that of the BT-cut resonator. Note that the frequency ranges in these two figures are different due to the difference in

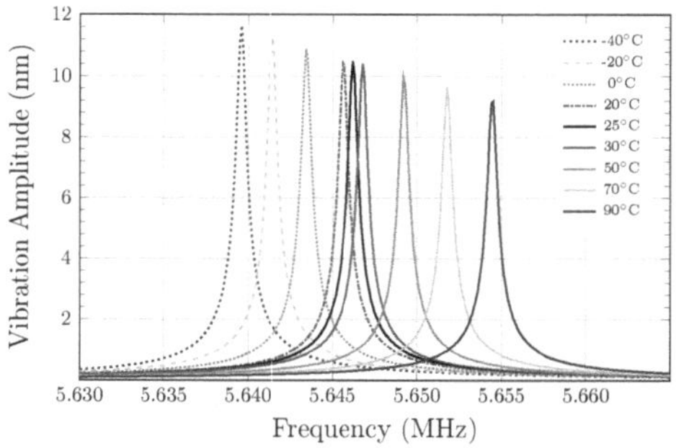

Figure 9.6 Temperature effect on the frequency response of a disc resonator made of BT-cut quartz.

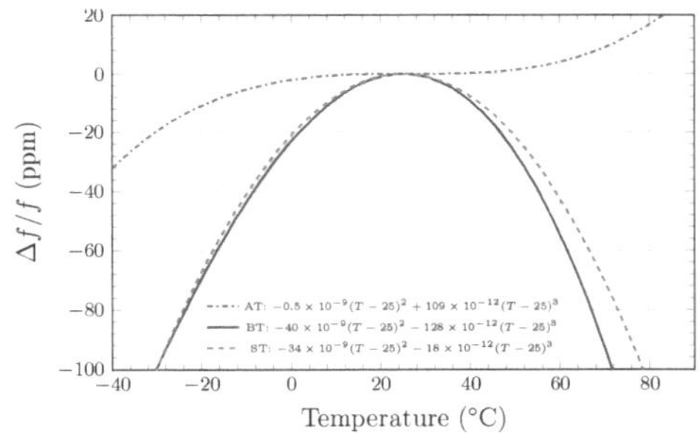

Figure 9.7 Temperature effect on the frequency shift in AT, BT abs ST cuts quartz.

the resonant frequency of the two resonators. In a close inspection, we can see that for the range of temperature change considered, namely, from −40°C to 90°C, the shift in the resonant frequency due to temperature change in the AT-cut resonator is much smaller than that in the BT-cut resonator. For example, for the BT-cut resonator, the overall shift in the resonant frequency from −40°C to 90°C is about 14.9 KHz, while for the AT-cut resonator, it is only 2.6 KHz, less than one fifth. Moreover, in both cases, the direction of the frequency shift is to the right as temperature increases.

It is worth noting that while the trend in which resonant frequency shifts due to temperature change for the AT quartz resonator is consistent with literature reported behavior, that for the BT quartz resonator seems different. Figure 9.7 shows the commonly referenced temperature-dependent frequency shift behavior of quartz in AT, BT and ST cuts based on the statistical regression studies reported by Ballato (1981). The variation of the AT curve indicates that as temperature increases from −40°C to 90°C, the resonant frequency is expected to increase. This is exactly the trend we can see in Figure 9.5. The variation of the BT curve, however, predicts a totally different trend. According to the BT curve, the resonant frequency is expected to increase as temperature increases −40°C to 25°C, and the rate of increase is higher than that in the AT curve. In this temperature range the predicted behavior seems consistent with the trend shown in Figure 9.6, but what happen at a temperature higher than 25° seems just opposite. The curve predicts a decreasing trend while the results in Figure 9.6 show a continued increasing trend in resonant frequency.

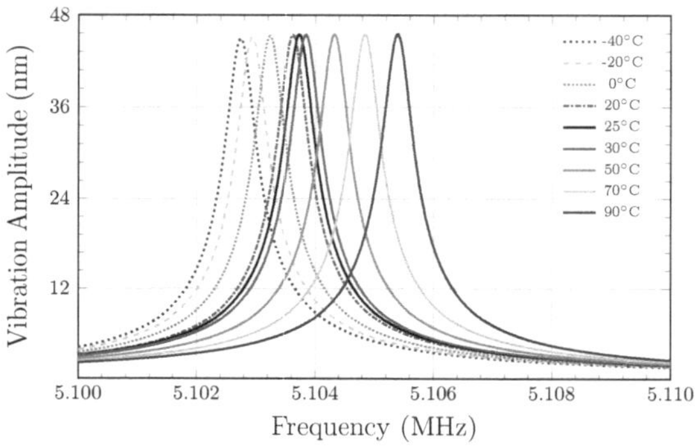

Figure 9.8 Temperature effect on the frequency response of a disc resonator made of ST-cut quartz.

Figure 9.8 shows the frequency response of a disc resonator made of ST-cut quartz. Aside from the difference in vibration amplitude, the span and shift in resonant frequency for the ST case seems almost the same as those for the At case (see Figure 9.5). Apparently, the frequency shifting behavior shown in Figure 9.8 is not consistent with the prediction by the ST curve in Figure 9.7. Reasons for these discrepancies may include (1) the effect of the presence of physical electrodes along with their materials and geometric shapes, which is technically ignored in the FE analysis, (2) inaccurate thermal constants and coefficients, and (3) other causes yet to be discovered. Judging from the fact that the same sets of thermal constants were used in Chapter 7 and produced reasonable TCD results, it is less likely these thermal constants are a major contributing factor. On the other hand, referring to Figure 7.4, the ST cut has zero TCD value but both the AT and BT cuts have negative TCD values. In a curious pursuit, we examine a case with a positive TCD value: 68° Rt.-Y. Figure 9.9 shows the obtained frequency responses. Interestingly, in this case, the resonant frequency shits in a different direction as temperature increases: Instead of increasing as we observed in the AT, ST and BT cases, the resonant frequency decreases as temperature increases, and the overall shift in the resonant frequency from −40°C to 90°C is about −14.5 KHz, close to that of the BT case.

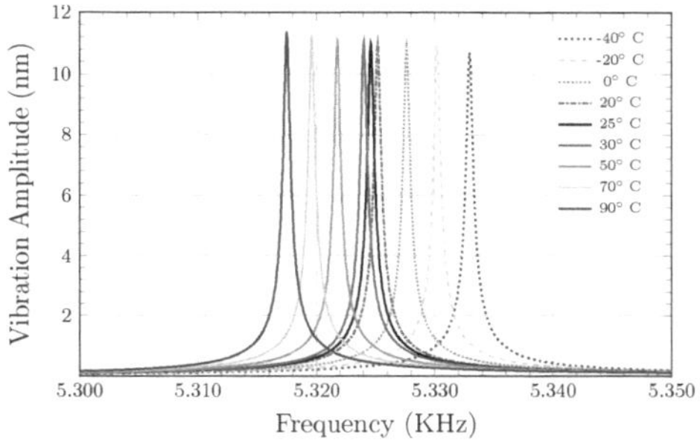

Figure 9.9 Temperature effect on the frequency response of a disc resonator made of 68° Rt.-Y.-cut quartz.

9.1.3 Disc Resonator as Sensors

As mentioned earlier, one common use of quartz resonators is quartz crystal microbalance, commonly known as QCM. The name "microbalance" suggests that the resonator is used for weighing the mass of an object at a micro level.

By using the inner circular region as highlighted in green in the lower right inset in Figure 9.10 as a sensing area for film deposition, the thickness of the deposited film can be detected by the resonant frequency shift in the QCM. In the case study presented here, we will see that the range of thickness detection can be even lower, from sub micro level to nano level. Figure 9.10 shows some selected results from an analysis in which a disc resonator made of AT-cut quartz is used as a QCM to detect the added mass in the form of uniform film deposition using COMSOL modeling. Note that the deposited mass has a density equal to that of gold. In this study we consider a control group without any film deposition and several testing groups with film deposition at various thickness values, ranging from 10 nm to 300 nm, and determine the frequency response curve of the disc resonator in each analysis condition.

From the results, we can see that thin film deposition with a thickness of mere 10s to few 100s nm will cause notable shifts in the resonant frequency of this QCM, and the thicker the film is, the

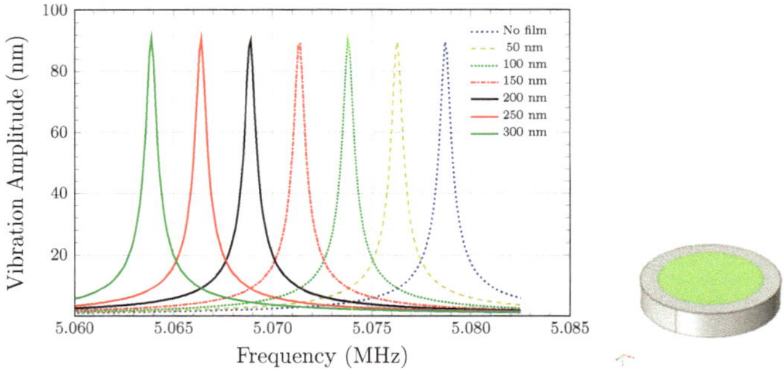

Figure 9.10 Effect of film deposition on the frequency response of a QCM thickness sensor made of AT-cut quartz.

more the frequency shift becomes. Specifically, compared with the control group, a shift by about 2.4 KHz, 7.3 KHZ and 14.8 MHz is seen when a film of 50 nm, 150 nm and 300 nm is deposited, respectively. Moreover, it is clear that a deposition of additional material onto the sensing area will cause the resonant frequency to shift to the left.

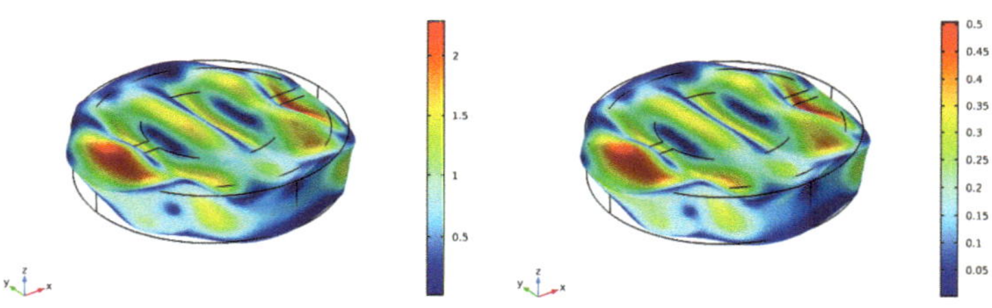

(a) No film (b) 300 nm film

Figure 9.11 Difference in induced stresses in the QCM with and without film deposition.

Figure 9.11 shows a comparison of the induced peak stresses in the QCM when there is no film deposition versus a deposition of 300 nm film in thickness. From the range of stresses indicated on the scale bars, we can see that the generated peak stresses in the QCM are reduced by more than a factor of four due to the attenuation effect of the added mass from the film deposition on wave generation and propagation.

9.1.4 Tuning-Fork Resonators

The second type of BAW devices we will examine is a tuning-fork resonator. As shown in Figure 9.12, in this finite element modeling case study, we consider a typical rectangular shape tuning fork with the following dimensions: overall length at 5 mm and thickness at 330 μm, and prong length, width and gap at 3 mm, 380 μm, and 200 μm, respectively. Again, to excite the piezoelectric effect, electrodes for positive and negative terminals are created on the various surfaces of the tuning fork according to the design of a commercial available device. Figure 9.12 shows several views of the developed model in which the positive terminal is highlighted in green and negative terminal in blue. In a nutshell, the

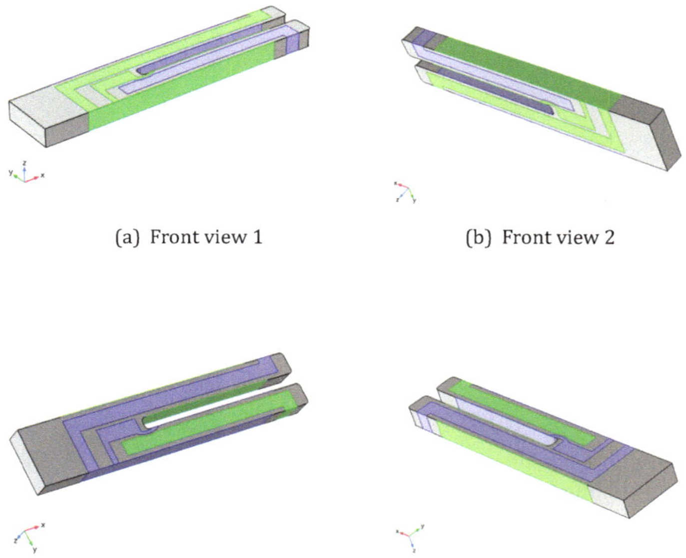

(a) Front view 1 (b) Front view 2

(c) Back view 1 (d) Back view 2

Figure 9.12 Illustration of a quartz resonator model. Green highlighted regions are for the positive terminal and blue highlighted regions are the negative terminal.

positive-terminal electrode runs from the upper and lower surfaces of one prong to the inner and outer side surfaces of another prong, and vice versa for the negative terminal electrode.

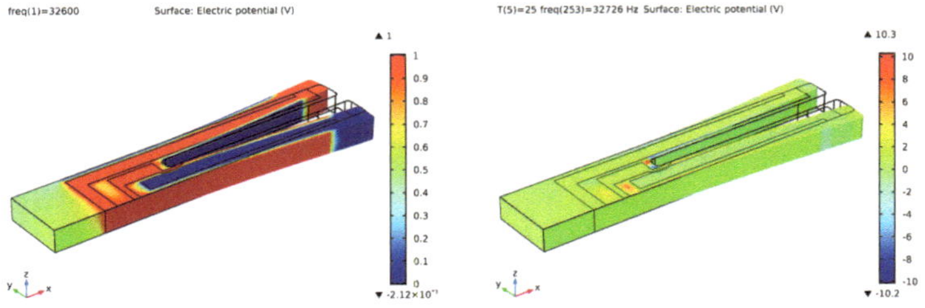

(a) Off resonance (b) At resonance

Figure 9.13 Tuning-fork resonator made of tuning-fork-cut quartz .

In this modeling analysis, the positive terminal is applied with a 1 V alternating potential signal and the negative terminal is grounded. For resonator materials, we use the commonly used tuning-fork and AT cuts as listed in Table 9.2.

Table 9.2 Quartz crystals used for tuning-fork resonators

Cases	Cuts	Euler angles (ϕ, θ, ψ)
1	Tuning-fork cut	$(0°, -178°, -90°)$
2	AT cut	$(0°, -54.75°, -90°)$

The vibrational mode can be seen from the deformed shape of the tuning-fork resonator made of tuning-fork-cut quartz as shown in Figure 9.13. The resonant frequency of this tuning-fork resonator is at about 32.727 KHz. Moreover, at the on-resonance situation (Figure 9.13(b)) certain areas on the tuning fork can have a potential that is about ten times higher than that at off-resonance (Figure 9.13(a)). Figure 9.14 shows the deformed shape of the tuning fork made of AT-cut quartz with a resonant frequency of about 35.488 KHz. Similarly, in certain areas of the tuning fork, the highest potential reached at resonance (Figure 9.14(b)) is more than double that at off-resonance (Figure 9.14(a)).

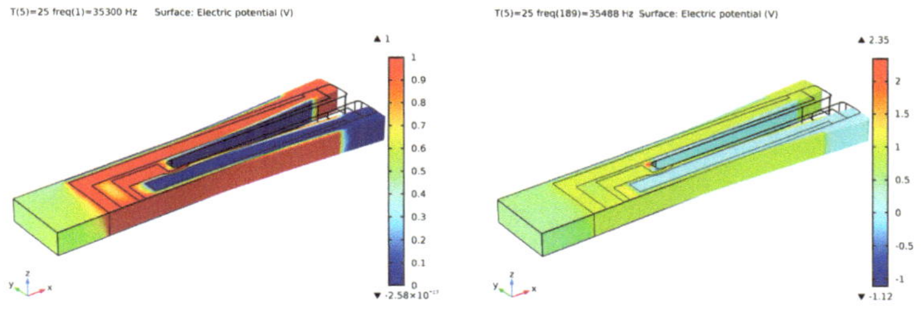

(a) Off resonance (b) At resonance

Figure 9.14 Tuning-fork resonator made of AT-cut quartz.

9.1.5 Temperature Influence on Quartz Tuning-Fork Resonators

In a similar way as we did for the disc resonators, by considering temperature-induced changes in quartz material properties, we evaluate the temperature influence on the tuning-fork resonators made of tuning-fork and AT cuts quartz. The temperature range considered in this case study is from −40°C, −20°C, 0°C, 20°C, 25°C, 30°C, 50°C, 70°C to 90°C. At each of these temperature point, we determine the frequency response curves of the tuning-fork resonators made of AT and tuning-fork cuts quartz, respectively.

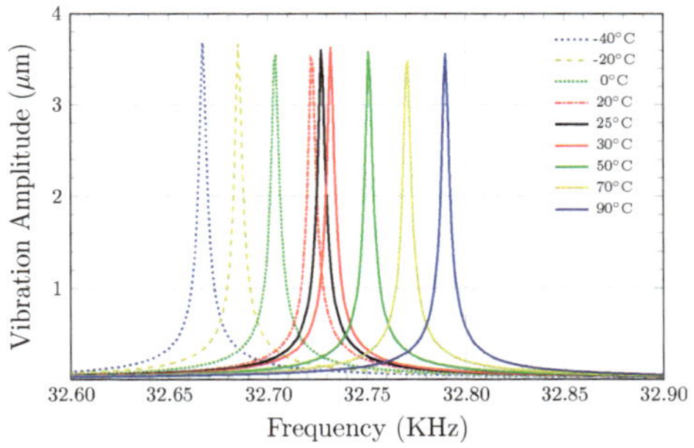

Figure 9.15 Temperature effect on the tuning-fork resonator made of tuning-fork-cut quartz.

Figures 9.15 and 9.16 show the results for the tuning-fork resonators made of tuning-fork and AT cuts quartz, respectively.

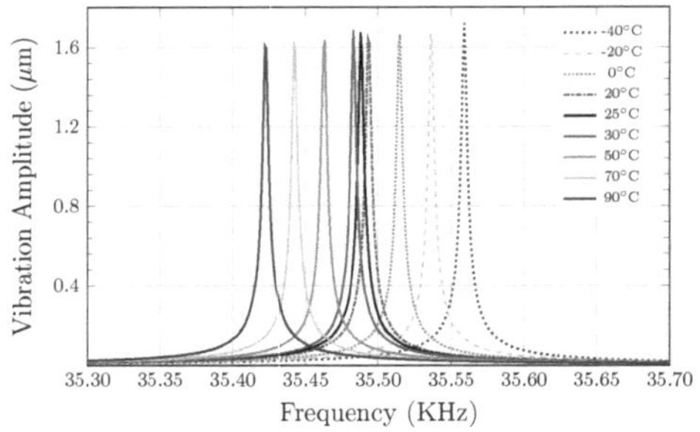

Figure 9.16 Temperature effect on the tuning-fork resonator made of AT-cut quartz.

Comparing between these two resonators, we can see that the direction of frequency shift due to temperature change is different. For the resonator made of tuning-fork-cut quartz, the resonant frequency increases, or shits to the right, as temperature increases, and for the resonator made of AT-cut quartz, it decreases, or shits to the left, as temperature increases. Moreover, of the range of temperature considered, namely, from –40° to 90°, the shift in the resonant frequency due to temperature change is slightly larger in AT resonator than that in tuning-fork one. For example, the shift is about 123 Hz and 137 Hz for the resonator made of tuning-fork and AT cuts quartz, respectively.

9.2 SAW Devices

For SAW devices there are many different types of electrode designs for converting electric energy to mechanical one. The simplest ones include one-port and two-port devices with interdigitated electrodes (IDEs) for piezoelectric excitation purposes. In this section, we will look at a two-port delay-line SAW device in the form of a transversal filter and a sensor, and a one-port SAW device in the form of a resonator.

As briefly discussed in Chapter 1, a two-port delay-line SAW device is commonly used for various applications. A typical two-port transversal SAW device can have an equal or unequal number of finger pairs for the transmitting and receiving IDEs as illustrated

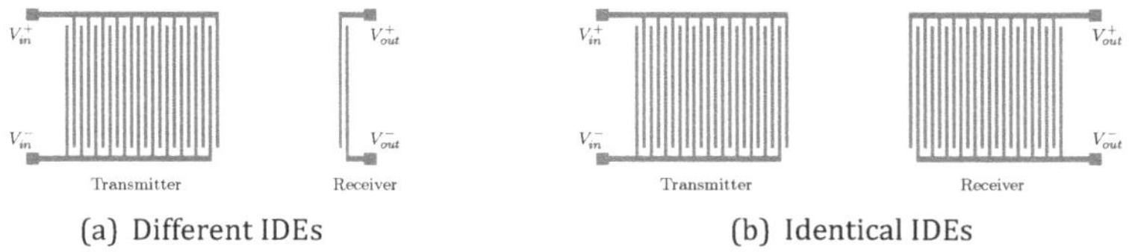

Figure 9.17 A two-port SAW device consisting of two interdigitated electrodes.

in Figure 9.17. In the illustration given in Figure 9.18, two identical IDE transducers are used, one is to generate acoustic waves as a transmitter and the other to detect the waves as a receiver. The transmitting IDE converts an alternating electrical signal into acoustic waves which will travel along the surface of the piezoelectric substrate, and the receiving IDE converts the acoustic waves back to an electrical signal for processing.

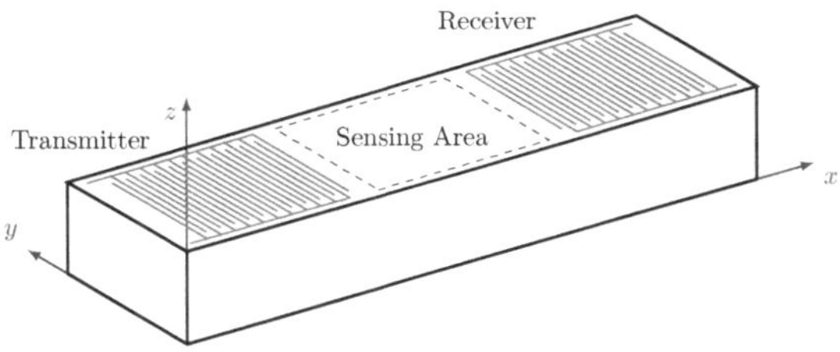

Figure 9.18 A two-port SAW device consisting of two interdigitated electrodes.

A two-port delay-line SAW device can be used either as a filter or as a sensor. For filter application, the input and output signals are used to assess the bandwidth, cut-off frequencies, attenuation and insertion loss of the device. For sensor application, a sensing area located between the two IDEs is often created to perturb the SAW waves traveling from the transmitting IDE to the receiving IDE and detection is quantified by the measurement of either frequency shift or insertion loss of the device. In the case studies for filter and sensor applications discussed in this section, we will consider SAW devices with identical transmitting and receiving IDEs.

9.2.1 SAW Transversal Filters

SAW filters can be designed to meet a specified need in terms of the desired center frequency and frequency bandwidth. The finger width and gap distance of an IDE are the parameters that can be adjusted to meet the design goals.

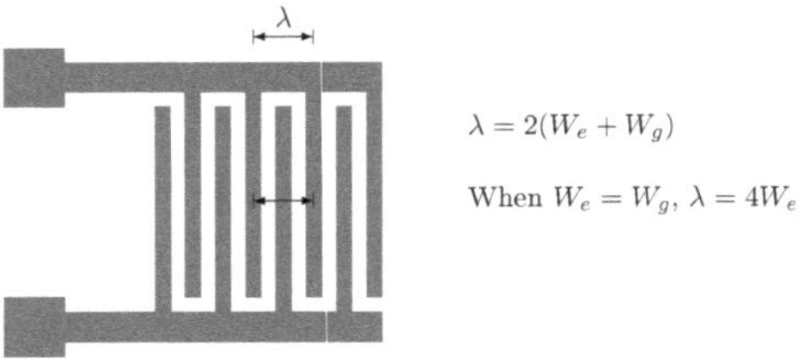

Figure 9.19 An IDE design with equal finger and gap width.

Since the fingers in an IDE serve as contact areas where electric signals are converted to mechanical (or acoustic) energy, it is important that they are placed properly to generate constructive rather than destructive addition for the acoustic waves in the piezoelectric material. It turns out that when the distance marked by the double-arrow line in Figure 9.19 equals the generated acoustic wavelength, additive construction of the acoustic waves will occur. For an IDE with equal-width fingers and equal-width gaps, this can be expressed as $\lambda = 2(W_e + W_g)$, where W_e is the width of each individual finger electrode and W_s is gap spacing between two adjacent fingers. When the finger width and gap spacing are kept the same, then we will simply have the following relationship to link the IDE design to the acoustics wavelength to be generated:

$$\lambda = 4W_e \tag{9.1}$$

We now examine a case study in which a transversal SAW filter is analyzed using FE modeling. As illustrated in Figure 9.20, for the geometric model, we use the following dimensions: overall length, width and depth at 320 μm, 80 μm, and 80 μm, respectively; and finger length, width and gap distance at 30 μm, 5 μm and 5 μm, respectively.

For electromechanical excitation of SAWs in the filter, IDEs are created on the upper surfaces of the SAW device. The transmitting

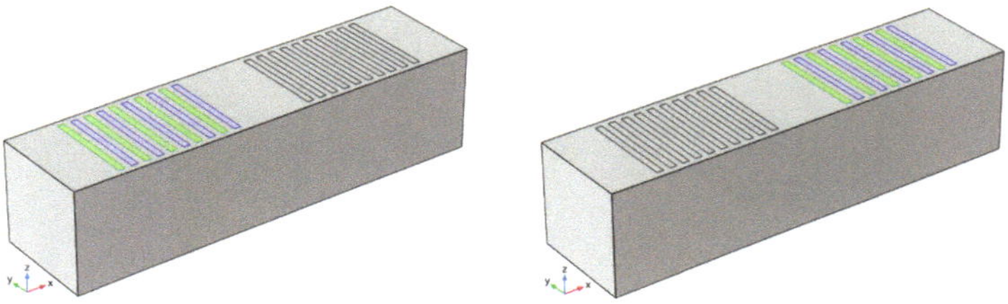

(a) V_{in}^{+} (green), V_{in}^{-} (blue) (b) V_{out}^{+} (green), V_{out}^{-} (blue)

Figure 9.20 A SAW filter with two sets of IDEs as input and output electrodes.

IDE on the left is used as the input end (see Figure 9.20(a)) where the positive and negative terminal are highlighted in green and blue, respectively, and the receiving IDE on the right is used as the output end (see Figure 9.20(b)) where the positive and negative terminal are also highlighted in green and blue, respectively. In the modeling analysis, the input end of the positive and negative terminals (see Figures 9.17(b) and 9.20(a)) are supplied with a bipolar impulse signal, defined as follows:

$$V_{in}^{+} = \begin{cases} 0.5\text{ V} & 0 \le t \le 1\text{ ns} \\ 0\text{ V} & t > 1\text{ ns} \end{cases}; \qquad V_{in}^{-} = \begin{cases} -0.5\text{ V} & 0 \le t \le 1\text{ ns} \\ 0\text{ V} & t > 1\text{ ns} \end{cases} \tag{9.2}$$

over a time period close to 70 ns. The output ends of two terminals are used to determine the output potential signal. For filter substrate materials, we use $LiNbO_3$ crystal in four different cuts listed in Table 9.3.

Table 9.3 $LiNbO_3$ crystals used for SAW filters

Cases	Cuts	Euler angles (ϕ, θ, ψ)	Cases	Cuts	Euler angles (ϕ, θ, ψ)
1	128° Rt.-YX(‖)	$(0°, 38°, 0°)$	3	41° Rt.-YX(‖)	$(0°, -49°, 0°)$
2	64° Rt.-YX(‖)	$(0°, -26°, 0°)$	4	YZ	$(0°, -90°, -90°)$

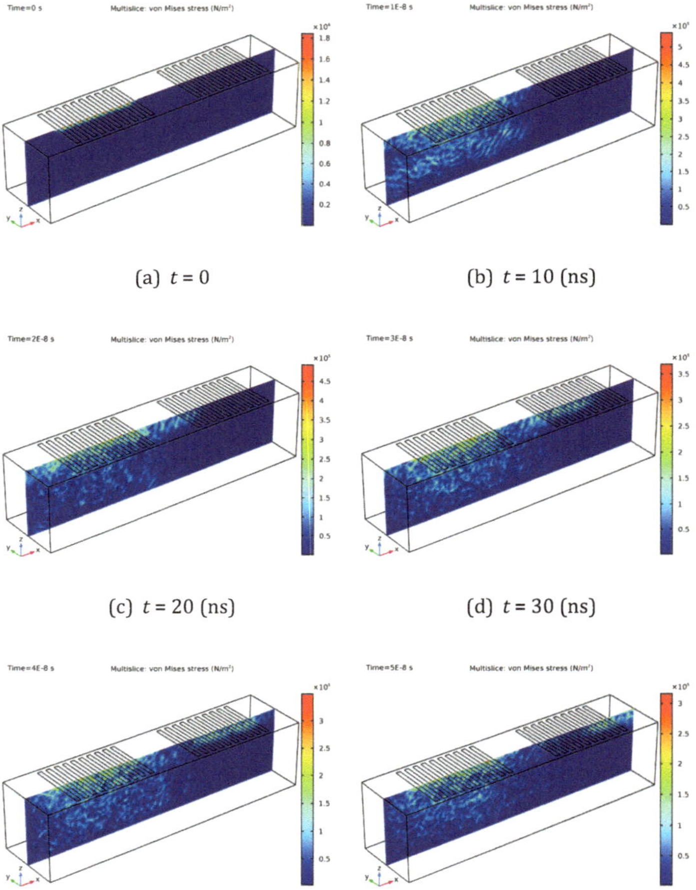

(a) $t = 0$ (b) $t = 10$ (ns)

(c) $t = 20$ (ns) (d) $t = 30$ (ns)

(e) $t = 40$ (ns) (f) $t = 50$ (ns)

Figure 9.21 Time-lapse section views of waves traveling (in the form of von Mises stress) in a transversal SAW filter made of 128° Rt.-YX(||)-cut $LiNbO_3$.

Figures 9.21 through 9.24 show at some selected time points section views of wave traveling in SAW filters made of 128°, 64° and 41° Rt.-YX(||) and YZ cuts of $LiNbO_3$, respectively. Apparently, the way SAWs travel in transversal filters made of $LiNbO_3$ in these cuts is different. In the 128° case (see Figure 9.21), aside from some dissipation of the acoustic energy into the bulk within the first 10 ns

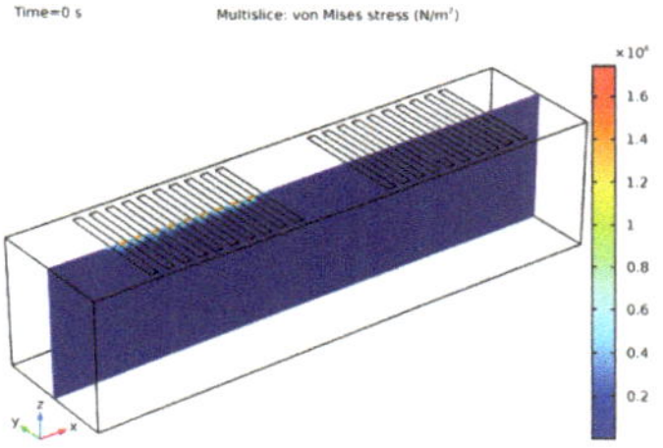

(a) $t = 0$

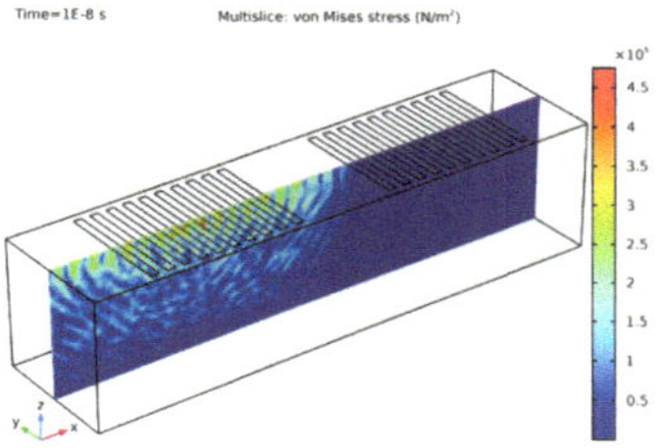

(b) $t = 10$ (ns)

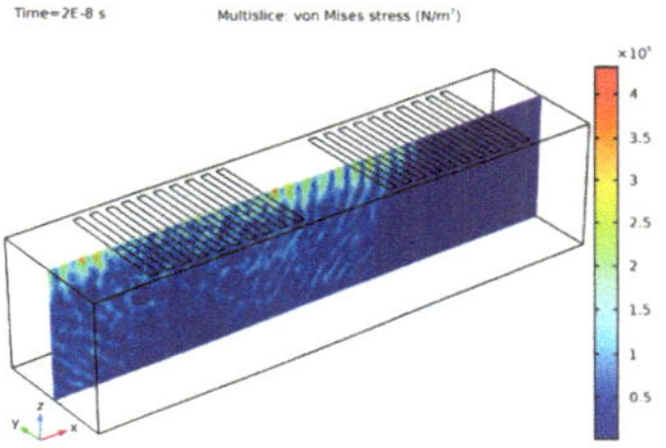

(c) $t = 20$ (ns)

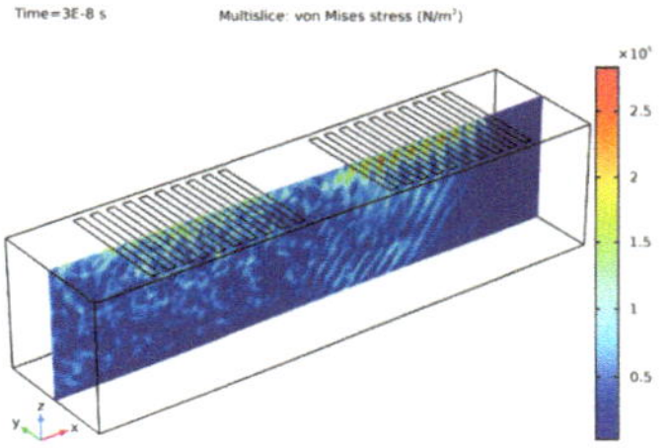

(d) $t = 30$ (ns)

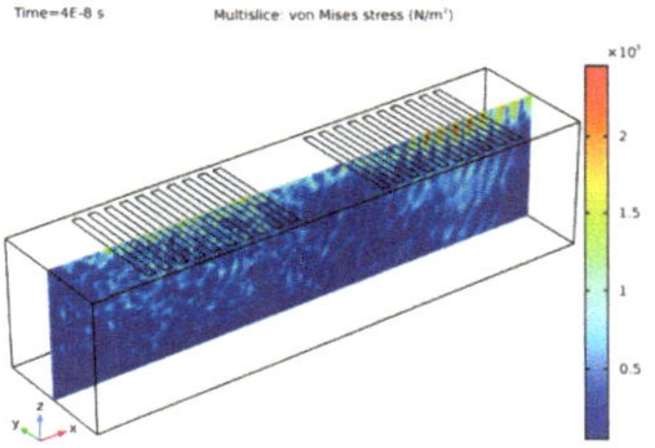

(e) $t = 40$ (ns)

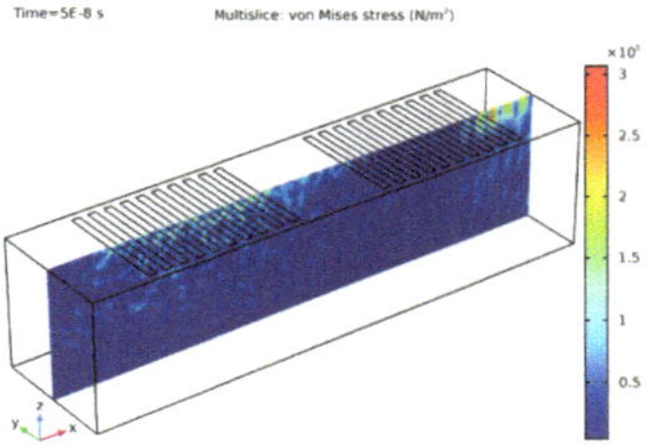

(f) $t = 50$ (ns)

Figure 9.22 Time-lapse section views of waves traveling (in the form of von Mises stress) in a transversal SAW filter made of 64° Rt.-YX(||)-cut $LiNbO_3$.

after the wave generation by the impulse signal, the waves are pretty much confined in a region near the surface as time goes by. However, in the 64° (Figure 9.22) and 41° (Figure 9.23) cases, in addition to the wave component that travels along the surface, the presence of a wave component propagating at an angle into the bulk part of the substrate is clearly visible. This type of surface waves are commonly known as surface leaky waves owing to the coupling of SAWs and

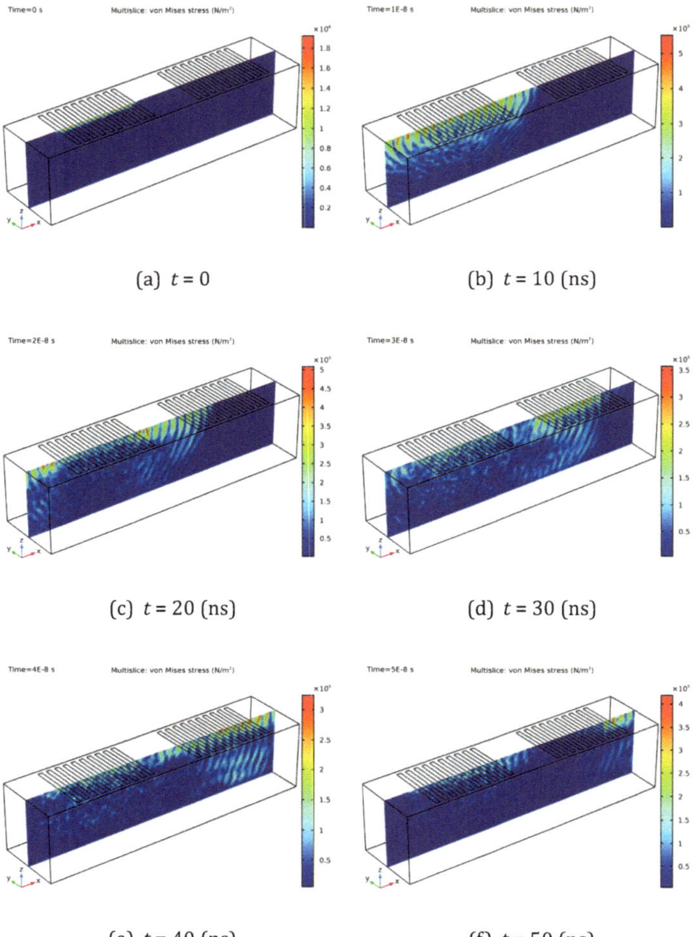

(a) $t = 0$ (b) $t = 10$ (ns)

(c) $t = 20$ (ns) (d) $t = 30$ (ns)

(e) $t = 40$ (ns) (f) $t = 50$ (ns)

Figure 9.23 Time-lapse section views of waves traveling (in the form of von Mises stress) in a transversal SAW filter made of 41° Rt.-YX(||)-cut $LiNbO_3$.

BAWs. Because of the coupling with a bulk wave component, leaky SAWs typically travel in a velocity much fast than that of conventional SAWs. In the YZ case (Figure 9.24), the situation is different. Although there exist a surface wave component and a bulk component, the bulk component seems different from those observed in 64° (Figure 9.22) and 41° (Figure 9.23) cases. The bulk component in the YZ case

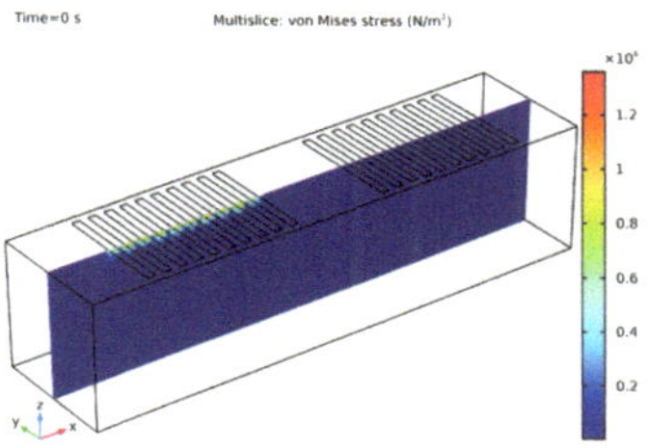

(a) $t = 0$

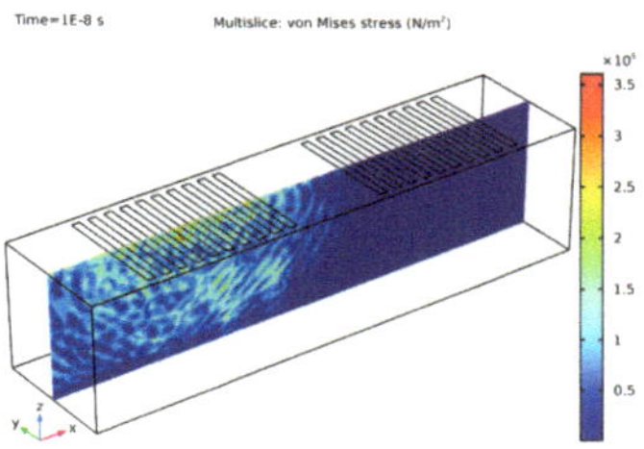

(b) $t = 10$ (ns)

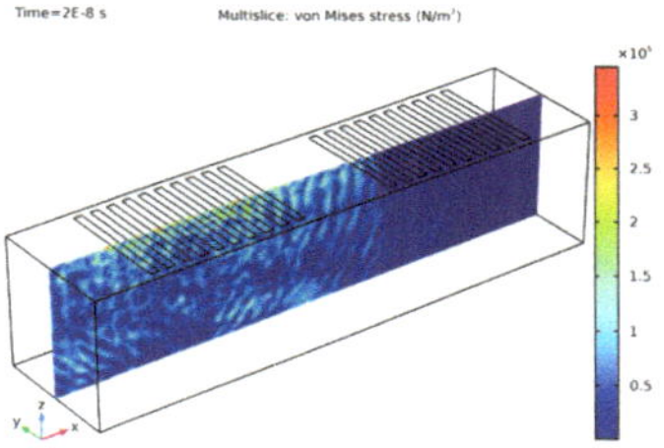

(c) $t = 20$ (ns)

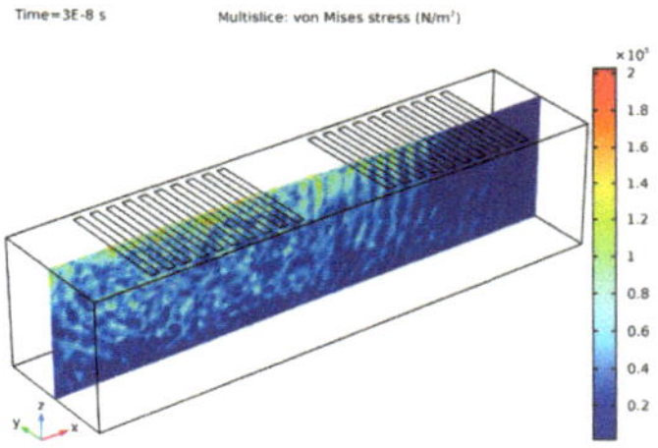

(d) $t = 30$ (ns)

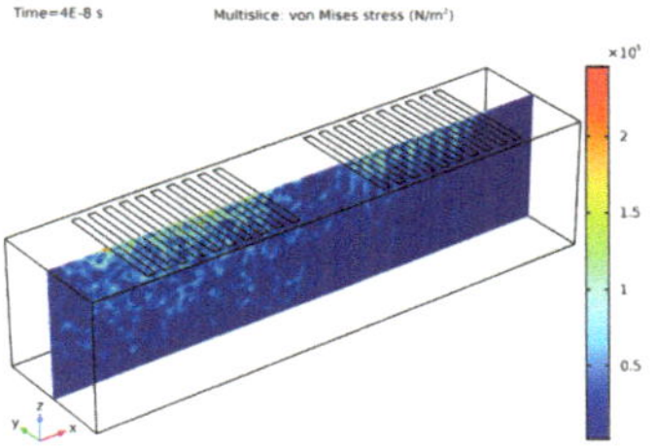

(e) $t = 40$ (ns)

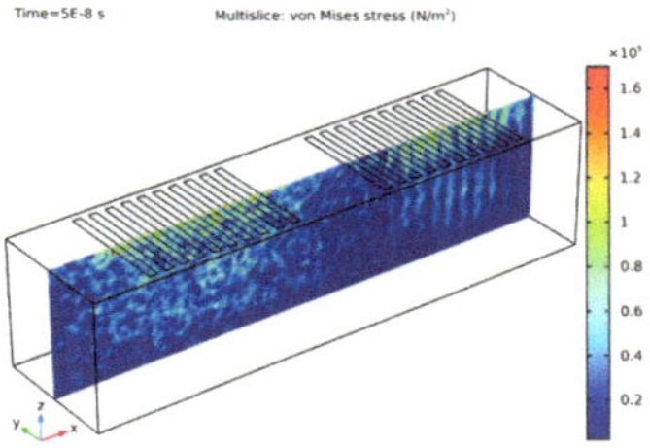

(f) $t = 50$ (ns)

Figure 9.24 Time-lapse section views of waves traveling (in the form of von Mises stress) in a transversal SAW filter made of YZ-cut $LiNbO_3$.

appears to travel along the *x* direction parallel with the surface rather than in an oblique angle. Recalling the discussion in Section 6.5.2, we know that for $LiNbO_3$, a Class 3m trigonal crystal, the YZ plane is its mirror plane, so both the YZ and ZY cuts will have their wave governing equations decouple into two separate waves. Referring to Example 7 in Section 6.6.4, we know that SAWs traveling in YZ $LiNbO_3$ are decoupled into piezoelectric stiffened Rayleigh waves and

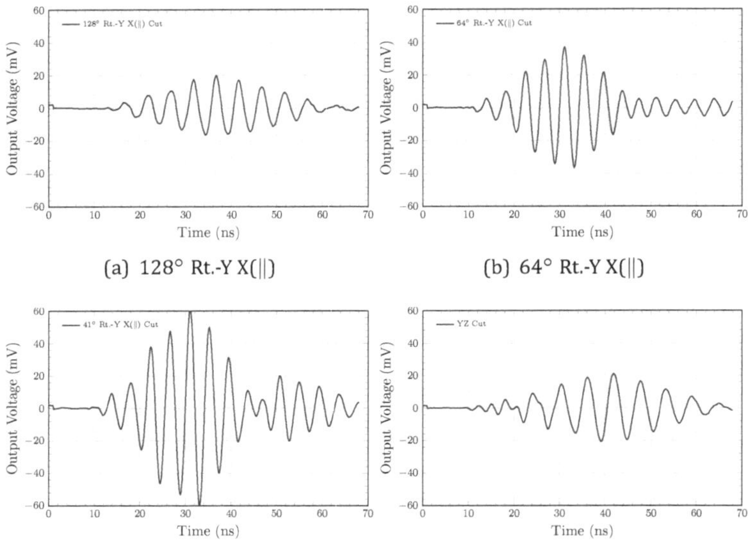

(a) 128° Rt.-Y X(||) (b) 64° Rt.-Y X(||)

(c) 41° Rt.-Y X(||) (d) Y Z

Figure 9.25 Output signals in response to a bipolar impulse signal determined from four transversal SAW filters made of $LiNbO_3$ crystals in different cuts.

a bulk TH wave. The captured views in Figure 9.24, especially the ones after 30 ns seem to confirm this fact: Aside from the Rayleigh waves propagating near the surface, a bulk wave is present in the deep part of the substrate material. Unlike the two leaky wave cases, because the surface wave and bulk wave components in the YZ case are not coupled, no increase in the phase velocity of SAWs will occur in this situation.

The output signals determined at the receiving IDE of the SAW filter in each of the four cases are shown in Figure 9.25. These results provide another look of the differences in these SAW filters. Of the four cases, the output signal seems the strongest and travels the fastest in the 41° Rt.-YX(||) case, followed by the 64° Rt.-YX(||) case. The YZ case appears to have slightly stronger signal than the 128° Rt.-YX(||) case, except that the waves travel a bit slower in the YZ case.

With the known input signal given in Eq. (9.2) and the output signal shown in Figure 9.25, by referring to the S_{21} parameter which

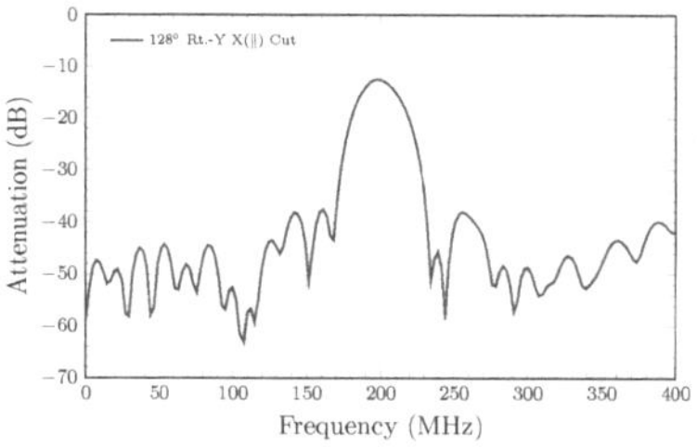

(a) 128° Rt.-Y X(||)

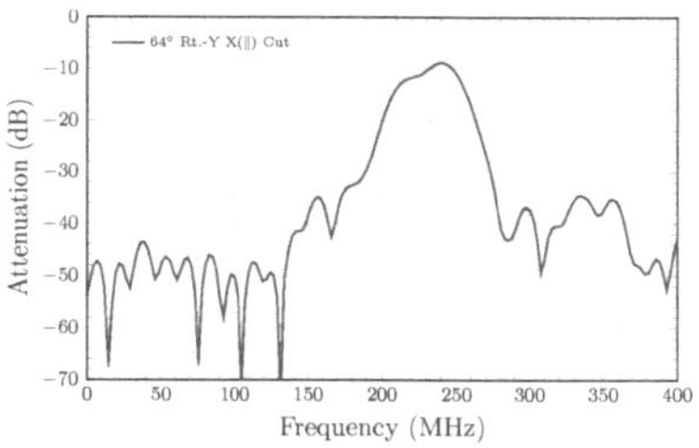

(b) 64° Rt.-Y X(||)

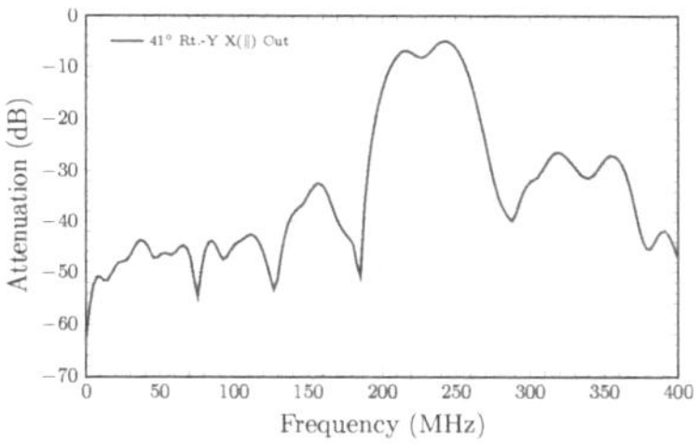

(c) 41° Rt.-Y X(||)

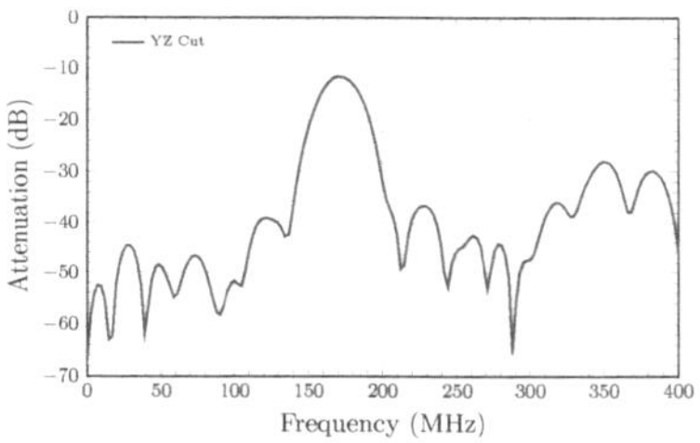

(d) Y Z

Figure 9.26 Frequency responses obtained from four transversal SAW filters made of $LiNbO_3$ crystals in different cuts.

is defined by the following expression:

$$S_{21} = 20 \log \frac{V^F_{out}}{V^F_{in}} \tag{9.3}$$

where V^F_{out} and V^F_{in} are the corresponding input and output signals in frequency domain after performing Fourier transform, we can determine the attenuation characteristics of these filter devices. The S_{21} parameter is often used to assess the attenuation and insertion loss behavior of the device and it is measured in decibel (dB).

Figure 9.26 shows the obtained frequency response curves for each of the four filters whose time-domain output signals are shown in Figure 9.25. The frequency at the peak of each of the response curves determines the center frequency of each SAW filter. For example, the center frequency (f_{center}) is found to be at approximately 199.59 MHz, 239.26 MHz, 241.70 MHz and 174.20 MHz for the 128°, 64° and 41° Rt.-YX(||) cases and the YZ case, respectively. Note that the double peak feature in both the 64° (Figure 9.26(b)) and 41°

Table 9.4 Wave velocity values

SAW filters	Case studies (m/s)	References (m/s)	SAW filters	Case studies (m/s)	References (m/s)
128° Rt.-YX(‖)	3992	3995	41° Rt.-YX(‖)	4834	4792
64° Rt.-YX(‖)	4785	4742	YZ	3484	3488

(Figure 9.26(c)) cases near the center frequency supports the fact that SAWs are coupled together with BAWs in these two situations.

Since all these filters have W_e = 5 μm, from Eq. (9.1) we know λ = 20 μm. With the familiar relationship, $v = \lambda f_{center}$, we can estimate the wave velocity of the SAWs traveling in these filters. For the four filter cases examined here, the estimated wave velocity values are provided in Table 9.4. For comparison, theoretical values from Chapter 6 for the 128° YX(‖) and YZ cases of $LiNbO_3$ and literature values for the other cases are provided as a reference. Apparently, the values obtained from the case studies presented here compare very well with the reference values. Figure 9.27 shows the same time- and frequency-domain results for these four SAW filters plotted together to provide another look of the differences in both the time and frequency responses of the SAW filters made of $LiNbO_3$ in four different cuts.

Aside from $LiNbO_3$ crystal, we also analyzed SAW filters made of made of $LiTaO_3$, SiO_2 and Langasite (LGS) at the cuts specified in Table 9.5. Figure 9.28 shows the frequency response results for these SAW filters.

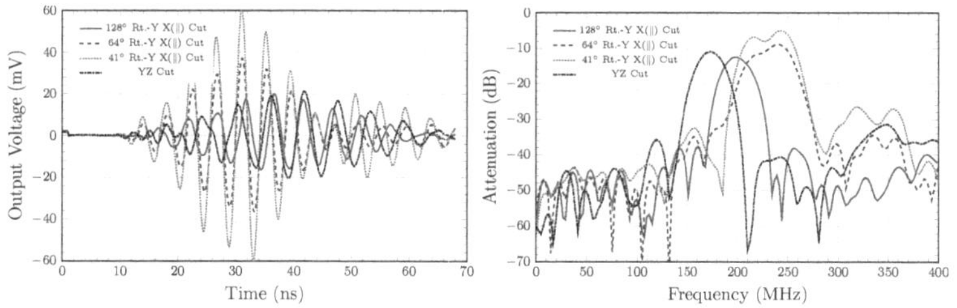

(a) Time-domain responses (b) Frequency-domain responses

Figure 9.27 Time and frequency responses of a SAW filter made of $LiNbO_3$ crystal in four different cuts.

Table 9.5 Other piezoelectric crystals used for SAW filters

Cases	Crystals	Euler angles (ϕ, θ, ψ)
1	$LiTaO_3$	$(90°, 90°, 112°)$
2	SiO_2	$(0°, -54°, -90°)$
3	LGS	$(0°, 138.5°, 26.6°)$

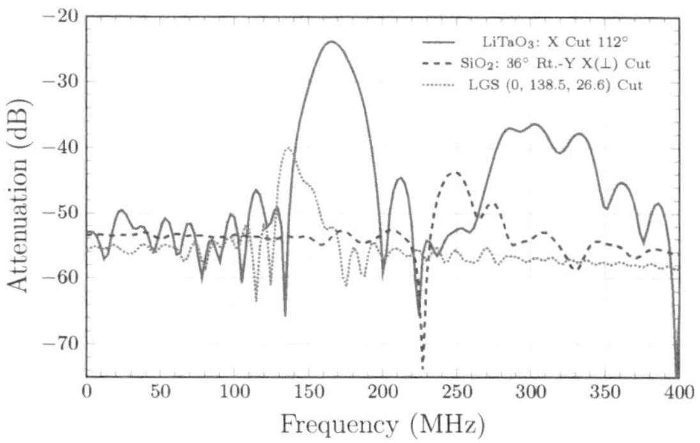

Figure 9.28 Frequency responses of a SAW sensor made of three different crystals.

From these frequency response curves, the center frequency is found to be approximately 3320 m/s, 4980 m/s and 2734 m/s for the X-cut 112° $LiTaO_3$, 36° Rt.-YX(⊥) SiO_2 and LGS $(0°, 138.5°, 26.6°)$ cases, respectively. The first value is very close to the theoretic value for X-cut 112° $LiTaO_3$ found in Section 6.7.1 (i.e., 3329 m/s). The other two are also close to the literature values, which are 5100 m/s and 2730 m/s, respectively. Moreover, SAW filters made of these crystals exhibit much high insertion loss as compared with those made of $LiNbO_3$ crystal. Here insertion loss is determined from the distance between the peak point at the center frequency and the 0 dB line. For instance, for the four $LiNbO_3$ cases, insertion loss is found to be between 4.89 dB and 12.59 dB, and here it is between 23.79 dB and 43.60 dB.

Figure 9.29 shows some selected views of acoustic waves traveling in these SAW filters made of $LiTaO_3$, SiO_2 and Langasite (LGS) at two different time points. Apparently, of these three cases, the SAW filter made of 36° Rt.-YX(⊥) SiO_2 has the highest wave velocity, and the one made of LGS has the slowest wave velocity. Moreover, both the SiO_2 and LGS cases exhibit leaky SAW behavior.

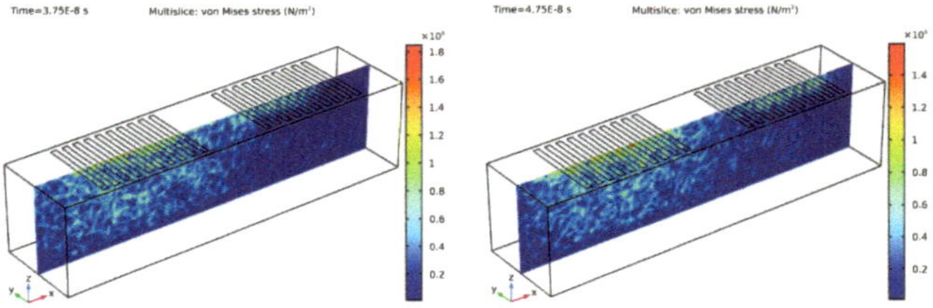

(a) $LiTaO_3$ at 37.5 (ns) (b) $LiTaO_3$ at 47.5 (ns)

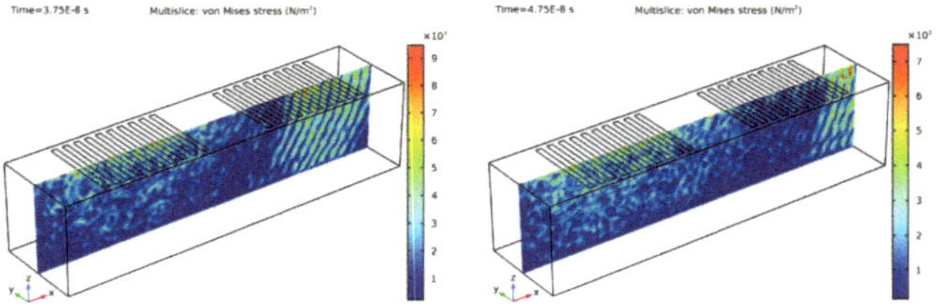

(c) SiO_2 at 37.5 (ns) (d) SiO_2 at 47.5 (ns)

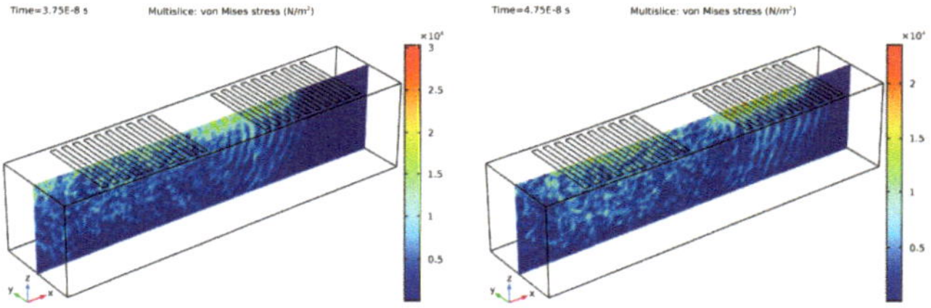

(e) LGS at 37.5 (ns) (f) LGS at 47.5 (ns)

Figure 9.29 Side-by-side comparisons of wave traveling in SAW filters made of three other crystals ($LiTaO_3$, SiO_2 and LGS).

9.2.2 Leaky SAWs

While a theoretical analysis of leaky SAWs is better discussed in an advanced future text, let us briefly highlight the essence of leaky SAWs. By the term "leaky," we already have a sense that leaky surface waves not just have components that travel along a surface, but also possess a wave component that propagates in the bulk part of the piezoelectric substrate of a SAW device. As we have seen in Figures 9.22, 9.23 and 9.29, because the bulk wave component in each of these cases travels at an angle into the substrate, it leaves an impression as if the surface waves are leaking into the bulk, hence the name *leaky surface waves*.

Recalling the discussion in Section 6.3.1, the existence of SAWs hinges on the fact that a real root exists for the boundary-satisfying matrix equation. As we have seen from all the examples discussed in Chapter 6, when a root does exist, the corresponding b values are either negative imaginary numbers or complex numbers with negative imaginary parts. This is so because only such b values will make all wave components disappear as the distance from the surface increases so that wave propagation is confined near the surface. In a leaky wave situation, at least one wave component will have a b value that is either a real number or a complex number with very small imaginary part. Such a b value will not make the corresponding wave component disappear as the distance from the surface increases. So strictly speaking, leaky SAWs are not like the SAWs in a conventional sense. But because the other b values are still complex numbers with negative imaginary parts, the propagation of other wave components is still confined near the surface. In this sense, acoustic waves that possess such b values are known as leaky SAWs or pseudo-SAWs. Typically, a real root may be found for the boundary-satisfying matrix equation when the wave propagation direction is tilted slightly downwards. This is the reason we see that in all the leaky wave cases (i.e., the 64° and 41° Rt.-YX(||) $LiNbO_3$, 36° Rt.-YX($\perp$) SiO_2 and (0°, 138.5°, 26.6°) LGS) a wave component is propagating at a downward angle into the bulk part of the substrates of the SAW filters.

Figure 9.30 provides a side-by-side comparison of the four leaky SAW cases and two non-leaky SAW cases at 50 ns to help the reader visualize the difference in wave propagation in these substrates. Clearly, in each of the four leaky SAW cases, namely the 64° and 41° Rt.-YX(||) $LiNbO_3$, 36° Rt.-YX($\perp$) SiO_2, and (0°, 138.5°, 26.6°) LGS, there exists a wave component propagating in the bulk part of the substrate at an angle. By contrast, in the 128° Rt.-YX(||) $LiNbO_3$ case all wave components are confined in a region very close to the surface. In the ZY $LiNbO_3$ case, although we see a bulk TH wave in addition to the surface waves as discussed earlier in Section

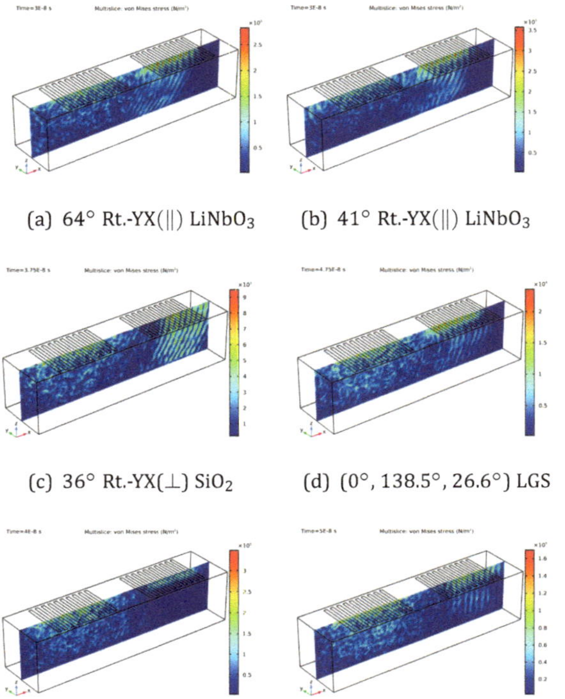

(a) 64° Rt.-YX(||) $LiNbO_3$ (b) 41° Rt.-YX(||) $LiNbO_3$

(c) 36° Rt.-YX(⊥) SiO_2 (d) (0°, 138.5°, 26.6°) LGS

(e) 128° Rt.-YX(||) $LiNbO_3$ (f) YZ $LiNbO_3$

Figure 9.30 Comparison of waves traveling in four transversal SAW filters at 50 ns.

9.2.1, because the bulk and surface waves are uncoupled, they travel independently along the *x* axis and the propagation direction of both waves is parallel with the surface.

Leaky SAWs exist in many elastic and piezoelectric materials. For example, of the many Rt.-YX(||) cuts of $LiNbO_3$ almost all of them except the 128° Rt.-YX(||) $LiNbO_3$ show leaky SAW behavior. To demonstrate that, we pick randomly six more cases of Rt.-YX(||)-cut $LiNbO_3$, namely, at 10°, 20°, 80°, 100°, 150° and 160°. Note that of these angles, two are less than 41°, two are in between 64° and 128°, and two are larger than 128°. Referring to Figure 6.27, we know that the phase velocity for the conventional SAWs at these angles is expected to be lower than that at 128°. However, from Figure 9.31 we can see that the center frequency obtained for each of the transversal SAW filters made of these Rt.-YX(||) cuts of $LiNbO_3$ is higher than that of the SAW filter made of the 128° Rt.-YX(||)-cut as shown in Figure 9.26(a). This fact indicates that SAWs in these six filters travel faster than in the filter made of 128° Rt.-YX(||) cut, confirming that they are leaky SAWs.

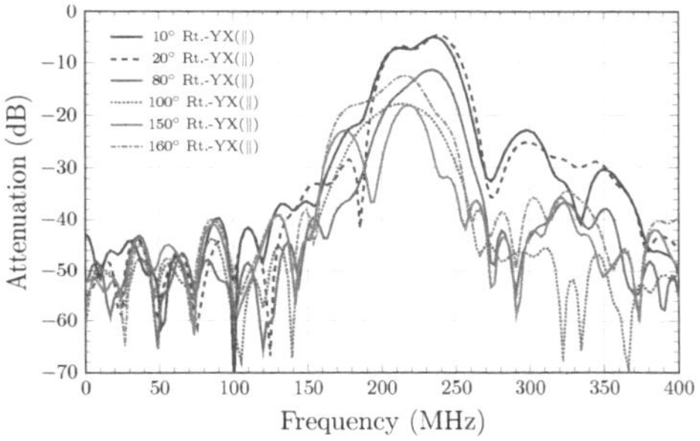

Figure 9.31 Frequency responses of a SAW sensor made of several Rt.-YX(||) cuts of $LiNbO_3$.

Typically, leaky SAWs travel faster than conventional SAWs with their phase velocity values falling in between the values of the phase velocity of the two T waves. Figure 9.32 shows the *xz* section views of the slowness curves of the three bulk wave components along with the corresponding Poynting angles obtained after the *xyz* coordinates system is rotated to the orientations defined by the 128°,64° and 41° Rt.-YX(||) cuts, respectively. An interesting observation we can make is that unlike the situations in the 128° case, in the 41° case the Poynting angle curves of the T(||) and T(⊥) waves almost overlaps with each other within a propagation angle ranging from approximately -20° to 20°, and in the 64° case they are very close. The slowness curves for these two waves also indicate near-zero normal in this propagation range, suggesting that the wave-front plane of both T waves is almost perpendicular to the *x* axis in the 64° and 41° cases. These insights seem to provide favorable conditions for the existence of leaky SAWs. Moreover, that the wave-front directions of the T(||) and T(⊥) waves are closer to each other for the 41° case than for the 64° case provides another piece of evidence to explain why leaky SAWs are stronger in the 41° case than in the 64° case as we have observed in Figure 9.27 earlier.

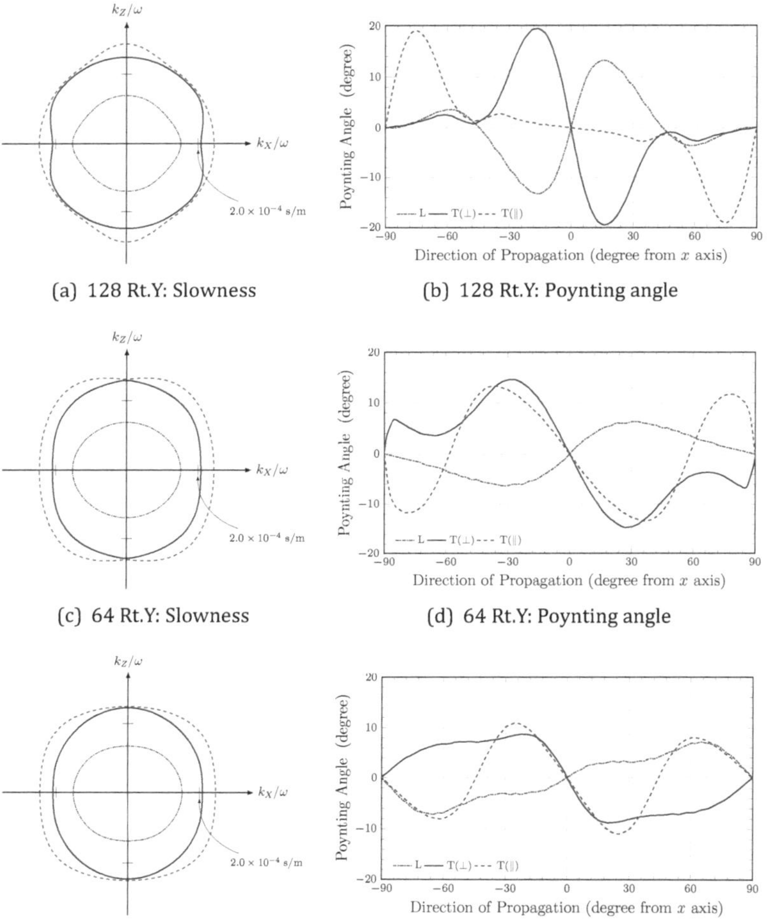

(a) 128 Rt.Y: Slowness (b) 128 Rt.Y: Poynting angle

(c) 64 Rt.Y: Slowness (d) 64 Rt.Y: Poynting angle

(e) 41 Rt.Y: Slowness (f) 41 Rt.Y: Poynting angle

Figure 9.32 Slowness curves and Poynting angle in *xz* section of Rt-Y cuts $LiNbO_3$.

9.2.3 SAW Sensors

The same SAW filter devices can be used as sensors when an area of sensing region is created between the transmitting IDE and the receiving IDE as illustrated in Figure 9.18. From a modeling point of view, we can simply expand the filter model by adding an area in

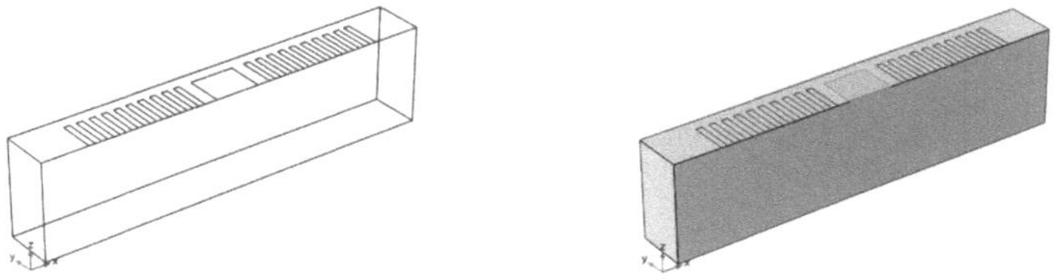

(a) Half model (b) Sensing area highlighted

Figure 9.33 A SAW sensor model developed and analyzed.

between the two IDEs as the sensing area to create a SAW sensor. For this case study, we use a half geometry model with addition of a sensing area as shown in Figures 9.33(a) and 9.33(b). With the same modeling considerations, we analyze the SAW device model as a sensor to determine the frequency responses of the device to an impulse signal as a function of mass loading in the form of uniform film deposition onto the sensing area. For substrate materials, we use 128° Rt.-YX(||) $LiNbO_3$.

Figures 9.34(a) through 9.34(d) show some side-by-side comparisons of SAWs traveling in this SAW sensor under two circumstances, without a film and with a 600 nm film, at two selected time points: 45 ns and 55 ns. With a close examination of the right and left figures at each time point, we can see that SAWs propagation in the cases with 600 nm film deposition is notably delayed as compared with the situation when there is no film deposition. This is a direct illustration of the effect of deposited film between the two IDEs on SAWs propagation.

Figure 9.35 shows the frequency response curves obtained for the SAW sensor in three situations: (1) no film, (2) 600 nm film deposition, and (3) 800 nm film deposition. Obviously, the deposition of a thin film in the sensing area has caused not only a left shift for the center frequency but also an increase in insertion loss. Clearly, the thicker the film gets, the larger the frequency shift and insertion loss become. For instance, for the three cases shown, namely, (1) no film, (2) 600 nm film, and (3) 800 nm film, the resonant frequency is found to be at approximately 197.75 MHz, 195.3 MHz and 192.9 MHz, respectively, and the insertion loss at approximately 11.4 dB, 13.6 dB and 15.7 dB, respectively.

(a) No film at 45 ns

(b) 600 nm film at 45 ns

(c) No film at 55 ns

(d) 600 nm film at 55 ns

Figure 9.34 Effect of film deposition on waves traveling in a SAW sensor made of 128° Rt.-YX(||) $LiNbO_3$ captured at various time points. Shown here are stress fields superimposed to a mechanically deformed crystal.

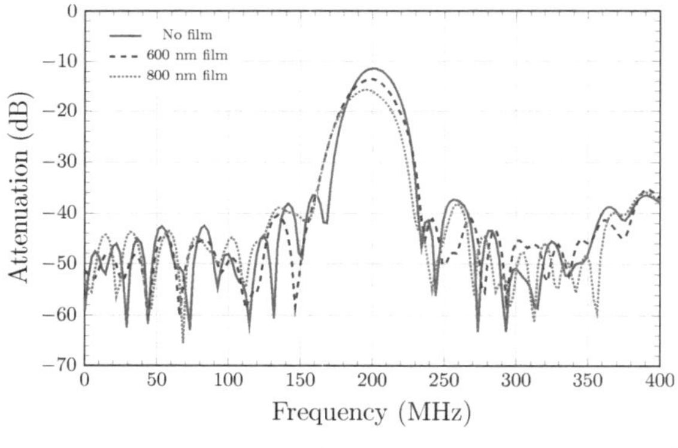

Figure 9.35 Frequency responses of a SAW sensor made of 128° Rt.-YX(||) $LiNbO_3$ crystal under different film deposition situations.

9.2.4 SAW Resonators

In this section, we will examine a SAW resonator. Figure 9.36 shows an illustration of a design for a one-port SAW resonator made of a single IDE placed at the center of the device with metal gratings on both sides to reflect and trap the acoustic energy within the resonator cavity.

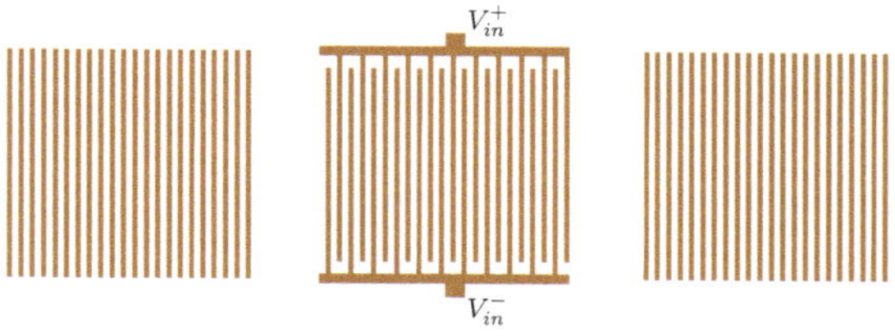

Figure 9.36 Electrode and grating design for a one-port SAW resonator.

For a case study, we analyze such a one-port SAW resonator by performing FE analysis. As shown in Figure 9.37, for the geometry of the model, we use the following dimensions: Overall length, width and depth of the resonator are set at 320 μm, 80 μm, and 80 μm, respectively; and finger length, width and gap distance of both the IDE and metal gratings are set at 60 μm, 5 μm and 5 μm, respectively. For electromechanical excitation and SAW generation in the resonator, an IDE and two sets of metal gratings are created on the upper surface of the SAW resonator with the IDE positioned at the center of the device and the gratings on both sides. Positive and negative terminals are assigned to the fingers of the IDE in an alternating pattern as highlighted in green in Figure 9.37 and the metal gratings on both sides are highlighted in blue. In the modeling analysis, the positive terminal is supplied with an

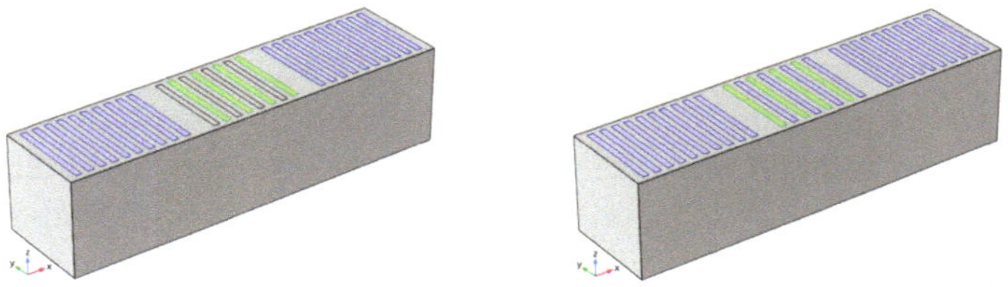

(a) Positive terminal (b) Negative terminal

Figure 9.37 A one-port SAW resonator with a single IDE in the middle (green) and metal gratings (blue) on two sides.

alternating potential signal with an amplitude of 10 V and the negative terminal along with the metal gratings is grounded. For substrate materials, we use $LiNbO_3$ crystal with two different cuts, 128° and 64° Rt.-YX(||) cuts, respectively.

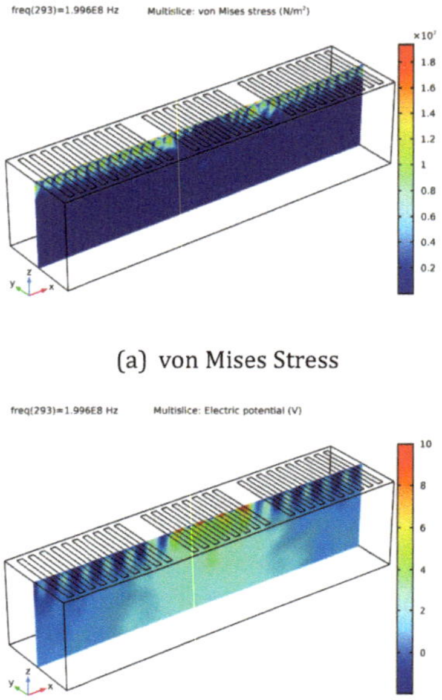

(a) von Mises Stress

(b) Potential

Figure 9.38 Stress and potential fields generated in a SAW resonator made of 128° Rt.-YX(||) $LiNbO_3$.

Figure 9.38 shows the induced stress field (Figure 9.38(a)) and potential field (Figure 9.38(b)) near the resonant frequency of approximately 199.60 MHz for the 128° case, Figure 9.39 shows the induced stress field (Figure 9.39(a)) and potential field (Figure 9.39(b)) near the resonant frequency of about 239.25 MHz for the 64° case, and Figure 9.40 shows the induced stress field (Figure 9.40(a)) and potential field (Figure 9.40(b)) near the resonant frequency of about 241.7 MHz for the 41° case. These resonant frequency values are almost the same as the respective center frequency values of SAW filters made of these cuts of $LiNbO_3$, thus suggesting that the SAWs generated in the 128° Rt.-YX(||) case

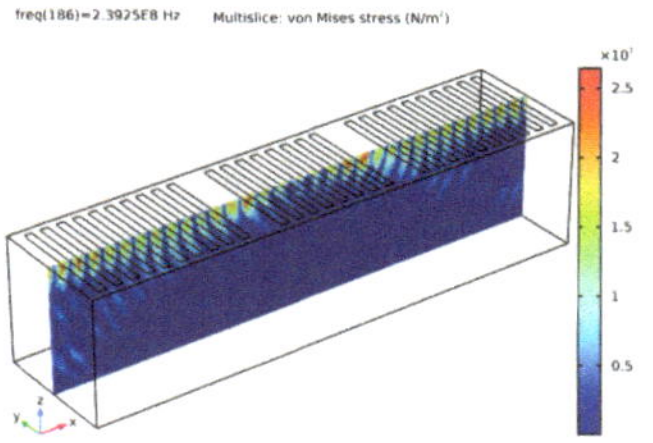

(a) von Mises Stress

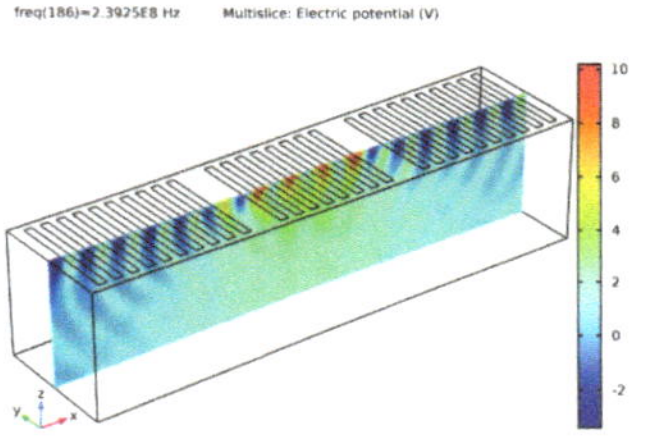

(b) Potential

Figure 9.39 Stress and potential fields generated in a SAW resonator made of 64° Rt.-YX(||) $LiNbO_3$.

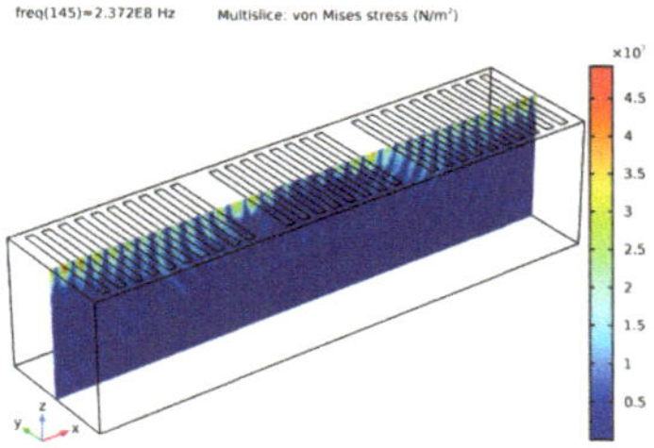

(a) von Mises Stress

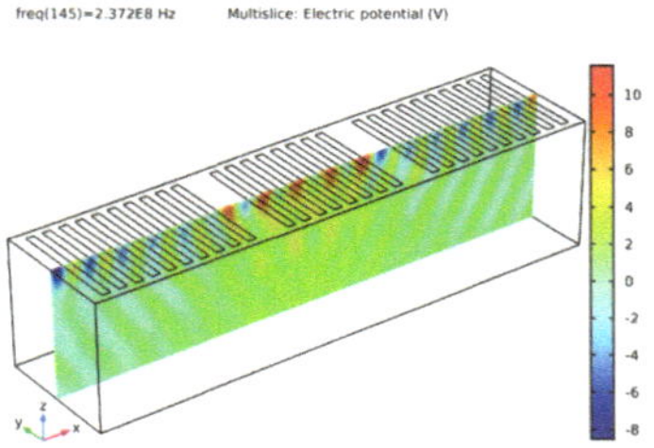

(b) Potential

Figure 9.40 Stress and potential fields generated in a SAW resonator made of 41° Rt.-YX(||) $LiNbO_3$.

are conventional SAWs and those in the 64° and 41° Rt.-YX(||) cases are leaky SAWs.

In a comparison of these obtained results, we can see that in the 128° case, the generated SAWs are confined within a very narrow region close to the surface. In the 64° and 41° cases, however, the generated acoustic waves are not only present near the surface, they also appear in the bulk part of the substrates. The surface waves in these two cases clearly possess at least one component that propagates into the bulk part of the substrate of the resonator. These results and visualization of wave propagation provide another illustration of the non-leaky nature of the generated SAWs in the 128° Rt.-YX(||) case and the leaky nature of the generated SAWs in the 64° and 41° Rt.-YX(||) $LiNbO_3$.

Appendix

Answers to Selected Problems

A.1 Chapter 1 Problems

Problem 4

Hint: use the expressions of $\sqrt{(\lambda_L + 2\mu_L)/\rho}$ and $\sqrt{\mu_L/\rho}$.

Problem 6

λ = 30 μm, so the finger width and gap are 7.5 μm

A.2 Chapter 2 Problems

Problem 2

$$10^{-10} \times \begin{bmatrix} 0.0487 & -0.0058 & -0.0125 & 0.0064 & 0 & 0 \\ -0.0058 & 0.0487 & -0.0125 & -0.0064 & 0 & 0 \\ -0.0125 & -0.0125 & 0.0436 & 0.0000 & 0 & 0 \\ 0.0064 & -0.0064 & 0.0000 & 0.1079 & 0 & 0 \\ 0 & 0 & 0 & 0 & 0.1079 & 0.0128 \\ 0 & 0 & 0 & 0 & 0.0128 & 0.1090 \end{bmatrix}$$

Bulk and Surface Acoustic Waves: Fundamentals, Devices, and Applications
Guigen Zhang

ISBN 978-981-4877-73-2 (Hardcover), 978-1-003-25662-5 (eBook)
www.jennystanford.com

Problem 4

$$[c]_{Plane-stress} = 10^8 \times \begin{bmatrix} 1.0321 & 0.4335 & 0 \\ 0.4335 & 1.0321 & 0 \\ 0 & 0 & 0.2993 \end{bmatrix}$$

$$[c]_{Plane-strain} = 10^8 \times \begin{bmatrix} 2.1699 & 1.5713 & 0 \\ 1.5713 & 2.1699 & 0 \\ 0 & 0 & 0.2993 \end{bmatrix}$$

Problem 5

Material stiffening will occur.

Problem 6

v_L = 6572.0 m/s

Problem 7

$v_{L(x)}$ = 6572.0 m/s, $v_{T(y)}$ = 3445.2 m/s, $v_{T(z)}$=4105.4 m/s

Problem 8

$v_{L(x)}$ = 6572.0 m/s, $v_{T(y)}$ = 4079.2 m/s, $v_{T(z)}$=4794.8 m/s

Problem 9

For $\{S\} = [s]\{T\} + [d]^T\{E\}$:

$$\begin{array}{cccc} \{S\} = & [s]\{T\} & + & [d]^T\{E\} \\ \downarrow & \downarrow & & \downarrow \\ 1 & m^2/N \times N/m^2 & & C/N \times V/m \rightarrow 1 \end{array}$$

For $\{D\} = [d]\{T\} + [\epsilon]\{E\}$:

$$\begin{array}{cccc} \{D\} = & [d]\{T\} & + & [\epsilon]\{E\} \\ \downarrow & \downarrow & & \downarrow \\ C/m^2 & C/N \times N/m^2 & & F/m \times V/m \rightarrow C/m^2 \end{array}$$

A.3 Chapter 3 Problems

Problem 2

$$[a] = \begin{bmatrix} 0.5783 & 0.7126 & 0.3971 \\ 0.0571 & -0.5210 & 0.8517 \\ 0.8138 & -0.4698 & -0.3420 \end{bmatrix}$$

$$[\mathbf{T}'] = \begin{bmatrix} 44.7464 & 4.5840 & 3.1721 \\ 4.5840 & 9.7840 & -2.3736 \\ 3.1721 & -2.3736 & 10.4696 \end{bmatrix}; \quad \{T'\} = \begin{Bmatrix} 44.7464 \\ 9.7840 \\ 10.4696 \\ -2.3736 \\ 3.1721 \\ 4.5840 \end{Bmatrix}$$

Problem 3

$$[a] = \begin{bmatrix} -0.6409 & 0.7071 & 0.2988 \\ -0.6409 & -0.7071 & 0.2988 \\ 0.4226 & -0.0000 & 0.9063 \end{bmatrix}$$

$$[\mathbf{S}'] = \begin{bmatrix} 11.2610 & -5.2512 & 3.6671 \\ -5.2512 & 10.0366 & -15.5903 \\ 3.6671 & -15.5903 & 9.0023 \end{bmatrix}; \quad \{S'\} = \begin{Bmatrix} 11.2610 \\ 10.0366 \\ 9.0023 \\ -31.1807 \\ 7.3343 \\ -10.5023 \end{Bmatrix}$$

A.4 Chapter 4 Problems

Problem 1

$$[a] = \begin{bmatrix} 0 & 1 & 0 \\ -1 & 0 & 0 \\ 0 & 0 & 1 \end{bmatrix}; \quad [M] = \begin{bmatrix} 0 & 1 & 0 & 0 & 0 & 0 \\ 1 & 0 & 0 & 0 & 0 & 0 \\ 0 & 0 & 1 & 0 & 0 & 0 \\ 0 & 0 & 0 & 0 & -1 & 0 \\ 0 & 0 & 0 & 1 & 0 & 0 \\ 0 & 0 & 0 & 0 & 0 & -1 \end{bmatrix}$$

$$[c] = [c'] = \begin{bmatrix} 27.5000 & 17.9000 & 15.2000 & 0 & 0 & 0 \\ 17.9000 & 27.5000 & 15.2000 & 0 & 0 & 0 \\ 15.2000 & 15.2000 & 16.5000 & 0 & 0 & 0 \\ 0 & 0 & 0 & 5.4300 & 0 & 0 \\ 0 & 0 & 0 & 0 & 5.4300 & 0 \\ 0 & 0 & 0 & 0 & 0 & 11.3000 \end{bmatrix}$$

$$[e] = [e'] = \begin{bmatrix} 0 & 0 & 0 & 0 & 21.3000 & 0 \\ 0 & 0 & 0 & 21.3000 & 0 & 0 \\ -2.7400 & -2.7400 & 3.7000 & 0 & 0 & 0 \end{bmatrix}$$

$$[\epsilon] = [\epsilon'] = \begin{bmatrix} 1744 & 0 & 0 \\ 0 & 1744 & 0 \\ 0 & 0 & 97 \end{bmatrix}$$

Problem 3

$$[a] = \begin{bmatrix} -0.5000 & 0.8660 & 0 \\ -0.8660 & -0.5000 & 0 \\ 0 & 0 & 1.0000 \end{bmatrix}$$

$$[M] = \begin{bmatrix} 0.2500 & 0.7500 & 0 & 0 & 0 & -0.8660 \\ 0.7500 & 0.2500 & 0 & 0 & 0 & 0.8660 \\ 0 & 0 & 1.0000 & 0 & 0 & 0 \\ 0 & 0 & 0 & -0.5000 & -0.8660 & 0 \\ 0 & 0 & 0 & 0.8660 & -0.5000 & 0 \\ 0.4330 & -0.4330 & 0 & 0 & 0 & -0.5000 \end{bmatrix}$$

$$[c] = [c'] = \begin{bmatrix} 8.7000 & 0.7000 & 1.2000 & -1.8000 & 0 & 0 \\ 0.7000 & 8.7000 & 1.2000 & 1.8000 & 0 & 0 \\ 1.2000 & 1.2000 & 10.7200 & 0 & 0 & 0 \\ -1.8000 & 1.8000 & 0 & 6.0000 & 0 & 0 \\ 0 & 0 & 0 & 06.0000 & -1.8000 & \\ 0 & 0 & 0 & 0 - 1.8000 & 4.0000 & \end{bmatrix}$$

$$[e] = [e'] = \begin{bmatrix} 0.1710 & -0.1710 & 0 & -0.04000 & 0 & \\ 0 & 0 & 0 & 0 & 0.0400 & -0.1710 \\ 0 & 0 & 0 & 0 & 0 & 0 \end{bmatrix}$$

$$[\epsilon] = [\epsilon'] = \begin{bmatrix} 4.4000 & 0 & 0 \\ 0 & 4.4000 & 0 \\ 0 & 0 & 4.6000 \end{bmatrix}$$

A.5 Chapter 5 Problems

Problem 1

With piezoelectric stiffening: v_L = 6304.5 m/s, v_{T1} = 3384.9 m/s, v_{T2} = 2345.2 m/s

Without piezoelectric stiffening: v_L = 5959.5 m/s, v_{T1} = 3384.9 m/s, v_{T2} = 2340.9 m/s

K^2_{BAW} = 10.64% (for the L wave)

K^2_{BAW} = 0 (for the T1 wave)

K^2_{BAW} = 0.36% (for the T2 wave)

Problem 2

With piezoelectric stiffening: v_L = 4452.4 m/s, v_{T1} = 1757.4 m/s, v_{T2} = 1757.4 m/s

Without piezoelectric stiffening: v_L = 4404.8 m/s, v_{T1} = 1757.4 m/s, v_{T2} = 1757.4 m/s

K^2_{BAW} = 2.12% (for the L wave)

K^2_{BAW} = 0 (for the T1 wave)

K^2_{BAW} = 0 (for the T2 wave)

Problem 3

With piezoelectric stiffening: v_L = 6029.8 m/s, v_{T1} = 3152.2 m/s, v_{T2} = 2748.0 m/s

Without piezoelectric stiffening: v_L = 5982.2 m/s, v_{T1} = 2962.0 m/s, v_{T2} = 2748.0 m/s

K^2_{BAW} = 1.57% (for the L wave)

K^2_{BAW} = 11.7%for the T1 wave)

K^2_{BAW} = 0 (for the T2 wave)

Problem 5

With piezoelectric stiffening: v_L = 5681.2 m/s, v_{T1} = 3883.5 m/s, v_{T2} = 3523.6 m/s

Without piezoelectric stiffening: v_L = 5554.0 m/s, v_{T1} = 3589.4 m/s, v_{T2} = 3523.6 m/s

K^2_{BAW} = 4.4% (for the L wave)

K^2_{BAW} = 14.6% for the T1 wave)

K^2_{BAW} = 0 (for the T2 wave)

Problem 9

With piezoelectric stiffening: v_L = 7128.2 m/s, v_{T1} = 4120.5 m/s, v_{T2} = 3857.3 m/s

Without piezoelectric stiffening: v_L = 6644.2 m/s, v_{T1} = 3978.5 m/s, v_{T2} = 3783.3 m/s

K^2_{BAW} = 13.1% (for the L wave)

K^2_{BAW} = 6.8%for the T1 wave)

K^2_{BAW} = 3.9% (for the T2 wave)

A.6 Chapter 6 Problems

Problem 1

$v_{Surface-Rayleigh} = 3017.3$ m/s, $v_{Bulk-Transverse} = 3195.7$ m/s,

Problem 3

$v_{Metallized} = 3254.5$, $v_{Free} = 3272.9$ m/s, $K^2_{SAW} = 1.12\%$

Problem 5

$v_{Metallized} = 2317.0$, $v_{Free} = 2406.5$ m/s, $K^2_{SAW} = 7.44\%$

A.7 Chapter 7 Problems

Problem 1

TCD=$\sim$ 21.1 ppm/C°

Problem 2

TCD=$\sim$ 36.8 ppm/C°

A.8 Chapter 8 Problems

Problem 1

$v_1 = 3895.4$ m/s

Problem 2

$v_2 = 5177.8$ m/s, $v_3 = 5707.6$ m/s

Problem 3

$v_1 = 3886.3$ m/s, $v_2 = 4111.6$ m/s, $v_3 = 4325.1$ m/s

Problem 4

$v_{asymmetric} = 3223.9$ m/s

Problem 5

$v_{symmetric} = 3441.1$ m/s

References and Further Reading

1. Adachi M., Karaki T., and Miyamoto W. (1999) Surface Acoustic Wave Properties of $La_3Ga_5SiO_{14}$ (LANGASITE) Single Crystals, *Japanese Journal of Applied Physics*, 38, 3283–3287.
2. Auld B. A. (1973) *Acoustic Fields and Waves in Solids*, Vols. I and II, John Wiley & Sons, Inc.
3. Ballato A., Lukaszek T., Williams D. F., and Cho F. Y. (1981) Power Flow Angle and Pressure Dependence of SAW Propagation Characteristics in Quartz. *Ultrasonic Symposium*, 346–349.
4. Bluestein J. (1968) A New Surface Wave in Piezoelectric Materials, *Applied Physics Letters*, 13(12), 412–413.
5. Campbell C. K. (1998) *Surface Acoustic Wave Devices for Mobile and Wireless Communications*, Academic Press.
6. Campbell J. J., and Jones W. R. (1968) A Method for Estimating Optimal Crystal Cuts and Propagation Directions for Excitation of Piezoelectric Surface Waves, *IEEE Transactions on Sonic and Ultrasonics*, Su-15(4), 209–217.
7. Royer D. and Dieulesaint E. (2000) *Elastic Waves in Solids, I and II*, Springer.
8. Engan H., Ingebrigsten, A., and Tonning. A. (1967) Elastic Surface Waves in α-Quartz: Observation of Leaky Surface Waves. *Applied Physics Letters*, 10(11), 311–313.
9. Hashimoto K. Y (2000) *Surface Acoustic Wave Devices in Telecommunications: Modeling and Simulation*, Springer.

Bulk and Surface Acoustic Waves: Fundamentals, Devices, and Applications
Guigen Zhang

ISBN 978-981-4877-73-2 (Hardcover), 978-1-003-25662-5 (eBook)
www.jennystanford.com

10. Inoue K. and Sato K. (1998) Propagation Characteristics of Surface Acoustic Waves on Langasite, *Japanese Journal of Applied Physics*, 37, 2909–2913.

11. Joshi S. G. and White R. M. (1969) Excitation and Detection of Surface Elastic Waves in Piezoelectric Crystals, *The Journal of the Acoustical Society of America*, 46, 17–27.

12. Kosinski J. A., Gualtieri J. G. and Ballato A. (1992) Thermoelastic Coefficients of Alpha Quartz, *IEEE Transactions on Ultrasonics, Ferroelectrics, and Frequency Control*, 39(4), 502–507.

13. Kozlov A. S., Tazier R. M., Vasiliev I. L., and Yakovkin I. B. (1995) Temperature Derivatives of Elastic Stiffness Evaluated from the SAW Delay-Temperature Behavior on Lithium Tantalate Plates, *IEEE Ultrasonic Symposium*, 393–396.

14. Lamb H. (1917) On Waves in an Elastic Plate, *Proceedings of the Royal Society of London*, 93(648), 114–128.

15. Lim T. C. and Farnell G. W. (1968) Character of Pseudo Surface Waves on Anisotropic Crystals. *Journal of the Acoustical Society of America*, 45(4), 845–851.

16. Love A. E. H. (1911) *Some Problems of Geodynamics*, Cambridge University Press.

17. Murota M. and Shimizu Y. (1991) Elastic Constants and Temperature Coefficients of $LiTaO_3$ for SAW Application, *Japanese Journal of Applied Physics*, 156–158.

18. Rayleigh (1887) On Waves Propagated along the Plane Surface of an Elastic Solid, *London Proceedings of Mathematical Society*, s1-17(1), 4–11.

19. Slobodnik A. (1972) The Temperature Coefficients of Acoustic Surface Wave Velocity and Delay on Lithium Niobate, Lithium Tantalate, Quartz, and Tellurium Dioxide, Physical Sciences Research Papers, No. 477. Air Force Cambridge Research Lab.

20. Takayanagi A., Yamanouchi Y., and Shibayama K. (1970) Piezoelectric Leaky Surface Wave in $LiNbO_3$. *Applied Physics Letters*, 17(5), 225–227.

21. White R. M. (1970) Surface Elastic Waves, *Proceedings of the IEEE*, 58(8), 1238–1276.

Index